Photovoltaic and Photoactive Materials –
Properties, Technology and Applications

NATO Science Series

A Series presenting the results of scientific meetings supported under the NATO Science Programme.

The Series is published by IOS Press, Amsterdam, and Kluwer Academic Publishers in conjunction with the NATO Scientific Affairs Division

Sub-Series

I. **Life and Behavioural Sciences**	IOS Press
II. **Mathematics, Physics and Chemistry**	Kluwer Academic Publishers
III. **Computer and Systems Science**	IOS Press
IV. **Earth and Environmental Sciences**	Kluwer Academic Publishers
V. **Science and Technology Policy**	IOS Press

The NATO Science Series continues the series of books published formerly as the NATO ASI Series.

The NATO Science Programme offers support for collaboration in civil science between scientists of countries of the Euro-Atlantic Partnership Council. The types of scientific meeting generally supported are "Advanced Study Institutes" and "Advanced Research Workshops", although other types of meeting are supported from time to time. The NATO Science Series collects together the results of these meetings. The meetings are co-organized bij scientists from NATO countries and scientists from NATO's Partner countries – countries of the CIS and Central and Eastern Europe.

Advanced Study Institutes are high-level tutorial courses offering in-depth study of latest advances in a field.
Advanced Research Workshops are expert meetings aimed at critical assessment of a field, and identification of directions for future action.

As a consequence of the restructuring of the NATO Science Programme in 1999, the NATO Science Series has been re-organised and there are currently Five Sub-series as noted above. Please consult the following web sites for information on previous volumes published in the Series, as well as details of earlier Sub-series.

http://www.nato.int/science
http://www.wkap.nl
http://www.iospress.nl
http://www.wtv-books.de/nato-pco.htm

Series II: Mathematics, Physics and Chemistry – Vol. 80

Photovoltaic and Photoactive Materials – Properties, Technology and Applications

edited by

J.M. Marshall

Electronic Materials Centre,
Department of Materials Engineering,
University of Wales Swansea,
Swansea, U.K.

and

D. Dimova-Malinovska

Central Laboratory for Solar Energy and New Energy Sources,
Bulgarian Academy of Sciences,
Sofia, Bulgaria

Kluwer Academic Publishers

Dordrecht / Boston / London

Published in cooperation with NATO Scientific Affairs Division

Proceedings of the NATO Advanced Study Institute on
Photovoltaic and Photoactive Materials – Properties, Technology and Applications
Sozopol, Bulgaria
9–21 September 2001

A C.I.P. Catalogue record for this book is available from the Library of Congress.

ISBN 1-4020-0823-6 (HB)
ISBN 1-4020-0824-4 (PB)

Published by Kluwer Academic Publishers,
P.O. Box 17, 3300 AA Dordrecht, The Netherlands.

Sold and distributed in North, Central and South America
by Kluwer Academic Publishers,
101 Philip Drive, Norwell, MA 02061, U.S.A.

In all other countries, sold and distributed
by Kluwer Academic Publishers,
P.O. Box 322, 3300 AH Dordrecht, The Netherlands.

Printed on acid-free paper

Table of Contents

Preface

The primary objective of this NATO Advanced Study Institute (ASI) was to present an up-to-date overview of various current areas of interest in the field of photovoltaic and related photoactive materials. This is a wide-ranging subject area, of significant commercial and environmental interest, and involves major contributions from the disciplines of physics, chemistry, materials, electrical and instrumentation engineering, commercial realisation etc. Therefore, we sought to adopt an inter-disciplinary approach, bringing together recognised experts in the various fields while retaining a level of treatment accessible to those active in specific individual areas of research and development.

The lecture programme commenced with overviews of the present relevance and historical development of the subject area, plus an introduction to various underlying physical principles of importance to the materials and devices to be addressed in later lectures. Building upon this, the ASI then progressed to more detailed aspects of the subject area. We were also fortunately able to obtain a contribution from Thierry Langlois d'Estaintot of the European Commission Directorate, describing present and future EC support for activities in this field.

In addition, poster sessions were held throughout the meeting, to allow participants to present and discuss their current activities. These were supported by what proved to be very effective feedback sessions (special thanks to Martin Stutzmann), prior to which groups of participants enthusiastically met (often in the bar) to identify and agree topics of common interest.

The content of these Proceedings is structured in a similar form to that of the ASI itself, presenting the contributions of the individual lecturers, followed by a selection of shorter papers arising from the poster presentations. Since the lecturers' articles have been designed to be accessible on an individual basis, there is necessarily some duplication of underlying concepts between them. However, we believe and hope that it will be of interest to the reader to compare and contrast the various approaches and viewpoints expressed. Papers arising from the lectures and poster contributions will also be published in a special issue of the Journal Solar Energy Materials and Solar Cells (many thanks to its principal editor, Carl Lampert, for supporting this initiative, and also for his revolutionary concept of a "solar-powered pub").

The ASI was held in the enchanting and historic Black Sea coastal town of Sozopol, Bulgaria, which dates back to 610 BC when Greek settlers founded the city of Apollonia. Its present status, combining many of its traditional (18th and 19th Century) architecture, cobbled streets, etc. with excellent beaches, bars and restaurants, makes it both an ideal venue for events of the present type and a haven for the Balkan artistic community. The ASI itself was held in the excellently appointed and self-contained complex "Izgrev" of the Ministry of the Affaire Interior of Bulgaria. We wish to express special thanks to the Manager, Mr. Georgy Davtchev, and his staff for the superb and highly effective support which they provided, both before and during the event. We would strongly recommend consideration of this venue to organisers contemplating the organisation of similar events in Sozopol. We also thank the ASI cat - Debbie (Ph.D.) - for maintaining an impressive interest in proceedings throughout the meeting (see the group photograph).

The meeting took place over a ten-day working period, with morning lecture sessions, followed by afternoon breaks and early evening lecture, poster, and tutorial/review sessions (with afternoons free to sample the delights of the local environment). There were also valuable opportunities for initial introductions and subsequent social interactions between participants, via a Welcome Party, an Excursion to the neighbouring historic town of Nessebar (plus an excellent subsequent meal in an ethnic restaurant), and a final Farewell Party. Photographs and a video - thanks Guiseppe [nobile@portici.enea.it] - *were* taken during these events, but are not included here, due to potential litigation issues.

The Editors are indebted to the authors for the scrupulous way in which they managed to avoid *almost all* of the camera-ready instructions and deadlines for preparing their manuscripts. Despite this, and after a mind-sapping process of proof reading, editing and formatting, we hope and believe that the resulting volume will be of value not just to the participants, but to other readers interested in the current status of this subject area. During the lectures and subsequent editing, some fascinating (and probably accurate, in practice) typographical errors were detected. These included *"dementional analysis"* (guess who!), and *"conduction by hoping"*.

A considerable debt of gratitude is due to our secretariat – Avgustina (Gusti) Rachkova, Ani Nedjalkova, Suzi Kavlakova, Mimi Kamenova and Krassimir Dimitrov - for the many hours of effort which they devoted to the smooth running of the event, both before and during it. However, we would strongly advise future collaborators not to allow Gusti to organise any more pub-crawls!

We wish to thank all lecturers and participants for making the event such a success, and for their kind expressions of appreciation at the end of the ASI. Best wishes to you all, and the best of luck in your future careers.

Before concluding, and by no means least, we wish to express our gratitude, on behalf of all participants, to the NATO Scientific Affairs Division (and in particular Dr. F. Pedrazzini) for having the foresight to realise that this event was a highly worthwhile and effective avenue for funding support, and for valuable advice during the organisational process. Finally, we thank Ms. Wil Bruins of the NATO Publishing Unit of Kluwer Academic Publishers for her help in the preparation of these Proceedings.

Joe Marshall (Swansea)
Doriana Dimova-Malinovska (Sofia)
April 2002

Organising Committee
D. Dimova-Malinovska
C. Lampert
J. M. Marshall
M. Palfy
M. Stutzmann

LARGE-AREA SMART GLASS AND INTEGRATED PHOTOVOLTAICS

C.M. LAMPERT
Star Science
8730 Water Road, Cotati, CA 94931-4252 USA
E-mail:cmlstar@juno.com

1. Introduction

The field of switchable materials is ever expanding, with many types of new market and the technology [1-4]. Switchable windows can be used for many applications including architectural and vehicle windows, aircraft windows, skylights and sunroofs. Related to glazing are applications for low-information-content displays for the presentation of information. Switchable windows rely on a variety of processes and materials. Several companies through out the world are developing dynamic glazing. University and National Laboratory groups are researching new materials and processes to improve these products. Switchable glazing for building and vehicle application is very attractive. Conventional glazing only offers fixed transmittance and control of energy passing through it. Given the wide range of illumination conditions and glare, dynamic glazing with adjustable transmittance offers the best solution.

Photovoltaic technology can be integrated with switchable glass, to give self-powering and possibly wireless features. This study covers selected electrical switching technologies, including electrochromics (EC), suspended particle devices (SPD), also known as electrophoretic media, and phase dispersed liquid crystals (PDLC). Currently, much development in switchable devices is aimed at deposition processes suitable for glass substrates. In the future, there will be a movement towards plastic substrates for weight savings. Cost of production and defects in large-area deposition are two of the largest issues. These exciting technologies create many challenges and opportunities for glass and plastic companies, along with excitement for building designers.

2. Markets

The use of flat glass is very widespread, with the global production running at about 3.8 billion m^2 per year, for all markets. The largest geographical producing regions are Asia (1.6 Bm^2), followed by Europe (1.2 Bm^2) and the Americas 875 Mm^2 [5].

1

J.M. Marshall and D. Dimova-Malinovska (eds.),
Photovoltaic and Photoactive Materials - Properties, Technology and Applications, 1–10.
© 2002 *Kluwer Academic Publishers.*

The ownership of glass float plants and coating facilities is widely distributed throughout the world, with considerable import and export of glass products. The North American market (1998) is split in to four usage sectors: 37% in residential architectural, 21% commercial architectural, 28% automotive, and 17% specialty glass.

Low-e glass accounts for 40% of the Insulated Glass market, with accumulated shipments of about 400 Mm^2 [6]. Low-e is still one of the strong growth products in coated glass. Indium tin oxide (ITO) and pyrolytic tin oxide are widely used as transparent conductors for switchable glazing. Antireflection coatings and self-cleaning coatings are other areas of development for glazing, displays and solar cell and solar thermal collector covers. Integrated photovoltaic glazing is a growth area for many glass makers and fabricators. A hybrid glazing could be made with both photovoltaics and a smart window. Photovoltaics used for residential and commercial buildings could deliver as much as 800 MWp annual production by 2010. The overall world annual photovoltaic production is predicted [7] to be at 1600 MWp by 2010.

2.1 ARCHITECTURAL GLASS

Architectural applications have dominated the research and development of switchable windows. The flat glass market for architectural glazing is one the most attractive. There are wide possible applications in a variety of buildings. Flabeg (Germany) has made the largest electrochromic architectural windows installed in buildings. They have at least four building projects in Germany using switchable windows. Switchable electrochromic skylights and light tube panels were made by Schott-Donnelly, (Tucson, AZ), but the Company ceased its operations in 2001. SAGE Electrochromics (Faribault, MN) have shown electrochromic skylights at the 2000 National Home Builders show (USA). PDLCs are made by Nippon Sheet Glass (Japan), Saint-Gobain (France) and Polytronix (Richardson, TX).

2.2 VEHICLES

In the automotive sector, there is considerable demand for glazing for the larger SUV (sport utility vehicle) type automobile in the United States. In Europe, there is growing demand for sunroofs in cars. Highly selective Low-e coatings have been introduced by Southwall Technologies (Palo Alto, CA) as laminate layers on plastic film for automotive safety glazing. If electrochromics could be made on plastic substrates, they could easily be introduced into the automotive market. Advances in electrochromics are driven by developments in electrochromic mirrors for cars. This follows the largely successful electrochromic automotive mirror. Switchable mirrors are available for most major makes of car. Electrochromics are used for automotive mirrors, made by Gentex (Zeeland, MI), Donnelly (Holland, MI), Toyota (Japan), Nikon (Tokyo, Japan), and Murakami-Kaimedo (Japan). Several groups are working on switchable sunroof glazing. Sunroofs represent the first automotive glazing to be switchable. In the near future, one can anticipate SPD vehicle sunroofs and aircraft windows from SPD, Inc. (Korea) working with Research Frontiers Inc. (Woodbury, NY).

2.3 AEROSPACE

In the aircraft market, there will be many changes in the future. Over the next 20 years there will be about 35% replacement of the aging aircraft on the world market. Also, the aircraft stock will grow from 13,000 planes in 1999 to 28,000 in 2018 [8]. The aerospace industry is interested in the development of visors and windows that can control glare for pilots and passengers. Research Frontiers, Inc. and InspecTech (Ft. Lauderdale, FL) have shown their SPD film on small jet aircraft at an aircraft show in 2001. Airbus Industrie has announced dimmable windows for their first class cabin of the new A380 Airbus airplane, to be on the market in about 2004-2006. Saint-Gobain (France) has shown its prototype electrochromic Airbus cabin windows [9]. These have a 40:1 contrast ratio, with the deeply colored visible transmittance of less than 1%. Chemically tempered and bent glass is used to satisfy the safety requirements for aircraft glazing.

2.4 DISPLAYS

There are certain segments of the display market that impact windows. For example, the UMU PDLC window made by Nippon Sheet Glass is also used as a display screen in building windows. The world market for displays is estimated to be about 55 billion US$ (2001) climbing to 100 billion US$ in 2005 [10-11]. The display market is divided into two major camps: the cathode ray tube (CRT) display and liquid crystal displays based on both active and passive matrix designs. These displays are seen in a variety of electronic products ranging from televisions to palm top computers. For flat panels, glass processing is moving towards larger sizes of glass. Soon the panel fabrication size will exceed 1 m^2

Within the category of flat panels, there are many new types being developed. The US Display Consortium (USDC) supports some of these technologies. Two of the most active areas of development are active-matrix liquid crystal and organic light emitting diodes (OLEDS). OLEDS can be designed to emit white light, making a glazing emitting light possible [12]. Other important areas of development are technologies such as electrophoretic systems aimed at electronic paper displays. The trend in displays is that they are going to flat panel configurations for many applications. Also, in future, flat panel displays will move to plastic substrates or extremely thin glass [13]. Another future display market is electronic paper, where information would be stored and displayed on a flexible substrate. Electronic paper could be integrated into glazing to give multifunctional properties. Most displays are currently based on glass. Displays are very dependent on high quality thin glass and transparent conductive coatings. They could be integrated with photovoltaics and film batteries to provide a self-powered display system.

3. Electrochromic Technology

Electrochromics are the most popular technology for large-area switching devices. Much of the electrochromic technology has been developed for building windows and automotive mirrors and windows. The major advantages of electrochromic materials are: (1) they only require power during switching; (2) they have a small switching voltage (1-5 V); (3) they are always specular; (4) they have gray scale; (5) they have low polarization; (6) many designs have adjustable memory, up to 12-48 hr. Typical electrochromic glazing devices can have an upper visible transmission of T_v = 70-50% and fully colored transmittance of T_v = 25-10%. Levels of transmittance as low as 1% are possible. The range of shading coefficients for electrochromic glazing (SC) is about SC = 0.67-0.60 for the bleached condition, and SC = 0.30-0.18 for the fully colored condition.

Electrochromic materials change their optical properties due to the action of an electric field, and can be changed back to the original state by a field reversal. Depending on the electrochromic material, various coloration ions can be used, such as: Li^+, H^+, Na^+, and Ag^+. The binary inorganic oxides that have gained the most research interest are: WO_3, NiO, MoO_3, V_2O_5 and IrO_x.

Tungsten oxide is the most commonly used of all the oxides [14-15]. The reversible reaction follows:

$$WO_3 \text{ (Colorless)} + xH^+ + WO_3 <----\text{(Voltage)}----> H_xWO_3 \text{ (Blue)}.$$

The viologens are the most used commercially of the organic electrochromics. Originally, organic electrochromics tended to suffer from problems with secondary reactions during switching, but more stable organic systems have been developed. An electrochromic device must use an ion-containing material (electrolyte) in close proximity to the electrochromic layer, as well as transparent layers for setting up a distributed electric field. Devices are designed in such a way that they shuttle ions back and forth into the electrochromic layer with applied potential. An electrochromic glazing can be fabricated from five (or less) layers, consisting of two transparent conductors, an electrolyte or ion conductor, a counter electrode, and an electrochromic layer. The construction is similar to that of a battery.

3.1 ELECTROCHROMIC GLAZING

The industry-projected selling price goals for electrochromic glazing are within the range 100-250 US$/m^2. However, most prototypes are a factor of ten higher in cost. Pilkington Deutchland/Flabeg companies have shown several "E-Control^{TM}" switchable glazing units of 0.80 x 1.6 m. installed in buildings. The window has a transmittance range of T_v = 50-15% [16]. This is in an insulated glass unit, using two panes of glass including low-emittance coatings. A group of E-Control^{TM} windows covering 8 x 17 m has been installed in the Stadtsparkasse Bank in Dresden, Germany. The glazing takes a few minutes to change color or to bleach. An example of a glazed electrochromic wall is given in Figure 1.

Figure 1. South facing view of 8 x 17 m of electrochromic E-Control™ glass made by Pilkington Deutchland/Flabeg. Installed in the Stadtsparkasse Dresden am Altmarkt. (Source: J. Cardinal, Flabeg GmbH).

In France, Saint-Gobain Glass is working to develop a range of electrochromic devices for a variety of applications. In Japan, Asahi Glass and Nippon Mitsubishi Oil have been steadily developing electrochromic windows of 1 sq. m based on Li_xWO_3/Li-polymer/Counter electrode for testing and evaluation. This smart glazing had a monochromatic transmittance, $T_{(633 nm)}$, of 60-19% [17].

AFG Glass (Petaluma, CA) is developing electrochromic glazing using the OCLI patents. SAGE Electrochromics, Inc., collaborating with Apogee Enterprises, has developed electrochromic glazing. SAGE showed a Sage Glass [R] switchable skylight product at the National Association of Home Builders 2000 Show (Dallas, USA). The skylight size is about 1 x 0.6 m. Prototype SAGE windows have a visible switching range of about 70-4%, and have survived 100,000 cycles of testing. SAGE has a new alliance with GE/Honeywell, directed at the development of switchable glass with an integrated control system. Gentex has been working with an organic based electrochromic window. The Company has combined photovoltaics with a smart electrochromic window. Staff found that for a 25 x 30 cm window, they could power it at 1V at 20 mA using 8 cm^2 of c-Si solar cells. Electrochromics are ideal for use with photovoltaics, since they are both low voltage high current devices.

3.2 ELECTROCHROMIC RESEARCH

Another type of electrochromic design is the nanocell structure (also known as organic PV). This type actually is fashioned from the nanocell photovoltaic cell, so the electrochromic can self-color when exposed to sunlight. The cell relies on a dye-sensitized anatase titanium oxide layer, which forms a distributed p/n junction. Its optical density can be regulated by resistively shunting the anode and cathode of the cell. Uppsala University (Sweden), NREL (Golden, CO) and NTera (Dublin, Ireland), working with the EPFL (Lausanne, Switzerland), have developments in this photoelectrochromic technology. Sustainable Technology Inc. (STI) (Australia) is developing electrochromic windows. They have research partnerships with the University of Technology-Sydney, Dept. of Physics, Monash University (Australia) and The Institute for New Materials (Saarbruken, Germany). STI uses solgel deposition to produce the films on glass.

Under the U.S. DOE (Dept. of Energy) Electrochromics Initiative, several organizations are funded to develop electrochromic glazing including SAGE and Eclipse Energy (Gainsville, FL). Eclipse Energy is developing PECVD deposited electrochromics using the National Renewable Energy Labs (Golden, CO,) (NREL) patents. The Lawrence Berkeley National Laboratory (LBNL) has developed building energy models to quantify energy saving of electrochromics. Phillips (NL) has been working on metal hydride materials that can switch from transparent to a reflective state [18]. NREL has been given the job of evaluating the lifetime and durability of electrochromic devices for the U.S. National Program.

4. Electrophoretic-Suspended Particles

RFI and their licensees are responsible for the commercial development of SPDs for goggles, eyeglasses and windows. Recent activities have been directed at polymer sheet development. Several companies have licenses with RFI for the development of specific products. Among other companies, Hitachi Chemical, Dainippon Ink and Chemicals, AP Technologies, among others have licensed the process to make ink and emulsion for devices. SPD Inc. has been very innovative in developing prototype products for flexible plastic and large panels of about $1m^2$. Figure 2 shows a prototype SPD window, in both the off and on condition. This glazing consists of 5 layers. The active layer has needle shaped dipole particles (< 1 mm long) suspended in a polymer. This layer is laminated or filled between two dielectric coated transparent conductors (ITO). In the off condition, the particles are random and light absorbing. When the electric field is applied, the particles align with the field, causing increased transmission. Typical transmission ranges are 6-75%, 15-60% with switching times of 100-200 ms [19]. The device requires about 100V a.c. to operate. With research, it is possible to get the operating voltage down to about 35V a.c. A SPD aircraft glazing was shown by InspecTech at the Aircraft Interiors Conference, June 2001 in Long Beach, CA, USA.

off   on

Figure 2. SPD Glazing (1 x 2 m) in the off (dark) and on conditions.
(Source: J. Harary, RFI. website: www.refr-spd.com).

Toyota Labs (Japan) has developed a new type of dispersed particle window. By modification of the particles, several colors (green, blue, red, and purple) can be achieved. New sub-micron dipole suspensions based on SiO_2 coated TiO_xN_y have been developed by Nippon Sheet Glass [20]. With this particle, up to 50% change in solar transmittance has been shown.

Related to SPD development are the Electrophoretic electronic paper displays being developed by E-ink (Cambridge, MA) and at NOK Corp. (Ibaraki, Japan). In the display, electrophoretic inks containing white particles are encapsulated in polymer bubbles and then embedded in a flexible polymer matrix. Depending on the applied field, the white particles move up or down in the bubble, giving a contrast effect. These displays operate at 90 V d.c and below, and have contrast ratios of at least 6:1. Also, they are bistable and exhibit a memory effect. E-Ink displays have been shown to be compatible with active matrix TFT polycrystalline silicon drivers [21]. It is possible to match the operating voltage and current requirements with SPDs and Electronic Ink to photovoltaics.

5. Liquid Crystal Glazing

The types of liquid crystals are nematic, semetic, twisted nematic, cholesteric, guest-host, and ferroelectric. For displays, twisted nematics are the most commonly used liquid crystals. The mechanism of optical switching in liquid crystals is to change the orientation or twist of liquid crystal molecules interspersed between two conductive electrodes, with an applied electric field. The orientation of the liquid crystal can also alter the overall optical reflectivity properties of the window or display.

One fairly unusual version of a liquid crystal system is to make an emulsion of a polymer and liquid crystal, to form a film. Such emulsions are called phase dispersed liquid crystals (PDLC), and have been commercialized for use in switchable glazing. The liquid crystal droplets (5 μm) are encapsulated within an index matched polymer matrix (22 μm) [22]. The polymer emulsion is fabricated between two sheets of transparent conductor coated polyester or glass, that serve as electrodes. The switching effect of this device spans the entire solar spectrum, up to the absorption edge of the glass. In the off-state, the device appears translucent white. When an electric field is applied, the liquid crystal droplets align with the field and the device becomes transparent. Typically these devices operate between 24-100 V a. c. Their power consumption is less than 5 W/m^2, but require continuous power to be clear. In general, compared to electrochromics, the power consumption is higher for liquid crystals because of the need for continuous power in the activated state. The typical integrated hemispherical visible transmission values for a PDLC device are T_v (off-on) = 50-80%. The shading coefficient changes by SC = 0.63-0.79. Pleochroic dyes can be added to darken the device in the off-state. The dyed film shows considerable control over visible transmittance, compared to an undyed one.

Raychem (Sunnyvale, CA) licenses PDLC processes to Saint-Gobain Glass (France), Nippon Sheet Glass (NSG) and Isoclima (Italy). NSG produces a PDLC product known as "Umu^{TM}" for specialty automotive and building applications. NSG has a 24V ac type with T_v = 12-69% change. This lower voltage PDLC window could be powered with photovoltaics, using inverted dc to make ac power. Saint-Gobain produces PRIVA-LITE in a variety of colors and curved and flat shaped glass, for various applications. Large-area PDLC glazing can be fabricated in 0.9 x 2.4 m sheets. New UV stable formulations allow exterior application of PDLC films on buildings. Open circuit memory is generally not possible with dispersed liquid crystals. However, by adding dipoles to the liquid, a memory effect can be achieved. Some of the issues that remain are UV stability and cost.

6. Conclusions

Electrochromics are favored for many glazing applications, because when they switch they remain specular, and non-scattering. Electrochromics have been commercialized for automotive mirrors. Fairly large windows have been shown and installed in buildings by Flabeg. Other companies are working on the introduction of glazing products for automotive sunroof applications. Production cost and process simplification are major issues for large-area electrochromics.

Suspended particle devices (SPD) are more absorbing in the off-state compared to the on-state. SPDs can be made into a flexible sheet form so they can be used in a variety of applications. Research Frontiers, Inc., SPD, Inc. of Korea are working hard to produce glass based and flexible products. SPDs have an advantage of having much lower scattering in the off-state compared to PDLCs, so they can be used for a variety of applications. PDLC windows are now available from NSG in Japan with operating voltages of 24V ac (compared to 100V ac), and can be used for exterior building applications.

The range of switchable glass products is very encouraging to provide several types of glazing with both energy and glare control. Electrochromics have power requirements that match photovoltaics well. SPD and PDLC technology may be better suited to lower voltage operation using lower voltage smart windows. For SPD and PDLC, the requirements are for a.c. power, so an inverter must be used.

7. Acknowledgements

I wish to thank Prof. J. Marshall, Univ. of Wales Swansea, and Prof. D. Dimova-Malinovska, Bulgarian Academy of Sciences, for inviting me to lecture at the NATO Advanced Study Institute on Photovoltaic and Photoactive Materials-Properties, Technology and Applications in Sozopol, Bulgaria. I thank NATO for their support of this valuable lecture series.

For technical information, I thank H. Wittkopf and J. Cardinal, Flabeg, for their E-Control photos and product information. Also, thanks goes to H. Kawahara, NSG, J-C. Giron, Saint-Gobain; J. Harary, RFI; P. van der Sluis, Philips; M. Myser, SAGE; and F. Pichot, NTera for information on their respective switchable technologies. I further wish to thank D. Mently, Stanford Resources and K. Brown, Display Search for market information on displays.

8. References

1. Lampert, C.M. (2001) Progress in Switching Windows, in C.M. Lampert, C-G. Granqvist and K. L. Lewis (eds.), *Solar and Switching Materials, Proc. SPIE* **4458**, 95-103.
2. Lampert, C.M. (2000) Switchable glazings for the new millennium, Proc. Eurosun, Copenhagen, Denmark.
3. Lampert, C.M. (2000) Smart switchable materials for the new millennium - windows and displays, 43rd Proc. Soc. Vacuum Coaters (USA).
4. Lampert, C.M. (2000) Functional coatings-Displays and smart windows, in H.A. Meinema, C.I.M.A. Spee and M.A. Aegertner (eds.), Proc. 3rd Intern. Conf. on Coatings for Glass, Maastricht, NL.
5. Cumpston, C. (2000) Primary leaders assess the global market, *Glass Mag.* **50-10**, 38.
6. Limb N., (2000) Industry challenges: past & future, *Glass Mag.* **50-1**, 34.
7. van Roedern, B. (2001) NREL, Golden, CO, USA and PV Energy Systems, Inc., private communication.
8. Boeing (1999), Boeing Aerospace Annual Report.
9. Betelle, F., Boire, Ph. and Giron, J-C. Highly durable all-solid state electrochromic glazing, in C.M. Lampert (ed.), *Switchable Materials and Flat Panel Displays, Proc. SPIE* **3788**, 70-74.
10. Mentley, D.E. (2000) Display Market Development and Perspective, Soc. for Information Display, Lecture Notes M-1/7.

11. Young, R. (2001) Display Market Overview (Display Search), *Soc. for Information Display*, Lecture Notes M-1

12. Mahon, J.K., Brown, J.J. *et al.* (2000) Recent Progress in Flexible OLED Displays *Proc. Display Works.*

13. Graff, G.L. *et al.* (2000) Fabrication of OLED devices on engineered plastic substrates, *43rd Proc. Soc. Vacuum Coaters.*

14. Granqvist, C-G. (1995) *Handbook of Inorganic Electrochromic Materials*, Elsevier, Amsterdam, NL.

15. Granqvist, C-G. (2000) Electrochromic tungsten oxide films: Review of progress 1993-1998. *Solar Energy Materials and Solar Cells* **60**, 201.

16. Becker, H. and Wittkopf, H., (1999) Variable solar control glazing-an outstanding application for electrochromics, Proc.3rd Int. Meeting on Electrochromics, London, UK *Proc. Electrochim. Acta.* **44**, 3268.

17. Kobo, T., Toya, T., Nishikitani, Y. and Nagai, J. (1998) Electrochromic properties and durability of large-area electrochromic window fabricated with carbon based electrode *Proc. Electrochem. Soc.* **98**, 86.

18. Van der Sluis, P. and Mercier, V.M.M. (2001) Solid state Gd-Mg Electrochromic devices with ZrO_2H_y electrolyte, Proc. 4th International Meeting on Electrochromism, Uppsala, Sweden, in *Electrochim. Acta* **46**, 2167-2172.

19. Yu, B.S., Kim, E.S. and Lee, Y.W. (1997) Developments in suspended particle devices (SPD), in C.M. Lampert, C-G. Granqvist, M. Gratzel and S.K. Deb, (eds.), *Optical Material Technology for Energy Efficiency and Solar Energy Conversion XV. Proc. SPIE* **3138**, pp. 217-226.

20. Saito,Y., Hirata,M., Tada, H., .Hyodo, M. and Kawahara, H. (1999) Electrically switchable glazing using a suspension of TiOxNy particles, in H. Pulker, H. Schmidt, and M.A. Aegerter (eds.), *Coatings on Glass 1998*, Proc. 2nd Int. Conf. on Coatings for Glass, Elsevier Science, Amsterdam, NL pp. 433-436.

21. Drzaic, P., Comisky, B., Albert, J.D., Zhang, L., Loxley, A., Feeney, R. and Jacobson, J. (1998) "A printable and rollable bistable electronic display" *Soc. for Information Display Int. Symposium Digest* **29**, 1131.

22. van Konynenburg, P., Marsland, S. and McCoy, S. (1989) Solar radiation control using NCAP liquid crystal technology, *Solar Energy Materials* **19**, 27.

PHOTOVOLTAIC MATERIALS, AN OVERVIEW OF HISTORICAL DEVELOPMENT, CURRENT STATE-OF-THE-ART AND FUTURE SCOPE

D. DIMOVA-MALINOVSKA
Central Laboratory for Solar Energy and New Energy Sources,
Bulgarian Academy of Sciences, Blv. Tzarigradsko Chaussee 72,
1784 Sofia, Bulgaria.

1. Introduction

Photovoltaics (PV) is a semiconductor technology that directly converts sunlight into electricity. PV power generation systems are clean, and utilize an inexhaustible and renewable energy source that shows great potential with respect to resource considerations. PV materials are abundant, and the Earth receives 6000 times more sunlight energy than humans consume. Also, because the "fuel" - sunshine - is available everywhere to everyone, any nation that builds a PV infrastructure will be less vulnerable to international energy politics and volatile fossil fuel markets.

Photovoltaics is a versatile technology that can be used for many applications from the very small to the very large. It is modular, enabling electricity generating systems to be built incrementally to match growing demands. Its systems are easy to install, maintain and use. PV equipment can be used anywhere there is sunshine, and can be mounted on almost any surface on the earth or in space.

The general trend in research and development of PV elements is directed increasingly towards low cost devices. This will not only bring down the cost of predicted energy (which is still the major target), but will also allow many applications of solar electricity in the consumer goods market.

High manufacturing costs have limited the sales potential for photovoltaic modules, but nevertheless a true photovoltaic industry has developed. Figure 1 illustrates the worldwide growth of photovoltaic cell and module shipments since 1986, including the "traditional" crystalline Si technology-based shipments and thin film technology (amorphous Si and CdTe) ones [1]. It is clear that an exponential pattern of growth for this industry has been established in recent years. A forecast [2] demonstrates (Figure 2) that if the future PV market grows by 20% or 30% annually, PV will surpass today's nuclear energy capacity within 2 or 3 decades, in terms of installed power.

11

J.M. Marshall and D. Dimova-Malinovska (eds.),
Photovoltaic and Photoactive Materials - Properties, Technology and Applications, 11–48.
© 2002 *Kluwer Academic Publishers.*

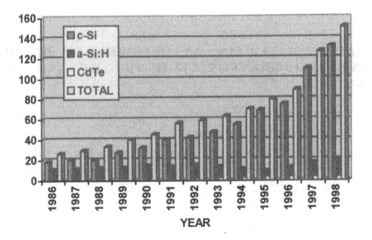

Figure 1. Annual worldwide PV shipments (MW) by various technologies [1].

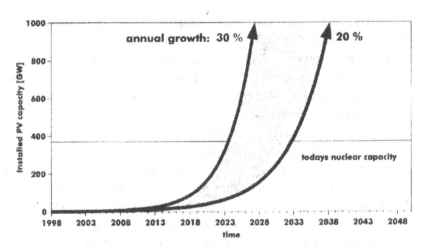

Figure 2. Forecast future growth of photovoltaics [2].

This paper will provide a review of the research and development of PV materials over the last 150 years. An overview of the present state of knowledge of the materials aspects of photovoltaic cells will be given, and new semiconductor materials with the potential for application in photovoltaic devices will be identified.

2. Historical Data

Regarding the development of photovoltaics, four district periods can be distinguished: incubation ("prehistoric"), research, demonstration, and industrial [3].

TABLE 1. Some dates of relevance to photovoltaic solar energy conversion.

Years	Development Steps
1817	Se discovered by Berzelius.
1821-1839	Si separated by Berzelius.
1840	Photovoltaic effect in an electrochemical cell (Bequerel).
1881-1900	Observation of photoconductivity in Se (Smith); Photovoltaic effect in Se (Adams and Day); Se photocells (Fritts, Uljanin)
1901-1910	Photosensitivity of the structure Cu-Cu_2O.
1911-1920	Photovoltaic effect in a barrier layer; Growth of Si crystals (Czochralski).
1921-1930	Cu-Cu_2O rectifier (Grondahl) and photocell (Grondal and Geiger); Solid state band theory (Brillouin, Kroning and Penney); Equivalent schema of the photocell (Schottky et al.).
1931-1940	Electron diffusion theory (Dember); Theory of the metal-semiconductor barrier (Mott; Schottky); Tl_2S photocells (1% efficiency) (Nix and Treptow).
1941-1950	Si p-n-junction photocell (formed during crystal growth) (Ohl); Theory of the p-n junction (Shockley).
1951-1955	Formation of p-n junctions by diffusion (Fuller); Si solar cell created (Person, Fuller and Chapin); CdS/Cu_2S solar cell developed (Reynolds et al.); Theory of solar cell (Pfan and Roosbroeck; Prince).
1956-1960	Improved theory of the p-n junction; Dependence of conversion efficiency on the band gap identified (Loferski, Rappoport, Wisocki); Theory of the spectral dependence of the photosensitivity, analysis of loss mechanisms (Wolf). Bell Telephone Labs: telephone pole power supply (Si); Si solar cells in space (Vanguard 1, Sputnik 3).
1960-1970	Ti-Ag contact deposited by evaporation; Si cells doped with Li; Model of CdS/Cu_2S solar cells developed.
1970-2000	PERL Si solar cells, GaAs, InGaAs, CdS/CdTe, $CdS/CuInSe_2$, Si ribbon and sheets, multicrystalline Si, a-Si:H, a-SiGe:H, a-SiC:H, μc-Si:H, pc-Si:H, organic semiconductors, C_{60}, TiO_2 dye photocells, nanomaterials, polycrystalline Si films.
21st Century	????

The first of these lasted from the discovery of the photovoltaic effect by Becquerel in 1839 [4] until the oil crisis in 1974, altogether about 135 years. One could call this the **incubation** or prehistoric period.

The most important events in the development of photovoltaic devices are given in Table 1. The discovery of the main effects took place over a long period - about 100 years [5]. In 1817, Berzelius discovered Se, and a few years later achieved the separation of Si. Smith observed photoconductivity in Se in 1837. In 1839, the photovoltaic effect was discovered, when Becquerel observed a photogalvanic effect in electrochemical cells. Such events resulted in a sharply increased interest in this scientific area, and subsequent activities included the study of the spectral dependence of Se photosensors, the development of the luxmeter, and the observation by Adams and Day in 1876 of the photovoltaic effect in a solid state Se structure.

Seven years later, the first Se photocell was designed by Fritts. Hallwacks reported photoconductivity in the system Cu-Cu$_2$O in 1904, and 10 years later the photovoltaic effect was related to the presence of a barrier layer. This discovery laid the foundations for the future development of the photovoltaic devices. However, not more then ten years later, a new active period started. The development of Cu-Cu$_2$O rectifier had attracted interest in its application as a photovoltaic device. As a result, the characteristics of this structure were studied in detail. Cells with front- and back-side barriers were created, and the first theoretical models were proposed. The model developed for the Cu-Cu$_2$O rectifier is still in use.

Later, the Se converter was developed and improved, and it replaced the Cu-Cu$_2$O device. A maximum conversion efficiency of 1% was reported, and the same value was observed for a Tl$_2$S photovoltaic device in 1941.

Following improvements in Si technology, a Si PV device was produced in 1941, by formation of a p-n junction during crystal growth. The possibilities for formation of a p-n junction by diffusion were unknown at this early stage. However, such a structure was developed 12 years later, and the crystalline silicon (c-Si) photocell was transformed into a device suitable for practical applications. Even in the first years of industrial production, the efficiency of c-Si photocells reached 6% [6]. Improvements in technology and design, plus the development of a theoretical framework, allowed progressive improvements in the operating parameters of this type of PV device, and in 1958 c-Si photocells demonstrated a 14% efficiency.

During this incubation period, researchers at the Bell Telephone Laboratories first used Si photovoltaic cells as practical sources of electricity. They installed a Si solar battery on a telephone pole, as a power supply for the repeater amplifier.

A more important subsequent application of solar cells was in spacecraft. In March 1958, the USA satellite Vanguard 1 was launched into orbit. It was the first to use solar cells to provide secondary electric power. Six small panels of crystalline Si solar cells delivered 50-100 mW to a Ge transistor oscillator, and provided a beacon signal throughout 6 years of operation. Analysis of the tracking data showed the Earth to be slightly pear-shaped. Two months later, soviet researchers launched the satellite Sputnik 3, using solar energy batteries to power the radio system - Mayak - which operated for 12,500 hours. In 1980, GaAs based solar cells were manufactured and used for critical space missions requiring higher operating power. For the Pathfinder mission to Mars, GaAs/Ge cells provided all the power for the Landers and the Sojourner Rover. In forty years, the power requirements per satellite have increased by more then six orders of magnitude. In addition to significant contributions to PV technology, space PV has provided a good example of the value of diverse niche markets in the widespread use of PV technology [7].

The efforts of many researchers were directed to developing the technology, and to studying and improving the properties of other semiconductor materials such as CdS, CdTe and GaAs. The viability of these materials for solar cell applications depended upon the development of thin film technology, which implied lower module prices. This incubation period lasted until 1974, when the oil crises occurred. For the first time in history, it was understood how fragile are the natural product foundations of world civilization. One of the consequences was intensive research activity in the field of renewable energy generally, and photovoltaics in particular. One could call this second phase the middle age or **research period**. During these years, many Institutes and Laboratories for R&D in the field of solar energy conversion were established.

Following the next shock - the Chernobyl accident in 1986 - efforts were directed towards demonstration of the productive power of PV and renewable energy on a viable scale: the **demonstration period** had started. In some countries, a number of enterprises extended their production capacities. In Germany, the USA and Japan, large programs for creating PV installations were funded. Thus, the long-term goal of large-scale production would finally be realized. In the 21st century, after the demonstration period, a new **industrial era** of photovoltaic development is beginning. However, this does not reduce the importance of research. On the contrary, it is well known that as relevant industrial activity increases, the role of research is also enhanced. For the industrial era of PVs to be successful, it is necessary to establish a strengthened relationship between research and industry, and to build up close international cooperation. **Let us hope that within the next hundred years, people can say: "We are living in the Solar Age".**

3. Solar Radiation

Solar radiative energy has its origin in a nuclear fusion reaction in the sun. The resulting energy is emitted mainly as electromagnetic radiation in the spectral range 0.2–3 μm (Figure 3). The intensity of solar radiation in free space at the average distance between the earth and the sun is called the solar constant, and has a value of 1353 W/m^2. The spectral distribution of the solar radiation which reaches the earth can be approximated by that of a black body at a temperature of 5800K. However, there is a sufficient departure from this idealized spectrum to make it desirable to use more exact data.

The sunlight is absorbed and scattered when it passes through the atmosphere on its way to the earth's surface. Basically, three sources of atmospheric absorption are important: atmospheric gases (O_2, N_2, etc.), aqueous vapor, and dust. The degree of attenuation is defined as the air mass (AM). This measures the atmospheric path length, m, relative to the minimum path length when the sun is at zenith and $m = 1/\cos\theta$, where θ is the angle between a line drawn through the observer and the zenith, and a line through the observer and the sun.

During the course of a day, θ varies from 90° to a noon minimum, θ_{min}. Furthermore, θ_{min} varies with the season of the year between the limits θ_{min} = latitude ± 23.50°. AM0 describes the solar spectrum outside the earth's atmosphere, AM1 (925 W/m^2) describes the situation at the earth's surface when the sun is at zenith, and AM1.5 that when the sun is 45° above the horizon. This spectrum (844 W/m^2) is used to characterize solar energy for terrestrial conditions. However, a standardized value of 1000 W/m^2 is used for simplicity in the characterization of solar cells and modules. The solar spectrum under different conditions is shown in Figure 3 [8].

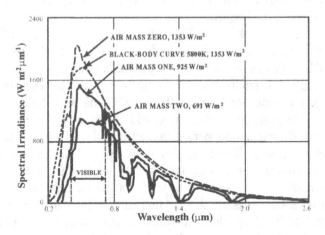

Figure 3. Solar spectral irradiance curves [8].

4. Requirements for Photovoltaic Materials

A solar cell can be considered the most challenging electro-optical device. The reason for this is that in order to produce efficient solar cells, the semiconductor materials used must have very good collection properties for both minority and majority carriers. In many other large area electronic devices, performance is predominantly governed by the majority carrier properties, which pose less stringent constraints on device optimization schemes.

A number of properties are required for candidate PV materials and device structures. The most essential ones concern photonic and electrical conditions:

i) Strong light absorption over a large spectral range. This property implies that a tunable band gap is desirable. The peak of absorption should be at 1.4–1.5 eV, for optimal efficiency.

ii) Good carrier collection properties for both minority and majority carriers, a low carrier recombination loss (in the bulk, at grain boundaries and at the front and back surfaces), and a large luminescence yield.

iii) Low cost, so that the thin film structures are preferable.

iv) Stability as functions of both time and illumination conditions (stable metal contacts, resistance to corrosion).

v) High abundance of the source materials (for large-scale production).

vi) Environment friendly technology.

PV materials and structures should meet as many of these desired properties as possible. However, it is a difficult task to meet all of them at once, and long range research efforts will be necessary.

A large number of theoretical considerations govern the choice of the optimal semiconductor for PV solar energy conversion [9-11]. The theory of the PV effect can be used to predict the characteristics of a semiconductor that will operate with optimum efficiency in a solar cell. The criteria for such an optimal material result from the interaction between the optical properties of the semiconductor, which determine the fraction of solar spectrum to be utilized, and its electrical properties.

Considerable attention is devoted to the effect of the forbidden energy gap (E_g) of the semiconductor. It has been shown by Loferski [9] that maximum efficiency, η_{max} (defined as the ratio of the maximum electrical power output to the solar power arriving on unit area), is given by:

$$\eta_{max} = Q\,(1-r)(1-\varepsilon^{-\alpha d})\,[\lambda\,V_{mp}/(1+\lambda\,V_{mp}]\,[en_{ph}(E_g)\,V_{mp}/N_{ph}E_{hv}]. \quad (1)$$

Here, Q is the collection efficiency (the ratio of the carriers passing through the circuit to those which have been generated by light), r the coefficient of reflection, α the absorption coefficient, d the thickness of the absorbing semiconductor, e the electronic charge, $\varepsilon^{-\alpha d}$ the fraction of the radiation transmitted, $n_{ph}(E_g)$ the number of photons/second/unit area of p-n junction whose energy is great enough to generate hole-electron pairs in the semiconductor, N_{ph} the total number of photons in the solar spectrum, V_{mp} the voltage at maximum power, and E_{hv} the average energy of such photons. Since $n_{ph}(E_g)$ decreases with E_g, while V_{mp} increases with E_g, it is evident that η_{max} will pass through a maximum as a function of this parameter [9].

Figure 4 shows the result of a calculation of the theoretical limit of conversion efficiency under AM1.5 global radiation, for several candidate solar energy materials. The theoretical limit means that the calculation has been made on the basis of 100% quantum efficiency of photo-carrier generation above the fundamental absorption edge [12]. Carrier recombination and ohmic loss factors have been neglected. The circle on each line specifying a particular material shows the experimentally-obtained maximum efficiency at room temperature. It is evident from the figure that there is still a large potential for improvement of efficiency, except in the cases of crystalline Si and GaAs. The differences between the values obtained in the laboratory and the theoretical values indicate the level of technological maturity of the respective semiconductor materials.

The progress made over the last 30 years is in the optimization for maximum efficiency of all the parameters of the PV materials and structures. Such advances are based on fundamental research, on computer simulations and on improved processing technology. Much has been learned from developments in microelectronics, using the same materials.

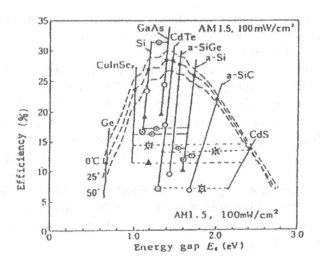

Figure 4. Theoretical limit of solar cell efficiency for various materials (solid circles). The triangles and open circles show the experimentally-obtained highest efficiencies on single junction cells and on mass production cells. The spotted and starred circles represent the highest values for tandem-type cells using the same materials [12].

A distinction has to be made between efficiencies obtained in the laboratory and in industrial production. For the former, there are no cost constraints on the choice of the materials or the fabrication process. In industrial production, other important factors include the large scale availability of feedstock material, the toxicity of the materials and the production process, and the costs of the material and process.

5. Current State-of-the-Art

Low-cost high-efficiency solar cells are the key to the large-scale applicability of a PV system. The cost of modules must decrease to less than 3 Euro/W_p in order to realize the goal of producing PV electricity (0.16 Euro/W_p). This goal can be achieved by various combinations of module cost reduction and increased efficiency. Research, therefore, is being conducted on various photovoltaic materials that can produce efficiencies in the range 10 -35%, at various costs.

PV technologies presently fall into three categories:

i) Crystalline silicon flat plates (single crystals and multicrystalline silicon);

ii) Si-based thin films (polycrystalline silicon, amorphous silicon and its tetrahedral alloys, protocrystalline and microcrystalline silicon) and polycrystalline films of CdTe, CuIn(Ga)Se$_2$, and

iii) GaAs and a number of concentrator approaches (including Si, GaAs, GaAlAs; and a multijunction approach using III-V alloy material combinations).

In the following sections, only the first two categories, which are important for terrestrial applications, will be reviewed. Additionally, the tendency to involve new materials, like nanomaterials, in PV applications will be presented.

5.1 CRYSTALLINE SILICON BASED TECHNOLOGIES

Crystalline silicon (c-Si) based technologies dominate today's PV production (Figure 1). c-Si is presently the most mature and best studied candidate for terrestrial PV applications. It offers several advantages over other PV materials, including abundance, an established technology base, high material quality and stability, and good surface passivation characteristics. The obvious disadvantages are the indirect band gap and the current high processing cost for silicon material and devices. Hence, research continues both in improving silicon and in studying and developing other Si-based technologies.

The vast majority of c-Si modules are fabricated on wafers sliced from Czochralski (Cz) single crystal ingots, or from polycrystalline ingots cast in a graphite or quartz crucible. Both growth techniques require relatively large amounts of energy to melt the silicon charge, and subsequent slicing into wafers results in a loss of some 70% of the silicon material. Most economical advantages were obtained by the application of techniques such as unidirectional wire-sawing cutting. This has increased the number of usable wafers per inch of ingot from 20-30 to 40-50, with a typical wafer thickness of 300 μm. A smaller number of crystalline Si cells are based on wafers grown as Si ribbons or as thick Si films deposited on conducting ceramic substrates.

A model calculation to specify various combinations of c-Si parameters (resistivity, minority carrier lifetime, surface texturing and recombination velocity) to reduce the discrepancy between present and practically-achievable silicon cell efficiencies has been advanced [10]. All other material parameters, such as carrier mobility and diffusivity, photon absorption, intrinsic recombination coefficients, etc. were kept constant, at acceptable values, in these calculations. Figure 5 shows the dependence of the calculated efficiency of a 200μm thick n^+-p-p^+ silicon solar cell (Si resistivity $\rho \geq 0.2$ Ωcm), assuming that the surface- and back-face recombination velocities are both zero. The conversion efficiency, η, is higher for a surface with a reduced surface recombination velocity, produced using a textured surface (inverted pyramids ($\eta = 26\%$)). It is a maximum (28%) if the light trapping is 100%. More recent models, however, predict higher values for η, as presented in Table 2 where the parameters for the best achieved solar cells are also shown.

The maximum reported efficiency for c-Si is for cells produced under laboratory conditions, where processing costs are not of major importance. The process implies high-quality materials and techniques to reduce energy-conversion losses

20

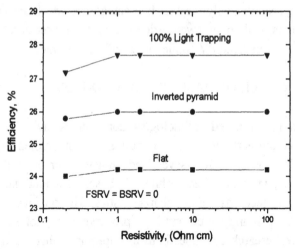

Figure 5. Importance of light trapping in high efficiency silicon solar cells [10].

TABLE 2. Parameters of the ideal silicon cell, and the best and achievable parameters
of a PERL (passivated emitter and rear locally diffused) cell (at AM1.5)

	J_{sc} (mA/cm^2)	V_{oc} (mV)	FF (%)	η (%)
ideal cell [13]	43.0	769	89	29
best Si crystalline PERL[14]	40.8	708	83.1	24.4
best Si multicryst. PERL [15]	38.1	654	79.5	19.8
achievable cell [16]	42.5	730	84	>26

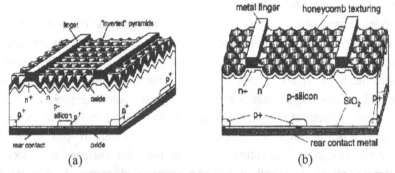

(a)   (b)

Figure 6. Structures of PERL cells: (a) c-Si with a pyramidal textured surface [17], and
(b) multicrystalline Si with a honeycomb [14] textured surface.

and to approach the fundamental efficiency limit. These cells usually have a small
(2-4 cm^2) area, and are produced on high-quality expensive float zone (FZ) silicon.
Today, the most efficient c-Si solar cell (4 cm^2) is a "passivated emitter and rear
locally diffused" (PERL) structure (Figure 6), showing an efficiency of about
24.4% with a value for a module (778cm^2) of 22.74% [14,17]. The main features of
this cell are: front surface reflection loss reduced by inverted pyramids, high quality

emitter-diffusion profile and high quality passivating thermal oxide on the front and back surfaces to reduce surface recombination losses, small front-contact fingers and localized rear p^+ diffusion to reduce the contact contribution to total recombination losses. A similar approach has been applied for a multicrystalline silicon solar cell, where "honeycomb" front surface texturing is performed (Figure 6b). As a result, multicrystalline Si solar cells with 19.8% efficiency have been produced [15].

Although the highest efficiency of the laboratory c-Si is observed for FZ material, very encouraging results have been achieved on Czochralski (Cz) Si wafers. An efficiency of 22% on 6.8 Ωcm Cz c-Si of 2x2 cm^2 area has been reported [18]. A 19.6% efficiency has been achieved in Cz 23.4 cm^2 c-Si cells, on a 10 Ωcm substrate with a randomly textured emitter [19]. These results show that high efficiency is possible for less expensive Cz crystalline silicon. They are thus very important for industrial production.

Despite significant progress in the slicing technique having been achieved, there is still about 200 μm of high quality silicon loss per wafer. This can be avoided in the ribbon and sheet silicon technologies. Although many technologies have been explored, only edge-defined film-fed growth (EDFG) polycrystalline silicon sheets have been used in high-volume production [20]. Thick films are prepared from Si "crystals" grown in the form of a hollow octagonal tube (5 m in length) with eight 10 cm wide faces and a tube-wall thickness of 300 μm. The faces are cut with a laser to appropriate lengths for cell production [20]. An efficiency of 9.1% has been announced by Texas Instruments, for a prototype developed on sheet Si layers [21].

5.2. THIN FILM PHOTOVOLTAIC MATERIALS

Thin film PV technology development is related to the advantages it proposes:

  i) lower usage of expensive semiconductor materials,

  ii) use of low cost substrates;

  iii) the possibility of deposition on large area cells and modules (without using soldering techniques for interconnecting them).

All this suggests a substantial reduction in solar cell manufacturing costs. Although a large number of materials have been explored for the preparation of the solar cells, only hydrogenated amorphous silicon (a-Si:H) cadmium telluride (CdTe) and copper indium diselenide (CuInSe$_2$ - CIS) thin film PV technologies have so far been commercialized. The maximum efficiencies obtained from the best thin film PV modules are presented in Table 3. A maximum efficiency of 12.1% is observed for a CdS/CIS alloy PV module produced by Siemens Solar [1]. According to the data compiled by the NREL Thin Film Partnership, CIS cells have the potential to achieve a performance level comparable to that of diffused Si wafer based cells [1].

TABLE 3. Best results for thin film solar cells (efficiency %, AM1.5 illumination) [1,22,23]

Material	Device	Best large area ($0.1m^2$)	Best R&D small area	Theoretical limit
Amorphous Si	single p-i-n	8.5-10	9-13	~20
Amorphous Si	dual junction	7.6	13-14.4	
Amorphous Si	triple junction	7.9	15.2	
CdTe/CdS	heterojunction	8.4 - 9.2	16	28
CIS	CdS/CIS	11-14	15-18	~45
Thin film Si		11	16	25

5.2.1. *Epitaxial Thin Silicon Films*

There is growing interest in thin silicon films, because of advantages in cost. The use of thin film Si for solar cells offers the potential for combining the high efficiency and stability of c-Si with the potential low cost of thin films. The requirements for thin film Si cell development can be summarized as:

i)   Silicon film thickness: 5 to 50µm.

ii) Light trapping.

iii) Planar film with single crystal grains at least twice as wide as thickness.

iv) Minority carrier diffusion length at least twice the thickness.

v) Back surface passivation.

vi) Passivated grain boundaries.

vii)Substrate to provide mechanical support and act as the back plane conductor, or to be removed for multi-deposition.

In this technology, low cost substrates can be used: metallurgical-grade silicon, stainless steel, graphite, ceramics, or even glass. One of the key technological challenges to achieving a commercially viable thin layer polycrystalline silicon solar cell technology is the development of a low cost supporting substrate. The requirements for the substrate material are severe:

i)   mechanical strength,

ii) thermal coefficient of expansion matching, to prevent the film from breaking or deforming during handling and high-temperature processing,

iii) the substrate must also provide good wetting and nucleation during the growth process, without contaminating the film,

iv) the substrate can be conducting or insulating, depending on device requirements,

v) the substrate-silicon interface must provide a high degree of diffuse reflectivity and surface passivation, for high performance.

Because of the lack of reliable cost data, there is not yet a well-established opinion about the most suitable deposition techniques and substrates to be used. The methods of thermally assisted Chemical Vapor Deposition (CVD) and Liquid Phase Solution Growth (LPSG) are applied, but require a high (>800°C) deposition temperature. Plasma ion-assisted deposition (PACVD), hot wire CVD and electron cyclotron resonance PACVD require some addition source of energy, but use temperatures below 700°C [24]. A separate group of techniques consists of genuine solutions, like Liquid Phase Epitaxy (LPE) - electrodeposition from molten salts, the epi-lift-off approach and the quasi-monocrystalline Si approach. These have in common that a high-quality monocrystalline Si substrate can be re-used several times, as a template or part of the active layer of a thin film crystalline Si solar cell.

Although these techniques are newly developed, promising results have been reported for the efficiency of a small-area Si thin film solar cell (14%) [24], and for large-area (25cm^2) n/p on p$^+$ c-Si substrate devices (13%).

Recently, an interesting new technique has been introduced to obtain directly textured monocrystalline silicon thin films for high efficiency thin film silicon solar cells on ceramic substrates [25]. Such structures can be produced using the following steps (Figure 7):

i) Thin silicon grid formation by photoelectrochemical etching, (Figure 8a).

ii) Grid release from the silicon bulk substrate.

iii) Transfer by pasting onto a ceramic substrate (mullite closely matches the substrate requirements); Pasting is realized at 1100°C with a silicate glass, allowing p$^+$ in-situ doping of the grid.

iv) LPE growth of the silicon active layer at 925°C (Figure 8b). Silicon crystals with (111) facets coalesce together, and fully cover the grid holes, leading directly to a pyramidally-textured thin film.

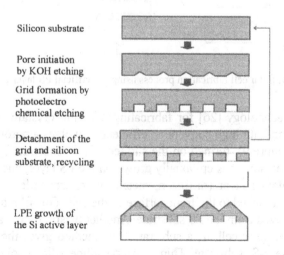

Figure 7. Process steps for pyramidally textured silicon thin film formation [25].

24

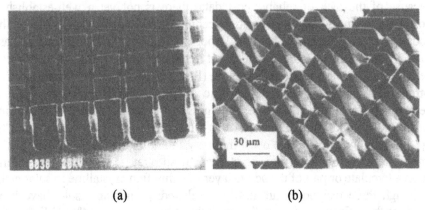

Figure 8. SEM of a silicon grid standing on a Si substrate with 30μm pore lateral dimensions [25] (a) and of the pyramidal Si epi-layer grown on such a Si grid (b).

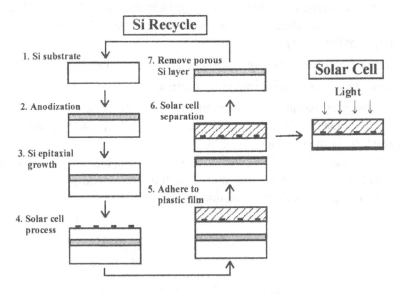

Figure 9. Solar cell fabrication process using a sacrificial PS layer [26].

Another new technology [26] for fabricating thin film crystalline silicon solar cells at low cost and with high conversion efficiency involves a porous silicon (PS) sacrificial layer (Figure 9), formed on the surface of a Si substrate by anodization. Thin film crystalline silicon is epitaxially grown on the PS layer, and a solar cell is formed in the epitaxial layer (diffusion, metallisation, anti-reflective coating). A plastic film is then attached to the front surface of the cell. This film and the formed cell are easily removed from the PS substrate. Another plastic film is attached to the back surface of the solar cell, as a substrate. The method gives the possibility of multiple use of the c-Si substrate. Thin film crystalline silicon solar cells having 12.5% efficiency (4 cm^2 area) have been prepared using this technology [26].

5.2.2. *Amorphous Silicon Thin Films*

Hydrogenated amorphous silicon (a-Si:H) thin films were proposed as advanced PV materials 25 years ago, after Spear and le Comber reported on successful n- and p- doping [27]. a-Si:H and related amorphous alloys, including those with Ge and C, allow the preparation of materials with different energy gaps in a large range - from 1.0 to 2.8eV. Amorphous silicon is a semiconductor with a direct energy gap, and consequently it has a large absorption coefficient. This advantageously allows very thin films (<1μm) to be used for photovoltaic devices.

a-Si:H is a non-crystalline solid, lacking long-range periodic ordering of its constituent atoms. However, it does have local order on an atomic scale. This short-range order is directly responsible for the observation of semiconductor properties such an optical absorption edge and an activated electrical conductivity. A distinction should be made between amorphous and polycrystalline materials. Polycrystalline semiconductors are composed of grains, each containing a periodic array of atoms surrounded by a layer of interconnective or boundary atoms. For progressively smaller grains, such as microcrystallites, the surface layer of a grain contains progressively larger numbers of interconnective randomly distributed atoms, relative to the periodically-arrayed interior atoms. Consequently, a semiconductor containing a large number of very small grains embedded in the amorphous phase of the material is called a microcrystalline semiconductor. It lies on the borderline between the amorphous and polycrystalline phases, and the difference between polycrystalline and microcrystalline semiconductors is in the amount of the amorphous phase.

A major disadvantage of a-Si:H has been the degradation of films and devices under illumination. In undoped films of very high quality, light exposure causes a reversible decrease of the dark- and photo-conductivities. The reason for this (Staebler-Wronsky) effect is an enhancement of the defect density to values of about 10^{17} cm^{-3} [28]. Most results are in agreement with a model in which recombination at weak bonds results in bond breaking. The broken bonds are stabilized by a mechanism involving the motion of hydrogen atoms. Many studies have been directed towards minimizing the metastability, by proper control of the deposition conditions, thereby influencing the microstructure and the density of weak bonds. Heavy hydrogen dilution of the process gases has been shown to lead to more stable material, with microcrystalline inclusions and a low density of micro-voids [29].

Amorphous Si films can be deposited by various methods, such as chemical vapor deposition (CVD), plasma enhanced CVD (PECVD), hot-wire CVD, electron cyclotron resonance (ECCVD), sputtering etc. The predominant approach for deposition of a-Si:H and related alloys is PECVD. Research related to other methods of deposition has a more fundamental character. It offers information about the control of the properties and stability of the material, and about the mechanisms limiting device performance.

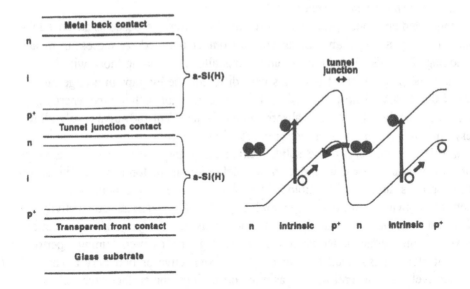

Figure 10. A double junction device structure (left), and the
band diagram of an a-Si:H stacked solar cell (right).

A p-i-n structure is standard for a-Si:H solar cells. It follows the classic recipe for a good cell: a well-chosen insulating absorber layer and contacts that are ohmic for electrons (holes) and blocking for holes (electrons). Another approach is proper device engineering. The proposal is to use stacks of p-i-n structures, where carriers are readily extracted by the high electric field in each thin i-layer (Figure 10).

The advantages of such structures are the following:

i) the carrier collection is enhanced through an increase in the internal electric field, since the thickness of the single cell is reduced in a tandem structure.

ii) a stack structure can be made, using materials with different band gaps, allowing better use of the solar spectrum.

The top cell, which absorbs the shortest wavelengths of the solar spectrum, is a-Si:H with a band gap of 1.8eV. p- and n-type a-Si:H thin layers serve as electrodes to collect generated carriers (Figure 11). An a-SiGe:H layer with a 1.55-1.6 eV band gap forms the i-layer of the middle cell. The bottom cell, using i-type a-SiGe:H with about 50% Ge and a bandgap ≈ 1.4 eV, absorbs the red part of the spectrum. As a back electrode, a textured Ag/ZnO back reflector is used, to enhance light trapping. The highest value reported for the efficiency of a laboratory solar cell (0.25 cm^2 area) based on a-Si:H is 15.2% initially, and 13% after stabilization [30]. Commercial stable efficiencies are about 7.5%. The large difference in efficiency is not surprising, because a higher deposition rate was chosen for production (0.3 nm/s instead of 0.1 nm/s), in order to reduce manufacturing costs.

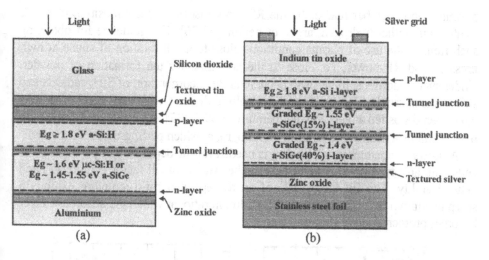

Figure 11. a-Si:H based single and triple-junction device structures.

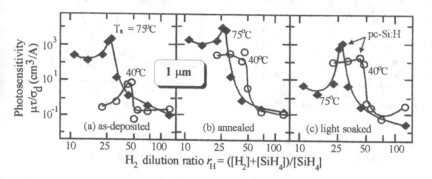

Figure 12. The ratio $\mu\tau/\sigma_d$ of pc-Si:H (1µm thick): as deposited (a), after 110°C annealing (b), and after AM1.5 light soaking (c) [37].

The principal technical difficulty with these cells is the formation of a tunnel junction between the n-layer of the upper cell and the p-layer of the lower cell. Good results can be achieved in the laboratory on small-area cells, but it is a major challenge to achieve similar levels of performance reliably and repeatable in commercial production on large-area modules.

The use of high hydrogen dilution during the preparation of the i-type a-Si:H layer results in the deposition of better quality films, with some medium range order or even monocrystalline inclusions [31]. Recently, many research groups have proposed such a microcrystalline material, µc-Si:H, as a bottom layer with a lower band gap - the so-called micromorphous tandem. Such structures show better stability, and an efficiency of about 12% [32-34].

Si:H with modified properties, called polymorphous (pm-Si) [35,36] or protocrystalline (pc-Si) [37,38] silicon, has been successfully used as the intrinsic layer in single junction p-i-n solar cells. The pm-Si:H was prepared in a multi-plasma mono-chamber reactor by the RF glow discharge decomposition of silane diluted with either hydrogen, argon or helium [35,36]. The guideline for obtaining such films is the use of plasma conditions close to a full dilution of silane at high pressure and RF power, i.e. close to the condition for the formation of powder. Under such conditions, the growth occurs via the incorporation of SiH_x radicals and silicon nanoparticles. Recently, protocrystalline Si films with very good properties have been deposited at very low temperatures $<80^{\circ}C$ [37]. As in Figure 12, the ratio $\mu\tau/\sigma_d$ exhibits a peak at the H_2 dilution ratio, r_H, at which pc-Si growth is expected.

At lower temperatures, a higher r_H is needed for the deposition of pc-Si:H. Solar cells with an i-type pc-Si:H layer have higher values of fill factor (Figure 13) and better stability. The fill factor, FF, of an as-deposited p-i-n structure exhibits a sharp maximum in a certain range of H_2 dilution ratio, at which the absorber i-layer becomes protocrystalline (Figure 13 a).

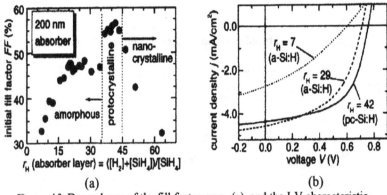

(a)   (b)

Figure 13. Dependence of the fill factor on r_H (a), and the I-V characteristic under illumination (b) of p-i-n solar cells with a pc-Si:H i-layer [37].

It has been shown that the film thickness at which the amorphous/microcrystalline transition occurs is, however, a function of the hydrogen dilution gas flow ratio, r_H (Figure 12). This material, applied as the i-layer in p-i-n structures, results in improved cell performance and stability (Figure 13). The use of this polymorphous material in a p-i-n solar cell with an optimized p-type a-SiC:H layer has resulted in a device with a stable efficiency (Figure 14). This is supposed to be the result of an equilibrium between the creation of a small number of defects and the activation of boron atoms in the p-layer.

Compared to a-Si:H, an enhanced photosensitivity and stability against light-soaking makes pc-Si:H a promised candidate for photovoltaic applications. However, the improved medium-range order and associated lower defect density of pc-Si:H are not well understood at present. Guha *et al.* [30] suggest that the electronic properties are related to the improved medium-range order and the associated lower defect density. Kamei *et al.* [39] and Tsu *et al.* [40] discuss nanometer sized silicon crystallites as recombination centers of excess carriers embedded in an amorphous matrix, to suppress light-induced defect creation.

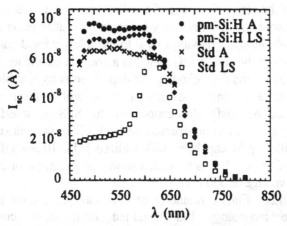

Figure 14. Spectral dependence of the short circuit current of a p-i-n solar cell with pm-Si:H and a-Si:H films (Std A) before (A) and after light soaking (LS). The pm-Si:H cell features excellent stability compared with a standard a-Si:H cell [35].

TABLE 4. Different types of device-quality thin silicon film, containing hydrogen

Material type Method of deposition	Substrate T, °C	Deposition rate, Å/s	Properties
Amorphous Various	100-300	>6	Good quality. DOS=10^{16}cm^{-3}. Short range order.
Protocrystalline and Polymorphous RFCVD	<200	1 H$_2$ dilution	Best quality. DOS<3×10^{15} cm^{-3} Short range order: 1 -2 neighbors
Microcrystalline Sputtering, hot-wire, MW HF PECVD	>200	>6 H$_2$ dilution 4-5 H$_2$ dilution	Lower density. Voids and (SiH$_n$)$_m$ complexes, 10-50 nm crystallites, 20%μc phase. Greater DOS. Longer range order: 2-3 neighbors. 5-10 nm crystallites. 50-70%μc phase
Polycrystalline hot-wire MW	>400		4% hydrogen in the film. 10-100nm crystallites. 90% polycrystalline phase.

Considering the structure and quality of a thin silicon film containing H, several types of material can be produced, using different methods and conditions of deposition (Table 4). The existence of protocrystalline layers is claimed to come from the indirect measures and physical continuity philosophy [38]. It is known that hydrogen dilution stimulates microcrystallisation. It can lead, through voids and big (SiH$_n$)$_m$ complexes, to large crystallites and a non-uniform layer, which is undesirable. Alternatively, at low sputtering rates, scattered point defects can induce high levels of nanocrystallisation, and such layers show excellent optoelectronic properties. With a relatively low hydrogen dilution, r$_H$, of 20-25, the growth of a-Si:H films not only becomes dependent on the substrate but also changes the growth, making the microstructure thickness dependent. As a result, during growth, the initially amorphous materials eventually become microcrystalline.

The reduction of the active thickness of a PV cell results in a decrease in short-circuit current. To avoid this, it is necessary to increase the absorption in the device materials. One possibility for achieving this is to use a lower band gap material in the active base region of the cell. A SiGe alloy is a possible candidate. Its band gap is smaller than that of Si, and the technology of deposition is compatible with the Si cell processes. Despite of these advantages, it is still unclear whether the unavoidable presence of misfit dislocations in the Si/SiGe interface can be overcome. Recently, a heterojuction structure with a CVD-grown relaxed layer on a p^+c-Si substrate with a p-Si cap layer was realized [41]. 10.3% efficiency was reported for a cell of 15 μm thickness, with a short circuit current of 24.2 mA/cm² and an open circuit voltage of 560 mV.

Hydrogenated silicon films continue to be important prospective materials for photovoltaics. The technology has benefited from many comprehensive fundamental research efforts, as well as from the use of a-Si:H in non-photovoltaic large-area thin film applications such as displays, sensors etc.

5.2.3. Polycrystalline thin films

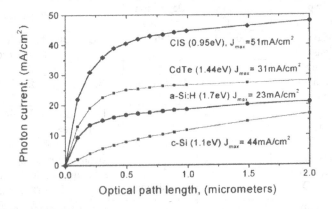

Figure 15. Photocurrent generated in CIS, CdTe, a-Si:H and c-Si solar cells at AM1.5, calculated from the optical absorption (no reflection losses) [42].

Polycrystalline solar cells are noted for their ability to absorb and collect most of the available solar spectrum. Typically, direct energy gap materials are considered, which strongly absorb the solar spectrum within a few micrometers depth. Figure 15 compares photocurrents generated in CIS, CdTe, thin film a-Si:H and crystalline silicon, as calculated from the optical absorption coefficient and assuming no reflection losses at 100 mW/cm² (AM1.5) [42]. The photocurrent increases as a function of optical path length (absorber layer thickness) and J_{max} is the maximum available photon current density for the material. CuInSe$_2$, with a 0.95 eV optical band gap (J_{max} = 51 mA/cm²) and CdTe with a 1.44 eV gap (J_{max} = 31 mA/cm²) both absorb 90% of the available photons in a depth of 1 μm, so film thicknesses of only 1-3 μm are sufficient for thin film PV applications.

Since common polycrystalline device types are heterojunctions, where a wide band gap window layer allows the light to be absorbed on or close to the space charge region, the minority carrier diffusion length can be 1 μm or less. The requisite minority carrier lifetime can be in the picosecond to nanosecond range, to achieve 15% efficiency. This contrasts with the case of crystalline silicon, which requires 10 μsec - 1 msec carrier lifetimes.

Thin film polycrystalline solar cells with a total device thickness of five micrometers or less contain significant levels of chemical, structural and electronic impurities, inhomogeneities and defects. Polycrystalline materials are compatible with several large-scale deposition technologies including vacuum evaporation, sputtering, electro-deposition, chemical spraying and sintering. However the spatial non-uniformities in thickness and composition, and the physical and chemical interactions at the interfaces, do impact upon the resultant device performance.

Cadmium telluride films.   Many of the basic properties of CdTe make it an ideal material for use in thin film solar cells. For example:

i) its energy gap is direct and its value of 1.45 eV supports a maximal conversion efficiency (Figure 4).

ii) it may be doped n- or p-type, and the preparation of the stoichiometric compound may be achieved easily in the production process.

iii) CdTe thin films can be deposited at high rates by various methods: sublimation/condensation (S), close spaced sublimation (CSS), chemical spraying (CS), electro-deposition (ED), screen printing (SP), chemical vapor deposition (CVD) and sputtering.

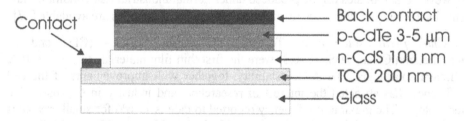

Figure 16. A typical CdS/CdTe solar cell structure.

The most commonly-used solar cell structure (Figure 16) employs a CdS/CdTe heterojunction. The n-type CdS is deposited onto glass coated with a transparent conductive oxide (TCO - In_2O_3, SnO_2, ZnO). It is preferable to deposit the CdS and CdTe using the same technique. The back contact is deposited over a p-CdTe film.

The carriers are generated in the field zone in the p-CdTe. Since there is no field in the highly doped CdS, and since the minority carrier lifetime is very low, the CdS layer is not photoelectrically active. Its thickness has to be minimized in order to reduce absorption losses.

The most critical point in CdTe cell technology is the formation of good ohmic contacts of high stability. Many different materials are used: Au, Cu/Ag, Ni, Ni/Al, ZnTe:Cu and graphite. The physical problems are that a metal of very high work function is required, and that CdTe cannot easy be doped p-type because of a strong tendency to self-compensation.

The quality of the CdTe/CdS heterojunction is a critical part of the solar cell. An important problem is the inter-diffusion of CdS and CdTe. It has been found that mixed CdS_xTe_{1-x} alloys are formed at the interface [43]. CdS diffusion is a limiting factor for both fabrication control and device performance. It has been observed that Cl treatment results in electrical passivation of the grain boundaries in the polycrystalline CdTe film, and enhances carrier lifetimes. Laboratory scale development has demonstrated a world record efficiency of 16% in a device deposited by the close space sublimation (CSS) technique [44]. For such solar cells, borosilicate glass substrates were used, because the deposition was performed at high temperature, 600°C. Using less expensive soda-lime glass, a lower efficiency of about 12% was obtained. On the basis of this technique, a commercial production line has been established for CdS/CdTe solar cells.

An interesting n-CdS/i-CdTe/p-ZnTe solar cell structure has also been proposed [45]. One advantage is that it uses each material in its natural form: e.g. CdTe prefers to be insulating, CdS - n-type, and ZnTe - p-type. It also avoids the problem of ohmic contacts to the CdTe, and increases the carrier collection efficiency due to the built-in field in the CdTe absorber. A higher short circuit current ($26mA/cm^2$) was achieved by the use of an optically-thin CdS layer, and an open circuit voltage of 840 mV was obtained, using a solution-growth CdS deposition process [46].

A disadvantage of CdTe solar cells concerns environmental and safety problems, since the product contains Cd, a toxic element. It has been shown, however, that modules can be produced under normal industrial risk conditions, and suitable techniques for waste treatment and recycling of modules are available [47].

Chalcopyrite solar cells. Ternary chalcopyrite $CuInSe_2$ (CIS) and its modification Cu(In,Ga)Se (CIGS) were the first thin film materials to achieve 10% efficient solar cells. The device stability, together with improvements of the cell efficiency, has attracted the interest of researchers and industry in CIS-based PV technology. The maximum efficiency reported to date is 17.6% for small area cells [48]. For large modules, the efficiency is 13.6% (for 50 cm^2) [49] and 13.9% (for 90 cm^2) [50].

A schematic cross section and the device parameters of a solar cell made using chalcopyrites are shown in Figure 17. Soda lime sheet glass with a thickness of 2-3 mm is used as a substrate. The Mo back contact is deposited by magnetron sputtering. The back contact is patterned by a Nd:YAG laser. Instead of CdS, other compounds like In(OH,S) can be applied to realize a Cd-free device [51]. This buffer layer adapts the absorber surface to the ZnO; this material having been found to be most suitable as the n-type transparent front electrode. The semiconductor patterning is realized by mechanical scribing. Modules are finished by bonding electrical contacts and glass encapsulation, using organic adhesives. In high

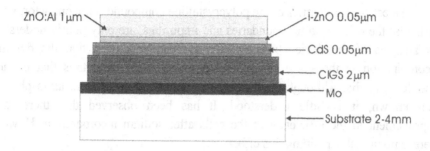

Figure 17. Cross section and device parameters of a chalcopyrite solar cell

efficiency cells, a CdS buffer layer is used, usually deposited by chemical bath deposition (CBD). This carries some disadvantages such as absorption losses, conduction band misalignment in higher band chalcopyrites and the toxicity of Cd. A preparation technology other than CBD is desirable, because of difficulty in integrating it into an in-line process. Proper band line-up at the interface is an important requirement for the potential use of the alternative buffer layer [52].

The most difficult step, and the key issue, in the fabrication of high quality and low cost CIS and SIGD based thin film modules is the deposition of the chalcopyrite layer. Several methods for its formation lead to devices with efficiencies exceeding 14%. These include simultaneous evaporation of the elements and selenization/sulfurization at high temperatures of sputtered or evaporated precursors in a H_2S/H_2Se atmosphere, or diffusion and crystal growth under atmospheric pressure. Other CIS and CIGS film deposition techniques like electro-deposition, screen printing, MOCVD etc. have led to moderate device qualities. The reproducibilities of the co-evaporated layers, and the homogeneity over large areas, are often poor. The stacked elemental layer method involves sequential deposition by vacuum evaporation of the elements comprising a compound or alloy. The elemental layers must have appropriate thicknesses, in order to fix the desired stoichiometry of the material. Normally, the deposition of a stack of sandwiches of the elements is required, in order to attain the required thickness of the film. Deposition is followed by a simple thermal annealing step in vacuum or inert gas, in order to synthesize the material and adjust the stoichiometry. The main advantages of this method are:

i) It is easier to deposit a uniform layer of a single element at the time than a combination of elements simultaneously, and to maintain the same composition across the substrate.

ii) If individual layers are deposited by the same method, using a similar geometry, then the resulting thickness variations are similar (this maintains the area uniformity of the final composition, which is critical, at the expense of overall thickness, which is not).

iii) There is design freedom to deposit individual layers by different techniques, if individual layer thickness control is not an issue.

The electronic properties of the polycrystalline compound semiconductor, and in particular the roles of grain boundaries and impurities, are only poorly understood. By analysis of the device behavior, it has been concluded that the dominant recombination in the space charge region involves the band tails that occur in disordered media. Although the influence of annealing in different atmospheres is well known, it is little understood. It has been observed that there is an improvement in the efficiency of the cells after sodium incorporation. However, there is no model explaining this effect.

Of particular interest is the development of materials with higher band gaps. The use of these will result in an increase in the open circuit voltage and a reduction of the current density. This will allow a reduction in the number of interconnects, and so ease the requirement for highly conductive transparent contacts. Widening the band gap of the chalcopyrite by introduction of Ga in CIS material is possible [53]. The CIGS films are influenced by chemical species in the aqueous solution used for the CBD process. The chemical species are a cadmium salt, ammonia, an ammonium salt, thiourea, and so on. The family of chalcopyrite semiconductors contains attractive materials with band gaps ranging from 1 to 2.4 eV. $CuInS_2$, with a band gap of 1.5 eV, has found interest for fabrication of cells free of the toxic component Se. The Cd-free buffer layers ZnInSe [54], ZnO [55] and ZnSe are used as alternatives for CdS, and efficiencies of 13-15.4% have been obtained.

If CIS technology turns out to be the lowest-cost PV option, a shortage of In and Ga may limit its growth. This is not of major importance for production levels of a few thousand MW per year. The commercial product would benefit from enhanced understanding of the mechanism limiting device parameters such as voltage. Device optimization is presently more of an art than a science.

CIS is a very promising photovoltaic technology, inducing several companies to become involved at various levels ranging from the development of new deposition processes to commercial manufacture. It promises to deliver near-term efficiencies close to that of crystalline Si.

5.2.4. *Transparent Conductive Oxides*

Semiconductors like SnO_2, F doped SnO_2 (FTO), $In_{1-x}Sn_xO_2$ (ITO) and ZnO, doped with Al or Ga, have found wide applications as transparent electrodes in thin film solar cells based on a-Si:H, CdS/CdTe, CdS/CIS etc. The requirements are the same as for the semiconductors in the cells, namely low cost large scale technology of preparation, high conductivity and optical transmission, and good stability. These oxides have energy gaps of 3-3.4 eV, depending on the deposition process. They have optical transmissions in the range 80-85% in the visible/near infrared range, and a sheet resistance of 5-10 Ω/square. Although ITO and FTO are generally used for practical applications, they are relatively expensive.

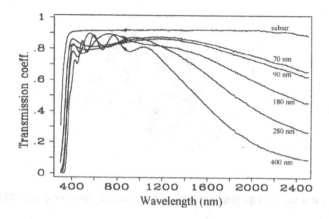

Figure 18. Optical transmission spectra of the system ZnO:Al film/glass substrate versus film thicknesses (RF power 100 W). The substrate transmission spectrum is also shown [56].

Recently, ZnO films have attracted interest as transparent conductive coatings. They offer various advantages:

i) They consist of cheap and abundant elements.

ii) They can be produced by large scale coating.

iii) They allow tailoring of the ultraviolet absorption.

iv) They have high stabilities in a hydrogen plasma.

v) They can be deposited at low growth temperatures. The most frequently used preparation methods are sputtering of ZnO doped with Al or Ga [56,57], reactive evaporation of Zn in an O_2 atmosphere, CVD etc.

Figure 18 shows the optical transmission spectra of ZnO:Al films deposited by magnetron sputtering to different thicknesses, at 100 W RF power. The average transmission in the range 400-900 nm is about 0.85, i.e. the transmission of the films is about 90% of that of the substrate, and does not appear to decrease with increasing film thickness. The resistivity of the films changes from $1.5.10^{-2}$ to 3.10^{-3} Ω.cm as the film thickness increases from 70 to 380 nm.

A very interesting approach has been proposed by Eisele *et al.* [58]. The authors use periodically-structured TCO (SnO_2 and ZnO) for light trapping structures in thin film solar cells. For the case of microcrystalline silicon cells, the indirect band gap leads to a low absorption, and the application of efficient light trapping structures is essential to achieve good efficiencies. Figure 19 demonstrates a typical surface topography for a periodically structured ZnO:Al layer. The periodic gratings are generated by holographic laser (Nd:YAG) interference illumination of photoresist deposited on a smooth ZnO:Al film. After 8ns exposure, (single shot of the pulsed laser), the photoresist is processed, and diluted hydrochloric acid is used to etch grooves into the ZnO:Al. On the top of this grating, a standard pin diode is deposited. The periodic structure is preserved after deposition of the solar cell.

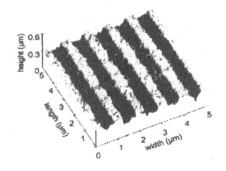

Figure 19. Typical surface topography (AFM) of a periodically structured ZnO:Al layer [58].

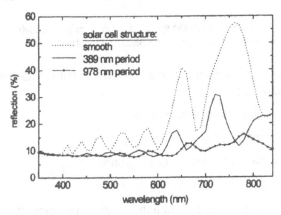

Figure 20. Reflection spectra of a-Si:H p-i-n solar cells with different grating periods [58].

Therefore, the Al back contact acts a periodic reflection grating for incident light. Figure 20 shows the reflection spectra of cells with different grating periods. A grating with a period of 987 nm provides the most effective light trapping structure.

5.2.5. *Ion Implanted Rare Earths in Anti-Reflection Films*

In the studies of solar cell efficiency improvements performed to date, almost all researchers have aimed to modify the structure of the solar cell itself, by fitting its spectroscopic sensitivity to the fixed sunlight spectrum. A reverse conception, however, is that the sunlight spectrum can be fitted to the solar cell sensitivity via a wavelength-shifting operation, using a rare earth ion. Rare earth complexes absorb light of short wavelength, and then emit light with longer wavelengths (Figure 21).

If the wavelength shifts for rare-earth ions are applied to solar cells, the higher photon energy component of sunlight can be shifted to a longer wavelength, at which cells can produce electrical power with high efficiency. Candidate rare earth ions for improvement of the conversion efficiencies are Ce^{3+}, Sm^{3+}, Eu^{2+}, Eu^{3+}, Yb^{3+}, Nb^{3+}, Tb^{3+} [59,60]. Results for the use of AR SiO_2 films containing Eu^{2+} ions implanted into commercial a-Si:H solar cells show an increased efficiency: the ratio of the relative efficiencies with and without implantation is 1.55 to 1.58 [60].

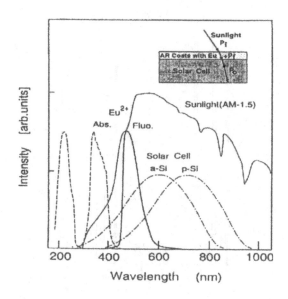

Figure 21. Absorption and emission spectra of the Eu^{2+} ion. The AM1.5 sunlight spectrum and the photocurrent spectra of a-Si:H and c-Si solar cells are also given [60].

5.2.6. Thermophotovoltaic Materials

To date, the photovoltaic community has focused on solar photovoltaic modules in which cells are illuminated by sunlight. However, there is a complementary concept known as thermophotovoltaics (TPV). A TPV system uses a heat source, e.g. a gas or wood burner, or the sun, to bring a thermal emitter to a suitable temperature (e.g. 1500K). The thermal infrared radiation is then converted to electricity using special infrared-sensitive photovoltaic cells. To increase the efficiency of energy conversion, the spectrum of the emitted light should be adapted to the spectral response of the solar cell used. TPV units both extend the potential market for photovoltaics and make traditional solar PV more attractive. The technology is complementary to that of solar photovoltaics, in that solar conversion works well in sunny climates and TPV in colder ones. It is also seasonally complementary, in that solar panels can supply electricity during the sunny summer months, while TPV can supply heat and electricity during the cold and cloudy winter ones.

An example of a fuel fired TPV converter assembly is presented in Figure 22 [61]. Here, GaSb infrared sensitive cells, which respond to wavelengths out to 1.8 μm, are used. At different laboratories, other cell types have been employed, including GaSb, InGaAs and Si,Ge. Recently, a monolithic interconnected module device was proposed [62]. It consisted of a small area of InGaAs photovoltaic cells, connected in series monolithically, on a semi-insulating InP substrate. This structure had a bandgap of 0.74 or 0.6 eV, depending on the semiconductor material. This resulted in the formation of a single high-voltage/low-current module with a peak quantum efficiency near unity for the most of the wavelength range of

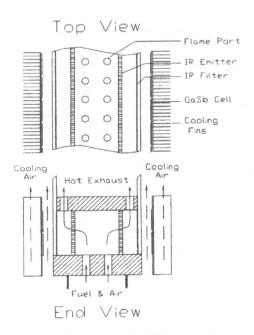

Figure 22. A fuel-fired TPV converter assembly [61].

interest, 0.8-1.8 μm. Under flashlamp testing, a module with 90 PV cells connected in series (1x1cm^2) demonstrated an open-circuit voltage of 39.6 V, and a fill factor of 71.2% at a short-circuit current density of 4.5 A/cm^2. Examples of the spectral response of GaSb and InGaAs TPV cells are presented in Figure 23 [63].

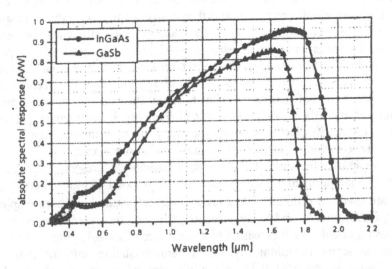

Figure 23. Absolute spectral responses of GaSb and InGaAs TPV cells measured at 25^{0}C [63].

In the near term, TPV holds the promise of high fuel-efficiency by co-generating electricity and heat in otherwise traditional furnaces at a reasonable cost. Fuel is converted to heat in a conventional furnace at 80-90% efficiency in homes, and TPV units can yield heat and electricity combined, at levels at least as high. The ability to generate electricity cost effectively from residential or industrial heating systems is a powerful new application of PV technology.

6. Nanocrystalline And Other New PV Materials

Nanocrystalline semiconductors are intriguing materials. Their optical properties are affected by light scattering and carrier confinement, and the carrier scattering lengths are of the order of the crystallite size. Because the cluster size is comparable to the period of the electron wave function, and to the scattering and diffusion lengths, nanocrystalline semiconductors are quite complex materials, which offer themselves as vehicles for highly tailored optoelectronic properties.

A novel and potentially important technology for the fabrication of structures with two or three degrees of carrier confinement has been proposed [64]. Two-phase materials with larger grains might also exhibit desirable combinations of optical absorption and carrier transport. The concept is based on the filling of natural or synthesized three-dimensional (crystalline or amorphous) cage templates. These are then used for the epitaxy of semiconductors. Zelolites are candidates for this approach, since the cage dimensions are in the nanometer range. They consist of wide band gap silica-aluminate or SiO_2. Such mesoporous structures are synthesized from a lipid micelle solution doped with Si. After annealing, a powder consisting of polycrystalline zeolite is obtained. High-resolution TEM reveals micron-sized grains composed of mesoscopic tubes with a regular hexagonal area. The tube diameter can be adjusted from 20 to 100 Å.

Such a zeolite cage can be loaded with semiconductor materials (Ge, Si, CdSe etc.). A method permitting epitaxial growth of the semiconductor inside the cage is desirable [65]. It is possible to use a vapor phase technique that permits a succession of loading, cracking and annealed cycles. The epitaxy is completed by an annealing cycle. Tomiya et al. [66] performed such experiments using a Y zeolite matrix and Ge as the prototype semiconductor. Scanning TEM data indicated that at least 10^{20} Ge atoms/cm^{-3} were present. The powder sample exhibited a strong luminescence band centered around 7500 Å, indicating the formation of Ge nanocrystallites [66].

The Ge semiconductor in this sample could easily be replaced by Si or other semiconductors such as ZnSe and other $A^{II}B^{VI}$ materials, GaP, GaAs etc. Proposing this approach for photovoltaic cells, the authors [64] hope to reach the very high absorption coefficient expected for a 0-dimensional semiconductor network formed in this manner. The close proximity of the nanocrystals could ensure that a tunneling process between cages will provide a transport mechanism for the photocarriers. Doping of these structures is also hopefully readily achievable.

Polymer photovoltaics also offer great technological potential as renewable, alternative sources of electrical energy. They can be regarded as a third generation of PV materials, after c-Si and inorganic thin films. Low cost, and easily applicable preparation techniques, are the main advantages. In the last couple of years, increasing efforts have been devoted to the development of solar cells based on organic molecules and conjugated polymers. The technological advantages for the fabrication of polymer based organic solar cells, like roll-to-roll production of large areas, lead to a possible reduction of the costs for large area production. Mechanical flexibility and tunability of the bandgap offer interesting prospects for polymer based solar cells, as compared to those based on inorganic materials. Because of these advantages, the development of polymeric solar cells would have a major impact, even though the efficiencies of such devices up to now are lower than those achieved in inorganic solar cells.

For the generation of electrical power by absorption of photons, it is necessary to spatially separate a photo-generated electron-hole (e-h) pair, before recombination occurs. In conjugated polymers, the stabilization of a photo-excited e-h pair can be achieved by blending the polymer with an acceptor molecule, with an electron affinity larger than that of the polymer, but still smaller than its ionization potential. In addition, the highest occupied molecular orbital (HOMO) of the acceptor should be lower than the HOMO of the conjugated polymer. Under these conditions, it is energetically favorable for the photo-excited conjugated polymer to transfer an electron to the acceptor molecule. The hole remains in the polymer valence band, which is the lowest available energy state for it.

Composites of organic polymers and inorganic nanocrystals are particularly interesting materials for PV application. Encouraging results have been obtained using mixture of polymers with different electron affinities that phase-separate on a length scale suitable to give effective charge separation, whilst providing efficient charge transport to the electrodes [67]. A photovoltaic effect has been observed in the heterostructure C_{60}/conductive polymer (poly(2-methoxy.5-(2^1-ethyl-hexyloxy)- p-phenylenvinylene) (MEH-PPV) [68]. In this composite, derivatization of fullerene molecules with a flexible alkyl group has been found to give optimum photovoltaic performance, although the detailed morphology of the composite has not been reported. Recently, large area photovoltaic devices based on inter- penetrating networks of donor (blends of poly(para-phenylenevinylene) (PPV) derivatives) and acceptor molecules (C_{60} derivatives) have been fabricated by using the doctor blade technique [69]. It was shown that among the various combinations of materials, short circuit currents are maximal (I_{sc} = 280 $\mu A/cm^2$, under illumination with white light at 6 mW/cm^2) for a blend of PPV and a highly soluble fullerene derivative. Open circuit potentials (V_o=0.72 V) stay almost the same for all devices. For the best combination of donor-acceptor materials, it has been shown that up-scaling to 10x15 cm^2 is possible by using the donor blade technique, without loss of efficiency.

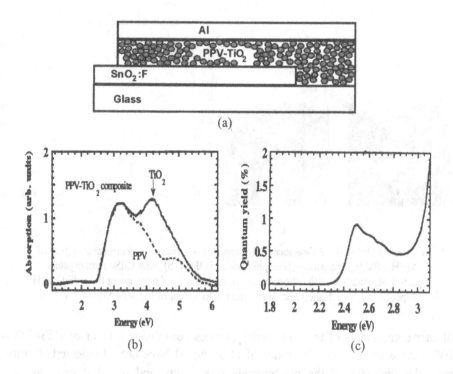

Figure 24. Structure (a), absorption spectra of the composite (b), and quantum yield (c) of a PV device using TiO$_2$ embedded in PPV [70].

Another potential next generation solar cell technology features devices based on dye-sensitized nanocrystalline titanium dioxide (Gretzel cells). To date, such cells have achieved conversion efficiencies of about 10%. While it has been suggested that a 15% efficiency level could be achievable, the exact potential for further improvement of this type of device is unknown. A major drawback is the utilisation of a liquid electrolyte that presently prevents the use of standard thin-film monolithic interconnection schemes. However, it has been proposed to replace the liquid electrolyte contact with gel or solid state ones.

Recently, a solid state polymer-nanocrystalline (TiO$_2$) photovoltaic device has been introduced [70]. The active layer is a composite mixture of two components, poly(p-phenilenevinylene) (PPV) and TiO$_2$ nanocrystals, spin-coated onto a conducting substrate (Figure 24). The collected current is due to absorption in the visible spectrum by the polymer, followed by electron transfer to the nanocrystals. Short circuit currents are between 20 and 100 μA/cm^2 under AM1.5 illumination, increasing with the weight % of TiO$_2$ nanocrystals in the composite. The open circuit voltage of 0.65V is nearly independent of the nanocrystal fraction.

Nanocrystals of A^{II}B^{VI} semiconductors, such as CdS and CdSe, have been used to prepare a composite with MEH-PPV [71]. It was observed that sufficient photoconductivity can be achieved at high nanocrystal concentrations, when electrons and holes can be transported, respectively, through the nanocrystal and

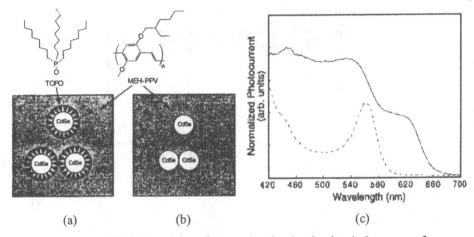

Figure 25. A MEH-PPV-nanocrystal composite, showing the chemical structure of MEH-PPV (a) and trionctyphosphinoxide (TOPO) (b), with CdSe nanocrystals. Spectral dependence of the short circuit current of a PV device using the CdSe/MEH composite (c). The dashed line shows the photocurrent of the pure MEH-PPV [71].

polymer components of the composite. Devices comprising a layer of CdSe/MEH PPV composite between electrodes of ITO and Al have been fabricated. In most cases, the diameter of the nanocrystals was 5 nm, and the first peak in their absorption spectrum was at 614 nm. Figure 25 shows the spectral dependence of the short circuit current from such a device. The spectrum shows a response in the region 600-660 nm, where there is no absorption in the polymer. This response corresponds to light which is absorbed in the nanocrystals, with subsequent hole transfer into the polymer. At wavelengths below 600 nm, the response is due to a combination of absorption in the polymer and in the nanocrystals. The pure MEH-PPV device has a peak in its spectrum at 560 nm. For high concentrations of CdSe, the spectrum follows the fraction of the incident light absorbed in the device. This indicates that both carriers are mobile within the composite material at high nanocrystal concentrations. A large improvement in the photovoltaic efficiency can be achieved by using a high concentration of nanocrystals, and further improvements have to be made to optimize the charge transport process.

Porous silicon (PS) is a form of nano-dimensional silicon that has been explored in PV device technology. It shows an extremely low reflection coefficient as an anti-reflection coating [72-74]. Additionally, PS has been used as the emitter in a solar cell [75]. However, the high value of the PS resistivity requires the use of a thin PS layer and optimisation of the contact grid to obtain a higher short-circuit current. An increase of about 30% in the short-circuit current and conversion efficiency of the heterostructure PS/(n^+-p)c-Si has been reported. The PS was created by stain etching between the Al grid electrodes, as in Figure 26 [72,73]. In such an approach, the Al grid is directly in contact with the n^+ diffusion layer.

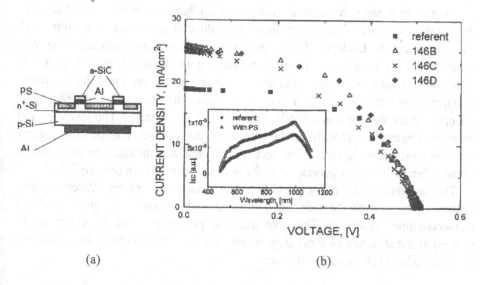

(a)   (b)

Figure 26. Structure (a) and I-V characteristics under illumination (b) of c-Si solar cells with PS between the contact fingers. The I-V curve of a reference cell without PS is shown for comparison. Inset: spectral dependence of the short circuit current for the two types of cell [73].

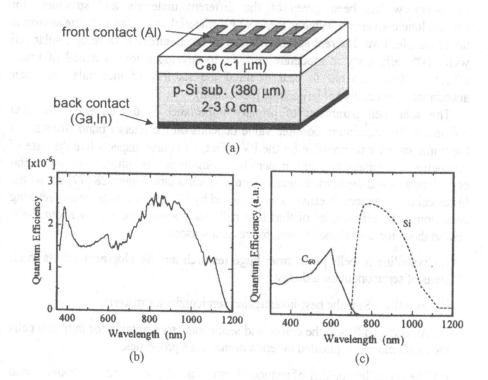

Figure 27. Structure (a), plus the measured (b) and calculated (c) responses of a C_{60}/p-Si heterojunction cell [76].

Another new nanosize material, fullerene (C_{60}), has been applied in PV devices. Polycrystalline films of C_{60} were deposited onto c-Si substrates in an ultra-high vacuum chamber. Undoped C_{60} acted as an n-type semiconductor. The c-Si substrates were p-type, so that a n-p-heterojunction was formed (Figure 27) [76]. Al front contacts were evaporated onto the C_{60} surface with a patterned shadow mask. A Ga_xIn_y alloy was used as a back contact. The short circuit current was 132 nA and the open circuit voltage was 0.306 V, under AM1.5 illumination. The low values of these parameters are thought to be due to a low built-in voltage (low carrier concentration in the C_{60} layer). The spectral response demonstrates the contribution of the fullerene to carrier generation in the region below 700 nm (Figure 27b,c).

The introduction of nanocrystalline semiconductors to photovoltaics requires further study of how to achieve controlled and reproducible growth of the nanocrystalline dispersion. The optoelectronic properties must be examined in relation to the structure of these dispersions, and the materials must be introduced into experimental photovoltaic devices.

7. Conclusions

An overview has been given of the different materials and structures for photovoltaic conversion. What is the ideal PV material and structure? The answer is not so simple. Two different approaches can be taken: **either** to develop a solar cell with >20% efficiency at moderate cost, **or** to develop a low cost cell of lower efficiency. One also has to bear in mind the stability of materials, and their abundances (especially for large scale production).

The solar cell parameter of primary importance is the energy conversion efficiency. The maximum possible value depends on the energy band structure of the semiconductor responsible for the PV effect. Of course, aspects like the ease of production, availability of materials, long-term stability, environmental considerations and the market situation are also important. Figure 28 [77] shows the historical development of efficiencies achieved by PV devices under manufacturing conditions. The efficiencies of thin film cells have always lagged by about 7-8% behind those for crystalline Si ones, for several reasons:

i) Crystalline Si cells profit from huge research and development efforts in all areas of semiconductor industry.

ii) Crystalline Si is the best-investigated semiconductor material.

iii) Amorphous Si and the compound semiconductors suitable for thin film cells are much more complicated in terms of materials properties.

iv) The crystalline quality of semiconductor films deposited non-epitaxially onto amorphous substrates cannot be as perfect as that of monocrystalline or even multicrystalline Si wafers.

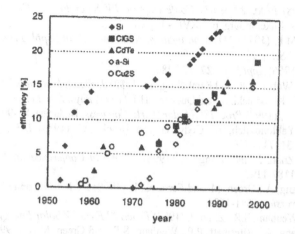

Figure 28. Development of the efficiencies of crystalline silicon wafer based cells, in comparison to thin film technologies [77].

The successful launch of commercial production lines promises an increased market share for thin film cells. However, they will not replace wafer-based silicon cells in the near future. In the mid term, they will complement crystalline silicon, which still has the potential to improve, e.g. by producing thin wafer cells on foreign substrates.

The history of photovoltaic materials is very long. Three generations of photovoltaic materials can be distinguished. The requirement for the first generation, mainly involving crystalline silicon, is a significant cost reduction by a factor of 4, to match the present cost of electricity - 0.16 Euro/W_p. The requirement for second generation PV materials and solar cells, based on thin films, is also a reduction in cost, to the level of the end-user price - 0.11 Euro/W_p. The third generation focuses on materials featuring easy production, but also a high conversion efficiency. The target cost has to be competitive with that for other energy sources - 0.05 Euro/W_p. Having in mind the parallel requirement for a high stability of device parameters, the ideal photovoltaic material remains a challenge for researchers in the 21st Century.

8. References

1. Von Roedern, B. (1999) *SPIE Conference on Solar Optical Materials XVI*, Denver, Colorado, **SPIE v.3789**, 104-115.
2. Schmid, J. (1998) *2nd World Conference on PV Solar Energy Conversion*, European Communities, Luxembourg, p.p. XLVIII-LII.
3. Sandtner, W. (1998) *Proc. 2nd World Conference on PV Solar Energy Conversion*, European Communities, Luxembourg, p. LX-LXI.
4. Bequerel, E. (1839) *Compt. Rendues* **9**, 561-566.
5. Wolf, M. (1972) *25th Power Sources Symp. Proc.*, Red Bank, N.J., 120-124.
6. Chapin, D.M., Fuller, C.F. and Pearson, G.L.J. (1954) *Appl. Phys.* **25**, 676-677.

7. Iles, P. (1998) *Proc. 2nd World Conference on PV Solar Energy Conversion*, European Communities, Luxembourg, p. LXVII - LXXII.

8. Thekaekra, M.P. (1974) *"Data on incident solar energy" in Suppl. Proc. 20th Ann. Meet. Inst. Environ. Sci.* p.21.

9. Loferski J. (1956) *Appl.Phys.* **27**, 777-784.

10. Rohatgi, A., Weber, E.R. and Kimbeling,, L. (1992) *J. Electr. Materials* **22**, 65-71.

11. Werner, J.M., Kolodinski, S. and Queisser, H.J. (1994) *Phys. Rev. Lett.* **72**, 3851-3854.

12. Hamakawa, Y., (1993) *Proc. ISES Solar World Congress*, Budapest, 83-90.

13. Tiedje, T., Yablonovitch, E., Cody, G. and Brooks, G. (1984) *IEEE Trans. Electron Devices* **ED-31**, 711- 716.

14. Green, M., Zhao, J. and Wang, A. (1998) *2nd World Conference on PV Solar Energy Conversion* 1187-1192.

15. Zhao, J., Wang, A., Green, M. and Ferrazza, F. (1998) *2nd World Conference on PV Solar Energy Conversion* 1681-1684.

16. Green, M., Wenham, S.R., Zhao, J. (1992) *Proc.11th Euro PV Solar Energy Conf.*, 35-40.

17. Zhao, J., Wang, A., Altermatt, P.P., Wenham, S.R. and Green, M.A. (1996) *Solar Energy Mat. Solar Cells* **41/42**, 87-99

18. Kobloch, J., Glunz, S.W., Biro, D., Warta, W., Schaffer, E. and Wettling, W. (1996) *Proc 25th IEEE PV Spec. Conf.*, 405-408.

19. Uebele, P., Tentscher, K.H., Kern, R., Mattes, S., Rasch, R.D., Schmidt, W., Schomann, F. and Strobl, G. (1994) *Proc. 12th European PV Solar Energy Conference*, 67-69.

20. Kardauskas, M. (1996) *6th Workshop: Role of Impurities and Defects in Silicon Device Processing*, p.172.

21. Schmit, R.R., Felder, B., and Hotchkiss, G. (1993) *Proc.23th IEEE PV Spec. Conf.*, 112-115.

22. "Solar Electricity" (1994) T. Markvart (ed.), (John Wiley & Sons), p.58.

23. Kolodziej, A., Wronski, C., Krewniak, P. and Nowak, S. (2000) *Optoelecron. Review* **8**, 339 -345.

24. Wang, L., Gu, M. and Reehal, H.S. (1998) *2nd World Conference on PV Solar Energy Conversion*, 1802 -1805.

25. Poortmans, J. and Dimler, B. (2000) European Research Conference "Photovoltaic Devices. Thin film technology", Teltow, Germany 17-18.

26. Tayanaka, H., Yamauchi, K. and Matsushita, T. (1998) *2nd World Conference on PV Solar Energy Conversion*, 1272-1277.

27. Spear , W.E. and Le Comber, P.G. (1976) *Phil. Mag.* **33**, 935-940.

28. Stutzmann, M. (1989) *Phil. Mag. B* **60**, 531-549.

29. Yang, J., Banerjee, A., Sygiyama, S. and Guha, S. (1997) *Proc.26th IEEE PV Specialist Conference*, 563-567.

30. Guha, S., Yang, J., Williamson, D.L., Lubianiker, Y., Cohen, J.D. and Mahan, A.H. (1999) *Appl. Phys. Lett.* **74**, 1860-1865.

31. Yang, J., Banerjie, A., Gitfelter, T., Sugiyama, S.S. and Guha, S. (1997) *Proc. 26th IEEE PV Specialist Conference*, 563 -567.

32. Meier.J., Keppner, H., Dubbail, S., Zuegler, Y., Feitknrcht, L., Torres, P., Hof, Ch., Kroll, U., Fischer, D., Cuperus, J., Anna Selvan, J.A. and Shah, A. (1998) *2nd World Conference on PV Solar Energy Conversion*, 375-380.

33. Rocca i Cabarrocas, P. (2000) *European Research Conference "Photovoltaic Devices. Thin film technology "*, Teltow, Germany, 19-20.

34. Longeaud, C., Kleider, J.P., Gauthier, M., Kaplan, R., Butte, R., Meaudre, R. and Rocca I Cabarrocas, P. (1998) *2nd World Conference on PV Solar Energy Conversion*, 680-683.

35. Roca I Cabarocas, P., Stahel, P., Hamma, S. and Poissant, Y. (1998) *2nd World Conference on PV Solar Energy Conversion*, 355-358.

36. Koch, C., Ito, M., and Schubert, M.B. (2000) *Proc. European PV Conference* Glasgow, (in press).
37. Wyrsh, N., Torres, P., Goetz, M., Dubail, S., Feitknecht, I., Cuperus, J., Shah, A., Rech, B., Kluth, O., Weider, S., Vetterl, O., Stiebig, H., Beneking, C. and Wagner, H. (1998) *2ⁿᵈ World Conference on PV Solar Energy Conversion*, 467-470.
38. Wronski,, C.R., .Jialo, L., Koval, R. and Collins, R.W. (2000) *Opto-electronic Review* **8**, 275-279.
39. Kamei, T., Stradins, P. and Matsuda, A. (1999) *Appl. Phys. Lett.* **74**, 1707-1709.
40. Tsu, D.V., Chao, B.S. and Ovshinsky, S.R. (1997) *Appl. Phys. Lett.* **71**, 1317-1319.
41. Said, K., Poortmans, J., Caymax, M., Nijs, J., Silier, I., Gutjahr, A., Konuma, M., Debarge, L., Christoffel, E. and Slaoui, A. (1998) *2ⁿᵈ WC PV Energy Conversion*, 36-39.
42. Bube, R., and Mitchell, K. (1993) *Electr. Mater.* **22**, 17-25.
43. Candless, B.E., and Birkmire, R.W. (1997) *Proc. 26ᵗʰ IEEE Photovoltaic Specialist Conference*, 307-310.
44. Ohyama, H., Aramoto, T., Kumazawa, S., Higuchi, H., Arita, T., Shuburani, S., Nishio, T., Nakajima, J., Tsuji, M., Hanfusa, A., Hibno, T., Omura, K. and Murosono, M. (1997) *Proc. 26ᵗʰ IEEE PV Specialist Conference*, 343-346.
45. Meyers, P.V., Ackerman, B. and Jordan, J.F. (1990) *IEEE Trans. Electron Devices* **37**, 434-439.
46. Chu, T.I., Chu, S.S., Ferekides, C., Wu, C.O., Britt, J. and Wang, C. (1991) *Proc.22ⁿᵈ IEEE PV Specialist Conference*, 962-965.
47. Bohland, J., Dapkus. T., Kamm, K., Smigielski, K. (1998) *Proc.2ⁿᵈ World Conference on PV Solar Energy Conversion*, 716-719.
48. Kohara, N., Negami, T., Nishitani, M., Hashimoto, Y. and Wada, T. (1997) *Proc. 14th European Photovoltaic Solar Energy Conference*, 2157-2160.
49. Gay, R.R. (1997) *Solar Energy Materials and Solar Cells* **47**, 19-26.
50. Dimller, B., Gross, E., Hariscos, D., Kesler, F., Lotter, E., Powalla, M., Springer, J., Stein.,U., Voorwinden., G., Gaen, M., and Schaffler, R. (1998) *2ⁿᵈ World Conference on PV Solar Energy Conversion*, 419-423.
51. Dimller, B. (2000) *European Research Conference "Photovoltaic Devices Thin film technology"*, Teltow, Germany, 45-46.
52. Ennaoui, A., Weber., M., Lokhande, C.D., Scheer, R. and Lewerenz, J. (1998) *2ⁿᵈ World Conference on PV Solar Energy Conversion*, 628-631.
53. Wada, T., Hayashi, S., Hashimoto. Y., Nishiwaki, S., Sato, T., Negami, T. and Nishitani, M. (1998) *2ⁿᵈ World Conference on PV Solar Energy Conversion*, 403-408.
54. Ohtake, Y., Okamoto, T., Yamada, A., Konagai, M. and Saito, K. (1997) *Solar Energy Materials and Solar Cells*, **49**, 269-275.
55. Olsten, Y., Lei, W., Addis, F.W., Shafarman, W.N., Contreras, M.A. and Ramanathan, K. (1997) *Proc. 26ᵗʰ IEEE Photovoltaic Specialist Conference*, 319-321.
56. Dimova-Malinovska, D., Tzenov, N., Tzolov, M. and Vassilev, L. (1998) *Mater. Sci. Technol. B* **52**, 59-62.
57. Jeong, W., Im, Y., Park, G., Park Chung, H. and Kim, C. (1998) *2ⁿᵈ World Conference on PV Solar Energy Conversion*, 747-750.
58. Eisele, C., Nebel, C. and Stutzmann, M. (1999) *Berich Annual Report of W. Schottky Institut, TU Muenchen*, 68-69.
59. Sendova-Vassileva, M., Nikolaeva, M., Dimova-Malinovska, D., Tzolov, M. and Pivin, J.C. (2001) *Mater. Sci. Engineering B* **52**, 185-187.
60. Kawano, K., Sado, T., Nishikawa, M. and Nakata, R. (2000) *2ⁿᵈ World Conference on PV Solar Energy Conversion*, 334-337.
61. Fraas, I., Samaras, J., Mulligan, W., Avery, J., Groeneveld, M., Huang, H., Hui, S., Ye, S., West, E. and Seal, M. (1998) *2ⁿᵈ World Conference on PV Solar Energy Conversion*, 25-29.

48

62. Fatemi, N., Hoffman, R., Mark, Jr., Stan, A., Weizer, V., Jenkins, P., Wilt, D., Scheiman, D., Brinker, D. and Murray, C. (1998) 2nd *World Conference on PV Solar Energy Conversion*, 345-348.

63. Beckert, R., Broman, L., Jarefors, K., Marks, J. and Bucher, K. (1998) 2nd *World Conference on PV Solar Energy Conversion*, 30-35.

64. Zinger, A., Wagner, S. and Petroff, P.M. (1993) *J Electr. Materials* **22**, 3-15.

65. Meier W.M. and Ison, D.H. (1992) *Atlas of Zeolite Structure Types*, 3rd Ed. Eng: Butterworth, Guilford.

66. Toniya, S., Petroff, P., Margolese, D., Srdanov, V. and Stucky, G. (1993) *Nanophase and Nanocomposite Materials. Mat. Res. Soc.* Pittsburg, 137-140.

67. Halls, J.J.M., Walsh, C.A., Greenham, N.C., Marseglia, E.A., Friend, R.H., Moratti, S.C. and Holmes, A.B. (1995) *Nature* **376**, 489-491.

68. Yu, G., Gao, J., Hummelen, J.C., Wudl., F. and Heeger, A.J. (1995) *Science* **270**, 1789-1790.

69. Padinger, F., Brabec, C.J., Fromherz, T., Hummelen, J.C. and Sariciftci, N.S. (2000) *Opto-electronics Review*, **8**, 280-283.

70. Safalsky, J.S. and Schropp, R.E.I. (1998) 2nd *World Conference on PV Solar Energy Conference*, 272-275.

71. Greenham, N.C., Peng, X. and Alivisatos, A.P. (1996) *Phys. Rev. B* **54**, 17628-17637.

72. Dimova-Malinovska, D., Sendova-Vassileva, M., Tzenov, N. and Kamenova, M. (1997) *Thin Solid Films*, **297**, 9-12.

73. Dimova-Malinovska, D. (1999) *J. Luminescence*, **80**, 207-211.

74. Yerohov, V., Melnyk, I., Tsisaaruk, A. and Semochko, I. (2000) *Opto-Electronics Review* **8**, 414-417.

75. Bastide, S., Strehlke, S., Cuniot, M., Bourty-Forveille, A., Le, Q.N., Sarti, D. and Levi-Clement, C. (1995) *Proc. 13th EPVECE*, H.S. Skephew and Ass. Publ., Bedford, U.K., 1280-1283.

76. Kojima, N., Ishikawa. N., Yang, M.-Ju. and Yamaguchi, M. (1998) 2nd *World Conference on PV Solar Energy Conversion*, 246-248.

77. Schock, H.W. and Pfisterer, F. (2000) *16th European PV Solar Energy Conference* Glasgow, UK (in press).

ELECTRONIC CHARACTERISATION AND MODELLING OF DISORDERED SEMICONDUCTORS

J.M. MARSHALL

Department of Materials Engineering, University of Wales Swansea Singleton Park, Swansea SA2 8PP, U.K.

1. Introduction

The viability of a semiconductor material for photovoltaic and related applications is critically dependent upon the mobilities and lifetimes of its charge carriers. The situation is complicated by the fact that many of the materials of interest are employed in forms (amorphous, nanocrystalline and polycrystalline films, etc.) which feature significant degrees of disorder. This can exert a great influence on the electronic properties, and can also make the interpretation of experimental data appreciably more complex and challenging.

This paper commences with a brief review of the conventional concepts of electronic transport in the solid state. It then describes the manner in which these concepts must be changed to incorporate the effects of disorder, how the carrier mobilities and lifetimes are influenced, and how some of the techniques used to determine these properties are affected. Finally, it is shown that computer-based modelling constitutes an extremely powerful tool in both predicting and understanding the experimental behaviour.

2. Electronic Transport in Crystalline and Disordered Semiconductors

2.1. CLASSICAL APPROACH TO ELECTRONIC TRANSPORT

The first attempt at a quantitative theory of electronic conduction in the solid state was made by Drude in the early 1900s. It was basically an adaptation of the kinetic theory of gases, with the significant modification that a "free" electron would be expected to experience very frequent scattering (with a mean free path, ℓ, of a few interatomic spacings) due to the rapid variations in potential within the lattice. The resulting analysis predicted that under an applied electric field F, such an electron would gain an average drift velocity $v_{drift} = Fe\tau/2m$, where e and m are the electronic charge and mass, and $\tau = \ell/v_{th} \leq 10^{-14}$ sec at room temperature, with v_{th} being the (randomly directed) thermal velocity. Thus, the carrier drift mobility, μ, (the average drift velocity per unity field) would be $e\tau/2m$, with an expected room temperature value of $1 - 10$ $cm^2V^{-1}sec^{-1}$.

49

J.M. Marshall and D. Dimova-Malinovska (eds.),

Photovoltaic and Photoactive Materials - Properties, Technology and Applications, 49–66.
© 2002 *Kluwer Academic Publishers.*

Assuming that each atom in a metal contributed one or two free electrons, this model gave reasonable values of electrical conductivity. However, it was unable to explain *why* only a small fraction of the electrons present were free to participate, or (for example) why some materials (e.g. Sn in Group IV of the periodic table) were metals, whilst others (e.g. C, Si and Ge in the same group) were not. Moreover, when measurements on such materials as good single crystal silicon became possible, room temperature mobility values of $\geq 10^3$ $cm^2V^{-1}sec^{-1}$ were observed, implying mean free paths of the order of 1000 inter-atomic spacings.

2.2. BAND THEORY OF SOLIDS

The above problems were resolved, at least for good single crystal materials, once it was realised that moving electrons had a wavelike nature, with the (time averaged) varying quantity being linked to the probability of an electron occupying a particular location (i.e. the Schrödinger equation).

Applying this concept to propagation in an infinite periodic lattice, it became easy to predict that the (classically continuous) spectrum of possible electron energies should split into bands of allowed energies, separated by forbidden energy gaps. Moreover, it became clear that an electron would only be scattered when it encountered some deviation from periodicity (thermal vibration, missing or impurity atom, grain boundary, etc. etc.). Also, it was readily shown that within a crystal, an electron could behave as if it possessed an "effective" mass, $m^* \neq m$.

Combining these factors, and employing a more sophisticated approach to the analysis of the effect of the electric field led to a mobility $\mu = e\tau/m^*$ which could easily be much higher than the classical Drude value. It also became easy to understand, in terms of the occupancies of the energy bands at zero temperature, why some materials could be metals (partially filled conduction band) and others potentially perfect insulators (full valence band, empty conduction band).

2.3. THE INFLUENCE OF DISORDER ON THE BAND STRUCTURE AND ELECTRONIC PROPERTIES

The above approach proved to be so powerful in analysing the properties of pure single crystal materials that for many years the existence of allowed energy bands was usually regarded as being intimately linked to presence of a perfectly periodic structure. However, this raised considerable problems with respect to non-crystalline solids. In a material such as vitreous SiO_2, the absence of a periodic structure was expected to invalidate the emergence of energy bands, so that the material should have been a metal, and thus opaque to light. Similarly, the existence of materials such as amorphous selenium, which was well known to exhibit semiconducting properties, could not be explained.

In order to understand the above situation, it is necessary to realise that materials loosely described as amorphous solids do not, in fact, usually feature the total lack of order implied by the term. Rather, as shown in figure 1, the short-range order typical of the single crystal material is mostly retained. However, relatively slight variations in interatomic bond length and angle rapidly combine to destroy the longer-range order, which typically disappears at distances greater than two or three interatomic spacings. Also, although most atoms retain the bonding configuration of the single crystal, a small fraction is unable to do so.

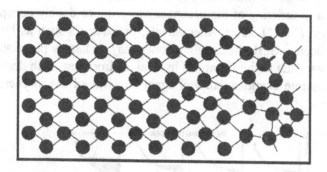

Figure 1. A periodic lattice (left hand side) can be transformed into a highly disordered one (right hand side) via relatively small variations in bond length and angle (exaggerated here). Most atoms retain their four-fold coordination, but a few are unable to do so.

In these circumstances, it is more appropriate to consider the emergence of energy bands via a tight binding approach. Under this (e.g. figure 2), the bands arise from the smearing out of the individual atomic (and molecular) allowed energy levels, as the interatomic (or intermolecular) separation decreases.

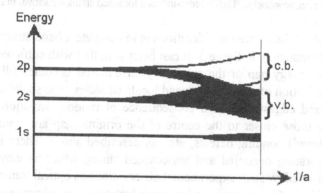

Figure 2. Tight binding approach to the emergence of energy bands in a solid (carbon), as the interatomic spacing, a, is progressively reduced.

This approach can be applied (although generally less quantitatively than that above) to a single crystal material. However, its advantage in the present circumstances is that it gives at least a qualitative approach to the case of a disordered one. Most of the outer electrons will remain bound in the same interatomic bonds as in the crystal (i.e. in the valence band (v.b.)). However, the variations in bond length and (mainly) angle can result in a modification of the energy required to break them (i.e. the energy gap). There will also be additional states within the gap, associated with the "dangling bonds" occurring at atoms for which the local coordination is not satisfied (see figure 1). Dangling bonds occupied by a single electron may not to be energetically favourable, in which case there is a re-ordering of the occupancy to produce doubly occupied (" D^- ") and unoccupied (" D^+ ") states. Alternatively, in materials such as hydrogenated amorphous silicon, a-Si:H, the dangling bonds may be largely passivated as Si=H bonds. Of course, due to the variations in local environment, such deep states will not exist at discrete energies, but will again be smeared out in energy to some degree.

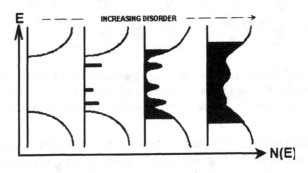

Figure 3. Influence of progressively increasing degrees of disorder upon the band structure of a semiconductor. The disorder-induced localised states are shown in black.

The net result of the above considerations is to generate a band structure of the form shown schematically in figure 3. It can been seen that with sufficient degrees of disorder, the energy gap of the single crystal material becomes filled with a continuous distribution of disorder-induced localised states. The shallow "tails" of states at the band edges arise as a consequence of random variations in local potential, while those closer to the centre of the original gap are mainly due to (suitably broadened) bonding defects, etc., as described above. There is then no energy gap separating occupied and unoccupied states, which is why theorists were initially puzzled by such experimental observations as optical transmission at long wavelengths, or the occurrence of semiconducting electrical properties.

To understand the above problems, we must examine the process by which electrons move around within such a material. In the extended electronic states within the conduction and valence bands, we would expect (iso-energetic) transport to be of a conventional form, as described in Section 2.2. However, as the boundary between the extended and localised states is approached, scattering will become increasingly frequent, and the mobility will fall. The lower limit

($\ell \sim$ inter-atomic spacing) yields a calculated room temperature mobility of about $10~cm^2V^{-1}sec^{-1}$, with a re-consideration of the transport as a Brownian diffusive process [1] yielding a limiting value of $\mu \sim ev_eR^2/6kT \sim 1~cm^2V^{-1}sec^{-1}$, where v_e is an electronic frequency and R the inter-site scattering distance).

Within the localised states immediately below the extended ones, transport between iso-energetic states must become a quantum-mechanical tunnelling process. The mobility will then be of the form [1]

$$\mu_{hop} = (ev_{ph}R'^2/6kT)~exp(-2R'/R_o)~exp(-W/kT) \tag{1}$$

where v_{ph} is a phonon vibrational frequency (often taken as $\sim 10^{12}$ Hz), R' is the inter-site separation, R_o the site localisation length, and W any activation energy associated with the tunnelling process. Even if we take R' to be the interatomic spacing, allowing one to assume $R_o >> R'$ and $W = 0$, the maximum value of this mobility will be about $10^{-2}~cm^2V^{-1}sec^{-1}$. Further into the localised states, the value will drop rapidly, primarily due to the effect of the $exp(-2R'/R_o)$ term. Thus, there will be an extremely rapid fall in iso-energetic mobility as the transition energy between extended and localised states is crossed.

It should be noted that the above analysis, like many aspects of the theory of disordered semiconductors, contains various questionable simplifications. In particular, the localised states are often treated as point defects, whereas a more accurate description should include the effects of longer-range variations in potential, particularly at energies close to the transition region. Similarly, in polycrystalline materials, it is unlikely that the localised states can be taken as being isotropically distributed, as opposed to (e.g.) being concentrated at grain boundaries. None the less, the general conclusion of a rapid fall in the iso-energetic mobility remains valid and is widely used.

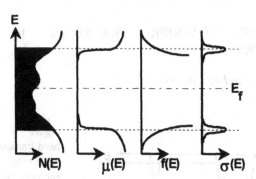

Figure 4. Expected energy variations of the density of states, N(E), mobility, μ(E), occupation probability for electrons (c.b.) and holes (v.b.), f(E), and resulting contributions to the electrical conductivity, σ(E) = e N(E) μ(E) f(E) in a disordered semiconductor.

The consequence of the above analysis of a disordered semiconductor is illustrated in Figure 4. Because of the rapid variation of mobility, the contributions of carriers to the electrical conductivity can peak (at sufficiently high temperatures) close to the energies separating the localised and extended states. For this reason, they are termed "mobility edges", separated by a "mobility gap".

These terms can, for many purposes, take similar roles to the band edges and band gap in single crystal semiconductors (and are sometimes even replaced by the latter terms). However, the distinction should be remembered, as should the fact that conduction must become dominated by hopping between localised states, at sufficiently low temperatures (where the energy variation of $N(E)f(E)$ overcomes that of $\mu(E)$).

Note also that the above approach explains the observation of low levels of optical absorption at sufficiently long wavelengths. Although there are unoccupied allowed localised states just above the occupied ones in the lower part of the mobility gap, the long intersite distances between states in mid-gap can yield very low transition probabilities. However, the optical gap will in general remain lower than the mobility gap. This is because a transition between an occupied localised state in the lower part of the gap and an extended state in the conduction band (or an extended valence band state and an unoccupied localised state in the upper part of the gap) does not involve tunnelling. Analyses based upon this concept have been advanced to explain the energy dependence of the optical absorption coefficient [1], yielding the so-called Tauc gap etc. However, we feel that these should be treated with caution since (a) they involve no consideration of the energy variations of the optical transition matrix elements, and (b) the underlying implication that the $N(E)$ variations feature simple (e.g. linear, parabolic or even constant) energy dependences, over ranges of 1 eV or more, seems very improbable. A common alternative empirical approach is simply to take the optical gap to be the energy at which the absorption coefficient attains a value of 10^4 cm^{-1}, yielding significant useable absorption in devices such as solar cells.

2.4 EQUILIBRIUM CARRIER TRANSPORT IN DISORDERED SEMICONDUCTORS

2.4.1. *Trap-limited Band Transport*

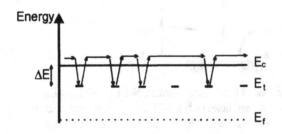

Figure 5. Trap-limited band transport in a material containing a single energetically discrete set of localised states

The existence of significant concentrations of localised states in disordered semiconductors suggests that it is unlikely that carrier motion in the extended states will persist for an extended period of time. Rather (ignoring recombination for the present), it is likely that such motion will be repeatedly interrupted by trapping into, and subsequent release from, localised states of varying depth.

Consider first, for simplicity, the case of a single set of shallow localised states, as shown in figure 5. An electron, moving in the extended states with a "free" mobility μ_o can become temporarily trapped in one of these centres. It will then stay there until it finds enough energy to escape back into the extended states and continue its progress. This mechanism is termed "trap-limited band transport". This periodic trapping/release will yield a reduced "drift mobility", μ_d. If the average time an electron spends free between trapping events is τ_f and the average time for re-release is τ_r, then the ratio of these two quantities is equal to the ratio of the concentrations of instantaneously free and trapped carriers, n_f:n_t, and

$$\mu_d = \mu_o \times \text{fractional time spent free} = \mu_o \, \tau_f / (\tau_f + \tau_r) = \mu_o \, n_f / (n_f + n_t). \qquad (2)$$

If N_t is the total density of trapping centres, and N_c the effective density of states at the conduction band mobility edge (i.e. the number in the bottom kT slice, where most of them are located), then, by Fermi-Dirac statistics:

$$n_f = N_c \exp(-(E_t - E_f)/kT); \; n_t = N_t \exp(-(E_t - E_f)/kT). \qquad (3)$$

At sufficiently high temperatures (with $N_t < N_c$), then the time between trapping events will dominate over that spent before release, and $\mu_d \rightarrow \mu_o$. However, at low temperatures the reverse will be the case, giving

$$\mu_d \rightarrow \mu_o \, \tau_f / \tau_r = \mu_o \, (N_c/N_t) \exp(-(E_c - E_t)/kT) = \mu_o \, (N_c/N_t) \exp(-\Delta E/kT). \qquad (4)$$

Thus, by measuring μ_d (e.g. via a time-of-flight technique - see for example [2] for extensive further detail) over a sufficiently wide temperature range, as in figure 6, then (ignoring secondary factors such as the temperature dependence of μ_o) valuable information can be obtained on the free carrier mobility, the depth of the traps limiting the actual drift mobility, and the ratio N_c/N_t (i.e. the trap density, if a reasonable value of N_c can be assumed).

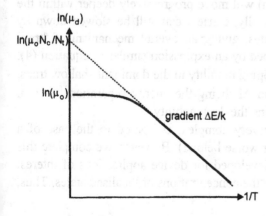

Figure 6. Temperature dependence of the trap-limited drift mobility, for the case of a single set of shallow traps.

In the case of a disordered semiconductor containing energetically-distributed traps, the situation will obviously be more complex. However, in principle, one can employ a modified version of equation (2), in which n_t now becomes the total equilibrium concentration of trapped carriers. For example, in the simple case of a linear bandtail of total depth ΔE, equation (4) becomes

$$\mu_d = \mu_o \, (\Delta E/kT) \, \exp(-\Delta E/kT). \tag{5}$$

Basically, for this and other rapidly-truncated energy distributions, the low-temperature drift mobility is controlled by carriers trapped in the bottom kT of the tail, and changes in the detail of $N(E)$ only influence the pre-exponential factor.

Another commonly-used model envisages an exponential bandtail: $N(E) = N_o \exp(-E/kT_o)$. Here, the high-temperature case is that for which $T > T_o$. At lower temperatures, computation would require use of the full Fermi-Dirac occupation statistics as the Fermi level is approached.

2.4.2. Conduction by Hopping in Band Tails

A competing alternative to trap-limited band transport is direct intersite hopping within a band tail. Whether or not a carrier is likely to make a hopping transition to a neighbouring site, before release to the extended states, obviously depends upon the inter-site separation, the value of R_o and the temperature. However, at low enough temperatures, hopping transport must eventually dominate. For a truncated distribution such as the linear band tail discussed above, this would tend to be dominated by hopping transitions within (or close to) the bottom kT slice of the tail. The expression for the resulting mobility would be similar to equation (1), with R' becoming the mean separation of sites in the bottom kT slice.

More complex variants are also possible. For instance, if the density of localised states just below the mobility edge is sufficiently high, then a carrier may make a transition from a deeper state to these, rather than to the extended states. In such a case, the effective transport path would be within the bandtail itself. For an exponential bandtail, it has been suggested [3] that the energy of the maximum contribution to the transport (i.e. the maximum product of the iso-energetic mobility and the carrier concentration) will move progressively deeper within the tail as the temperature falls. Additionally, carriers can still be slowed down by trapping events involving deeper states, giving an overall mechanism of "trap-limited hopping". This can be described by an expression similar to equation (4), but with μ_o being replaced by the hopping mobility in the dominant shallow traps, N_c becoming their concentration, and ΔE being the energy separation between these states and the deeper ones limiting the drift mobility.

The above situation is obviously very complex, compared to the case of a single crystal material (and it will get worse below!). However, we complete this Section by noting that in materials developed for device applications of interest here, every effort is made to minimise the concentrations of localised states. Thus, it is *hopefully* possible, at normal temperatures, to assume that iso-energetic transport is dominated by states close to the mobility edge, and that hopping between deeper states can be neglected in comparison to trap-limited band transport.

2.5 TRANSIENT CARRIER TRANSPORT IN DISORDERED SEMICONDUCTORS

A further significant complication to the above, already torturous, situation is that the properties of interest are usually those applicable *on a finite timescale*. This is obviously the case in a measurement of the drift mobility (see below), and also applies in device applications (e.g. in solar cells, where photo-induced carriers take a finite time to drift to the collection electrodes).

The basic significance of this is that carriers (assumed to be) initially generated in the extended states may not have time to interact with the deep states which would ultimately limit their motion. In this case, the measured drift mobility (or that applicable for device purposes) will be significantly dependent upon the appropriate timescale. In the limiting case in which carriers could be extracted before any trapping, the free mobility would obviously apply. At longer times, for a localised state concentration which decreases continuously from the mobility edge, carrier trapping would first tend to occur into the shallowest states, by virtue of their high concentration. Subsequently, "thermalisation" (i.e. a transition in the energy of maximum occupation) to progressively greater depths would occur. Thus, the trapped carrier distribution might be expected to evolve with time as indicated schematically in figure 7. Obviously, the ratio of the concentrations of free and trapped carriers will decrease continuously with time, as therefore will the effective drift mobility.

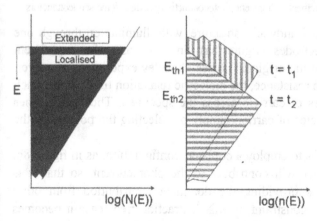

Figure 7. Time (t) evolution of the energy distribution of initially free carriers, in an exponential bandtail.

The above phenomenon is often described in terms of a "thermalisation energy", $E_{th}(t) = kT \ln(vt)$, where v is the attempt to escape frequency for the trapped carriers. This represents the energy at which the carrier release time is equal to the time since initial generation. It is argued that for shallower traps, sufficient time will have elapsed to allow the establishment of a quasi-thermal equilibrium between the trapped and free carrier populations. For deeper centres, the occupation will be determined simply by trapping considerations, and will thus follow the form of N(E).

Although the above concept allows a very convenient simplification of the analysis, we do not consider it valid in the general case. A more fundamental approach is to regard thermalisation as being primarily controlled by *trapping*, rather than *release*, considerations [4]. Even so, the thermalisation energy concept can be employed, with appropriate caution, in a variety of circumstances of practical interest, as illustrated below.

2.6. MEASUREMENT AND INTERPRETATION OF TRANSIENT CARRIER TRANSPORT

The thermalisation of initially-free excess charge carriers is most readily examined via transient photoconductivity experiments. In these [2], excess carriers are typically generated by a short flash of strongly-absorbed light from a pulsed laser.

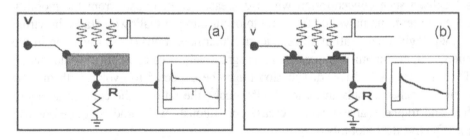

Figure 8. Systems for the measurement of transient photoconductivity in disordered semiconductors

Specimens may feature a "sandwich" structure, with illumination through one of the two (blocking) electrodes (figure 8a). In this case, the experiment constitutes a modification of the original Haynes-Shockley experiment. However, for materials of suitably high resistance, the dielectric relaxation time can be much longer than the transit times of carriers across the specimen. Thus, it becomes possible to study either species of carrier, simply by selecting the polarity of the applied field.

An alternative approach is to employ a coplanar configuration, as in figure 8b. Here, both species of carrier will contribute to the photocurrent, so that it is necessary to infer which is providing the dominant contribution from other experiments. Subject to this constraint, in this "extraction free" case it becomes possible to study the photocurrent over an extended time period (until recombination forms a limitation, or even thereafter, provided the photocurrent remains detectable (see below)).

From Section 2.5, it can be seen that the thermalisation of a set of initially free excess carriers is intimately related to the localised states with which they interact. Thus, transient photoconductivity data should contain valuable information on the nature and energy distribution of the shallow trapping centres. Various techniques, featuring different levels of complexity, have been advanced to extract such information (e.g. see [2] for reviews of the topic), and there is still significant activity in seeking to refining these. Below, we will only attempt a brief overview of some of the underlying concepts.

The simplest approach [5] to the interpretation of transient photoconductivity data arises from the thermalisation energy model of the relaxation process [6-8]. Here, it is argued that for a suitable (i.e. an exponential or similarly continuous and slowly-decaying) trap distribution, most of the trapped carriers will be situated within kT of $E_{th}(t)$, as in figure 7 (note the logarithmic $N(E)$ and $n(E)$ scales). Moreover, this population is assumed to be in quasi-thermal equilibrium with that of the free carriers providing the photocurrent at this time. Thus, adapting equation (4), one can write the time-dependent effective drift mobility as

$$\mu_d(t) = \mu_0 n_f/n_t(E_{th}) = \mu_0 \ (N_c/(kT \ N_t(E_{th}))) \ exp(-(E_{th})/kT). \qquad (6)$$

Rearranging this, and remembering that $E_{th} = kT \ ln(vt)$, we obtain $N_t(E_{th}) \ kT \ /N_c = (\mu_0/\mu_d(t)) \ (vt)^{-1}$, so that

$$N_t(E_{th}) = const \ / \ (I_{ph}(t) \ t), \qquad (7)$$

since the photocurrent, $I_{ph}(t)$, is obviously directly proportional to $\mu_d(t)$. Thus, a plot of $1/(I_{ph}(t).t)$ against $E_{th}(t)$ should map out the $N_t(E)$ distribution. Care is needed in this procedure, to ensure that measurements are genuinely being performed in the "pre-transit" (i.e. no carrier losses by extraction at the electrodes, or by recombination) regime.

At the opposite extreme, consider the "post-transit" regime of the photocurrent in a sandwich cell specimen. Here, most of the carriers have already been lost by extraction, and the residual current is controlled by emission of those carriers trapped in untypically deep centres. It can then be shown [9] that equation (7) should be replaced by the expression

$$N_t(E_{th}) = const' \ (I_{ph}(t) \ t), \qquad (8)$$

i.e. (other than for a different constant term on the r.h.s.) its complete inverse!

The above "pre-transit" and "post-transit" procedures are now quite commonly used in exploring materials of interest in the present context (particularly thin film amorphous and polycrystalline semiconductors). They naturally tend to broaden out sharp features in $N(E)$, because of the grouping of the states into kT slices, and also because of the basic assumption that $n(E)$ always peaks at E_{th}. They can thus seriously distort such distributions, making them appear closer to a simple exponential bandtail [10]. Also, they obviously depend critically on the assumption of a trap-limited band transport mechanism, with no contributions to thermalisation or to the photocurrent via hopping etc. However, subject to these caveats, they can prove extremely valuable and relatively easy to perform experimentally. Indeed, they and their variants may well constitute the only genuinely novel form of spectroscopy to arise from the study of amorphous semiconductors.

Before leaving this topic, it worth mentioning some of the more recent variants of the above interpretive techniques; in particular those based upon Fourier or Laplace transformation of the photocurrent data (e.g. [11]). Such transformations firstly convert the raw $I_{ph}(t)$ data into the complex frequency domain ($I_{ph}(\omega)$). A further manipulation yields a discrete representation of $N(E_n)$, where $E_n = kT \ln(v/\omega_n)$. This expression is formally equivalent to that for the thermalisation energy, but does not imply the same underlying assumptions concerning the thermalisation process. Impressively, such techniques are applicable to data from either the pre-transit or post-transit regimes, or where significant recombination losses are present (or any combination of these). The most recent variants are also able to yield a significant reduction in the "kT broadening" effect, producing impressively accurate reproductions of even sharply structured energy distributions.

3. Modelling of Transient Carrier Transport in Disordered Solids

As noted at various points above, the understanding of the properties of non-crystalline solids is greatly hindered by the difficulty of treating the effects of disorder in a rigorous manner. There is currently no equivalent of the Bloch theorem and associated concepts which allowed precise and elegant analyses of the properties of single crystal materials.

In the above situation, as far as the author is aware, *all* techniques so far advanced for interpreting experimental data for disordered semiconductors involve important assumptions and simplifications. The idea of a thermalisation energy, as defined above, is one common example. This concept is employed not just in the interpretation of the transient photoconductivity, but also in many other related experiments such as deep level transient spectroscopy.

It is also common to use highly simplified models for the energy distribution of localised states. The assumption of an exponential distribution is particularly frequently encountered. This is primarily, we believe, because it can greatly simplify calculations, rather than because its presence is predicted theoretically. However, another factor is that the broadening effect outlined above can distort even highly structured energy distributions towards an apparently exponential form [10].

In these circumstances, it becomes extremely valuable to assess the validity of the assumptions being made in any interpretive technique. Here, computer-based numerical modelling has proved to be particularly effective. Also, in addition to allowing checks to be made of underlying assumptions, it can provide valuable new insights into the details of transport processes etc.

The first attempts to undertake such modelling (e.g. [12]) utilised a Monte Carlo procedure. This remains an extremely useful approach, since any underlying assumptions can be clearly identified, and their consequences explored. For example, to examine the decay of the trap-limited transient photoconductivity for any required form of localised state distribution, the following basic steps are involved:

(a) use an appropriately-weighted random number, in conjunction with the overall free carrier trapping time, to determine the time before an initially free carrier becomes trapped,

(b) use a second random number, together with the N(E) being studied, to chose the depth of the centre into which trapping occurs,

(c) use a third random number, together with the trap depth and attempt-to-escape frequency, to decide the time before release to the extended states,

(d) loop back to (a), to continue following the progress of the carrier until the required conditions (e.g. arrival at an electrode or expiry of a specified time limit) have been reached,

(e) repeat the above procedures for a sufficiently large number of carriers, combining the results to yield the required data (current vs. time, trapped carrier energy distribution vs. time, etc.).

The Monte Carlo approach can be extended in various ways, for example to explore transport by hopping as opposed to trap-limited band transport (see below). Its only significant drawback is that data for significant numbers (typically 10^4) of carriers must be combined to reduce the noise level associated with the individual random events. Thus, it can require considerable amounts of computation time, and sometimes (e.g. for the modelling of hopping in a random spatial array of sites) computer memory.

An alternative approach for the modelling of trap-limited band transport is to divide the N(E) distribution into thin (e.g. kT) slices, and generate a matrix containing the rates for trapping into and release from these. This can then be employed to obtain similar information to that extracted via the Monte Carlo approach. This procedure allows much more rapid computation, but does have some limitations. For example, because the matrix elements are average values, it is not possible to generate and study noise data for various trap distributions. Also, because of problems associated with the choice of slice widths (see below), it has only recently become possible to consider using it to study transport by intersite hopping.

3.1 EXAMPLES OF THE NUMERICAL MODELLING OF TRAP-LIMITED BAND TRANSPORT

Here, we here provide some illustrative examples to demonstrate how the above procedures can be employed, and to indicate the insights they can provide. A more detailed treatment may be found in the literature [2,10].

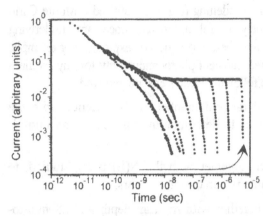

Figure 9. Monte Carlo simulation of the room-temperature transient photocurrent for a sandwich-configuration specimen of a material featuring a linear band tail of depth 0.25 eV. The arrow indicates the direction of increasing free carrier transit time (i.e. decreasing applied field).

Figure 9 displays the results of a Monte Carlo simulation of the (recombination-free) transient photocurrent for a specimen (sandwich configuration) with a linear band tail of depth 0.25 eV. The currents in the pre- and post-transit regions clearly do not have the simple power-law forms which occur for an exponential trap distribution [7,8]. Rather, the pre-transit current tends to saturate at low applied fields, since carriers have had sufficient time to reach quasi-thermal equilibrium with states at the bottom of the tail.

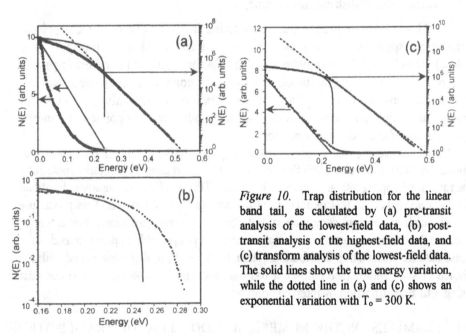

Figure 10. Trap distribution for the linear band tail, as calculated by (a) pre-transit analysis of the lowest-field data, (b) post-transit analysis of the highest-field data, and (c) transform analysis of the lowest-field data. The solid lines show the true energy variation, while the dotted line in (a) and (c) shows an exponential variation with $T_0 = 300$ K.

The results of applying various analytical techniques to some of the data in figure 9 are shown in figure 10. The pre-transit technique clearly distorts the true N(E) into an approximately exponential form, with a characteristic temperature equal to the experimental one, over much of the energy range, as emphasised by the exponential plot in figure 10a. Obviously, the thermalisation energy $E_{th}(t) = kT \ln(vt)$ *must* increase indefinitely with time, even though the band tail here has a finite depth. This clearly illustrates an important fundamental limitation of the procedure. It also demonstrates that non-exponential energy distributions will tend to be distorted towards an exponential form, possibly explaining why so many measurements tend to suggest such an approximate distribution.

The post-transit procedure (figure 10b) again attempts to compute N(E) for energies beyond the tip of the linear tail. However, the distortion is rather less than in figure 10a, as can be understood by a full consideration of the approach [10].

As shown in figure 10c, the transform procedure provides a considerably improved regeneration of the true linear form, but still yields a false exponential region (with $T_0 = T$) beyond the tip of the tail.

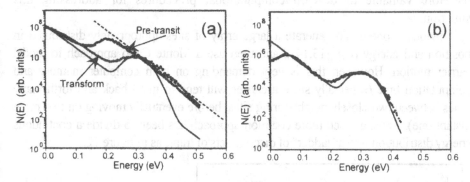

Figure 11. Trap distribution for a composite exponential + Gaussian N(E), as calculated by (a) pre-transit and transform analysis of low field data, and (b) post-transit analysis of high field data. The solid lines indicate the true N(E), while the dotted line in (a) is an exponential with $T_0 = T$.

Obviously, an N(E) as sharply structured and terminated as the linear one is not to be expected in reality. Thus, figure 11 shows similar data for a more realistic distribution, comprising an exponential tail plus an additional Gaussian feature. Here, the pre-transit technique again gives generally poor results, and even generates a peak in N(E) close to 0.2 eV, where the true distribution goes through a minimum. The transform technique again gives a much better reproduction of the actual N(E), but once more is unable to regenerate the correct distribution beyond about 0.4 eV. In this case, the post-transit technique is the most effective, as shown in figure 11b. It does not completely regenerate the minimum at about 0.2 eV, but does prove more effective in reproducing the rapidly-decaying regions. The reason for this is explained in the literature [10].

It can be seen from the above illustrations, that computer-simulation can provide valuable insights into the effectiveness of the various procedures which have been advanced for interpreting transient photoconductivity data in disordered semiconductors. A particularly important conclusion is that if any procedure yields an approximately exponential N(E), with a characteristic temperature similar to the measurement one, *the results must be regarded with extreme suspicion.* Measurements over a range of temperatures should then be made, to assess the accuracy of the conclusions.

3.2 EXAMPLES OF THE NUMERICAL MODELLING OF HOPPING TRANSPORT

The procedures examined in the previous Section all feature the assumption that transport occurs by trap-limited band transport. This is probably reasonable for measurements at sufficiently high temperatures in materials which do not feature very high degrees of disorder. However, when these conditions are not met, hopping transport will obviously emerge as a competing mechanism. It is therefore valuable to develop computational procedures for addressing this situation.

Obviously, one could generate a large array of sites appropriately distributed in position and energy (e.g. [13,14]), and then use a Monte Carlo approach to model carrier motion. However, this is very demanding on both computer memory and computation time (especially since a carrier will tend to jump back and forth many times between two closely neighbouring sites, before eventually moving on to a more distant one). Thus, a much more common approach has been to divide a continuous energy distribution into a "ladder" of discrete sets of traps, as in figure 12.

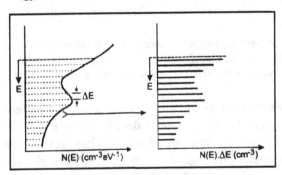

Figure 12. Division of a continuous energy distribution of localised states into a ladder of discrete sets of centres.

One can then compute average separations of sites within and between these slices, and use this as the basis for an analysis. However, there is a critically serious problem with such a procedure. This is that these average separations are obviously dependent on the slice widths, ΔE, chosen. This is vitally important, in view of the dominant role of the term $\exp(-2R'/R_o)$ in equation 1. One *could* chose $\Delta E = kT$, but this is entirely arbitrary (and iso-energetic hopping rates within a slice would then become significantly temperature dependent, which is illogical).

To surmount this problem for transitions to *iso-energetic* or *deeper* sites, the author has evolved two alternative new procedures [15].

3.2.1. *Fractional Probability Approach*

(i) The total number of sites within the band tail which are deeper than an energy E_m (assuming $N(E)$ to become negligibly small before the Fermi level is approached) is $N_{deep}(E_m)$, and can be obtained by straightforward integration for any model distribution.

(ii) Thus, the average separation of these is $r_{deep}(E_m) = N_{deep}(E_m)^{-1/3}$, and the net rate for hopping from a site at E_m into some deeper or iso-energetic centre is $\rho_{deep}(E_m) = \nu \exp\{- r_{deep}(E_m)/R_o\}$. Note the assumption here, as in many other analytical approaches, that random variations in intersite separation can be neglected for a sufficiently large population of carriers.

(iii) This transition rate can be employed (e.g. in a Monte Carlo approach) to calculate whether this process occurs prior to release into extended states or hopping into shallower traps. If so, since there is no energy term in the expression for $\rho_{deep}(E_m)$, the band tail can now be divided into thin slices, with the rate for hops to some arbitrary site in some other slice n simply being proportional to the fractional concentration of sites within it, i.e. $\rho_{m,n}/\rho_{deep}(E_m) = N(E_n)\Delta E/N_{deep}(E_m)$. Note that, as required, $\rho_{m,n}$ is now linearly dependent upon ΔE, provided that $N(E)$ does not vary appreciably across a slice.

3.2.2 *Differential Probability Approach*

Again consider hopping iso-energetically or downwards from slice m to slice n. Let $N_{deep}(E_m)$ be reduced by the number of sites in slice n (i.e. by $N(E_n)\Delta E$ for narrow slices). This yields a new increased intersite separation, corresponding to that of all *remaining* states in slice m and below. The hopping rate thus falls by an amount which should correspond to the *effective* rate $\rho_{m,n}$ for hopping from a site in slice m to the total collection of sites in slice n.

3.2.3 *Effective Transition Probabilities for Hopping to Shallower States*

The above approaches allow computation of the total effective rates of hopping to sites in an iso-energetic or deeper slice. To obtain corresponding effective rates for hopping to shallower slices, a relatively straightforward extension of the treatment is possible. Consider two slices at energies E_m and E_n, with $E_m > E_n$. The effective rate $\rho_{m,n}$ for hopping downwards can then be calculated as above. Detailed balance then requires that, in equilibrium, the fluxes in both directions must be equal. These are $n_{n,m} = N(E_n)\,\Delta E\,f(E_n)\,\rho_{n,m}$ and $n_{m,n} = N(E_m)\,\Delta E\,f(E_m)\,\rho_{m,n}$, where $f(E)$ is the Boltzmann occupation probability. Thus, one obtains $\rho_{m,n} = \rho_{n,m}(N(E_n)/N(E_m))\exp(-(E_m-E_n)/kT)$. The total rate for hopping to shallower states from slice m can then easily be calculated by summation.

3.2.4. *Contributions to the Hopping Current*

The procedures outlined above constitute a relatively straightforward procedure for calculating effective transition rates. Moreover, evaluation [15] has shown that they give very similar results when applied to an exponential bandtail (see below). To use them in a (e.g. Monte Carlo) simulation of the hopping current, it is necessary to couple them to the appropriate hopping distances. This, will not be independent of the choice of ΔE (or at least we have not yet found a way of making it so). However, the dependence is quite weak (as $\Delta E^{-1/3}$ for the hopping distance and $\Delta E^{-2/3}$ for the hopping current), in comparison to that of the $\exp(-2R'/R_o)$ term dominating the "ladder" approach. Thus, we can ignore it for many purposes.

3.2.5. *Illustration of the use of the hopping simulation procedures*

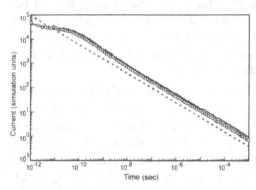

Figure 13. Transient photoconductivity for hopping conduction in an exponential bandtail. ($kT_o = 0.025$ eV; $kT = 0.01$ eV; for other parameters see [15]), determined by Monte Carlo simulation using the differential (solid line) and fractional (open circles) techniques. The dotted line is a power law of gradient 0.6 ($= 1 - (T/T_o)$).

Finally, in figure 13 we provide a brief illustration of the use of the above procedures. The functional form of the photodecay is very similar to that which would be obtained for trap-limited band transport with an identical exponential bandtail (i.e. a power law of gradient $1-(T/T_o)$). Thus, the simulation demonstrates that it is very difficult to distinguish between the two different transport processes from transient photoconductivity data. It is also possible to study various details of the thermalisation process, such as the time evolution of the trapped carrier distribution. So far, we have only done this for the zero-temperature case [16], but more detailed examinations are now in progress.

4. References

1. Mott, N.F. and Davis, E.A. (1979) *Electronic Processes in Non-Crystalline Materials*, 2nd Edition, Oxford: Clarendon.
2. Marshall, J.M. (1983) *Rep. Prog. Phys.*, **46**, 1235; (1985) *J. Non-Cryst. Sol.*, **77-78**, 425.
3. Monroe, D. (1985) *Phys. Rev. Lett.*, **54**, 146.
4. Marshall, J.M., Berkin, J. and Main, C. (1987) *Phil. Mag. B.*, **56**, 641.
5. Marshall, J.M. and Barclay, R.P. (1985) in D. Adler, H. Fritzche and S. Ovshinsky (eds.) *Physics and Chemistry of Disordered Solids*, New York: Plenum, p. 567.
6. Arkhipov, V.I. Iovu, M.S., Rudenko, A.I. and Shutov, S.D. (1979) *Phys. Stat. Sol.* (a), **54**, 67.
7. Tiedje, T. and Rose, A. (1981) *Solid State Commun.*, **37**, 44.
8. Orenstein, J. and Kastner, M. (1981) *Phys. Rev. Lett.*, **46**, 1421.
9. Seynhaeve, G., Barclay, R., Adriaenssens, G. and Marshall, J. (1989) *Phys. Rev. B*, **39**, 10196.
10. Marshall, J.M. (2000) *Phil. Mag. B*, **80**, 1705.
11. Main, C., Webb, D.P. and Reynolds, S. in J.M. Marshall, N. Kirov and A. Vavrek (eds.) *Electronic Optoelectronic and Magnetic Thin Films*, Taunton: Research Studies Press, p. 12.
12. Marshall, J.M. (1977) *Phil. Mag.*, **36**, 959.
13. Marshall, J.M. (1978) *Phil. Mag.*, **38**, 335; (1981) *Phil. Mag. B*, **43**, 401; (1983) *Phil. Mag. B*, **47**, 323.
14. Marshall, J.M. and Sharp, A.C. (1980) *J. Non-Cryst. Sol.*, **35-36**, 99.
15. Marshall, J.M. (2000) *Phil Mag. Lett.*, **80**, 691.
16. Marshall, J.M. (2000) *Phil Mag. Lett.*, **80**, 777.

MEASUREMENT METHODS FOR PHOTOACTIVE MATERIALS AND SOLAR CELLS

GEORGE S. POPKIROV

Central Laboratory for Solar Energy and New Energy Sources,
Bulgarian Academy of Sciences, 72 Zarigradsko Chaussee,
1784 Sofia, Bulgaria

1. Introduction

Solar cells and modules have been used for over four decades in space and terrestrial applications. During this time, the realisable energy conversion efficiency has increased markedly, due to a better understanding of the operating principles, the development of new design schemes, the application of new materials and a steadily improving manufacturing technology. This technological advance has been continuously supported by various new measurement and test methods for characterisation of the materials properties and device parameters, related to solar cells and modules. A well equipped laboratory may have facilities for: surface structure observation, structural and crystallographic analysis, compositional analysis, analysis of optical properties, resistivity measurements, determination of carrier density and mobility, solar cell performance testing, characterisation of solar cell inhomogeneities by mapping techniques, etc. Knowledge and experience from different fields of science and engineering should be united in order to use properly the existing, and develop new measurement and investigation methods.

The aim of this article is to introduce briefly some of the many different measurement techniques applied to solar cell characterisation. It is, however, beyond the scope of this work to present these methods in full detail, or to give literature references. The book by Schroder [1] can be recommended here. In some cases, a link will be given to websites, where an interested reader will find useful additional information.

The paper is divided into two parts. The first is concerned with the investigation and measurement methods applied to materials used in the production of solar cells. The second presents measurement techniques used to test or characterise ready-to-use solar cells.

2. Measurement Methods for the Characterisation of Photoactive Materials

We may consider materials used in the production of solar cells or modules as being photoactive if they contribute to the process of conversion of solar energy into

67

J.M. Marshall and D. Dimova-Malinovska (eds.),
Photovoltaic and Photoactive Materials - Properties, Technology and Applications, 67–92.
© 2002 *Kluwer Academic Publishers.*

electrical energy. The solar cell semiconductor and the antireflective coating are examples of photoactive materials. The metal electrodes used for front and rear contacts, the transparent conducting oxides used (in some cases) as front contacts, even the materials used for the encapsulation of the cells in modules and photovoltaic arrays are also related to the photoactivity of the solar cell.

A variety of properties of such photoactive materials are relevant to solar cell operation. Among these are:

- chemical composition, structure, grain size and grain boundaries,
- band-gap, light absorption,
- lifetime, mobility, diffusion length of current carriers,
- doping concentration profile in homojunctions,
- crystal lattice mismatch in heterojunctions,
- concentration and distribution of impurities, point defects, dislocations, defect clusters, metal precipitates,
- ageing, irradiation resistance, etc.

Here, we briefly present some selected measurement methods for the characterisation of photoactive materials in solar cells. We will try (but not precisely) to follow a very pragmatic classification of these methods, based on the way we (as experimentalists) disturb the materials under investigation, by:

- applying and/or measuring mechanical forces,
- applying or extracting heat (heating or cooling),
- applying and/or measuring electrical/magnetic fields,
- applying and/or measuring electromagnetic radiation (X-ray, ultraviolet, visible, infrared light),
- applying and/or investigating the influences of electron or ion beams,
- using combinations of the above (which further complicates the classification).

2.1. SURFACE PROFILOMETRY

Surface profilometry is used to measure minute physical surface variations as a function of position. Topographic maps of a sample surface, i.e. the surface texture (roughness, waviness, step heights and widths) can be obtained. Three types of profilers have been developed: the stylus profilometer, the optical profilometer and the AFM-profilometer. Scanning electron microscopy can also be used for surface profilometry, as described later in this paper.

Stylus profilometry is a contact-method, using a diamond stylus contacted to the sample surface with a force of $0.1 - 50$ mg. The sample under test is mounted on an X-Y stage, and scanned to obtain a 3-dimensional topographic map. An electromagnetic or laser interferometric transducer is used to precisely measure the vertical motion of the stylus as a function of its position (X,Y). The resolution in the Z-range is typically 50 Å $- 800$ μm.

Optical profilometry is a non-contacting method. It can use: (i) dynamic focusing of a laser beam, whereby a focus error signal is used to determine the surface topography; (ii) a fiber optic confocal laser scanning microscope; (iii) digital holography microscopy; or (iv) a reflection light microscope, equipped with an interferometer, the reference surface of which is mounted on a piezoelectric transducer. In all these cases, a CCD-camera is linked to a frame-grabber in a computer, for software control and evaluation. The Z-range resolution can extend from 0.1 nm to several mm.

Atomic force profilometry utilises an Atomic Force Microscope (AFM) head, mounted on an X-Y drive to extend the AFM measurement range in the X and Y directions. The precision obtained can reach 5 Å for 1 μm profiles and 10 nm for 10 mm profiles.

2.2. THERMAL HYDROGEN EFFUSION [2,3]

Hydrogen plays an important role in modern silicon technology. In the fabrication of amorphous (a-Si), microcrystalline (m-Si) and single crystal (c-Si) silicon solar cells, the incorporation of atomic hydrogen essentially decreases the bulk recombination, because it can be trapped at unsaturated covalent bonds, point defects and extended lattice defects. Thermal effusion measurements and SIMS (to be described later) can yield valuable information on the partial content and diffusion of weakly and strongly bonded H and H_2.

The sample to be tested is mounted in a quartz tube in an open vacuum system. The temperature is raised to 950°C at constant rate, and the effusion of H and H_2 is monitored using a quadrupolar mass spectrometer. To avoid errors due to traces of water, D_2 can be used instead of H_2 (the difference in binding energies Si-H and Si-D is < 1%).

2.3. FOUR-POINT-PROBE MEASUREMENT OF RESISTIVITY

The 4-point-probe method for the measurement of resistivity ρ and sheet resistivity R_s for wafers and films was originally introduced by van der Pauw [4]. In the usual configuration, four spring-loaded electrodes are aligned in a row and pressed onto the sample surface. A current I is driven between the outer two electrodes. The resulting voltage developed between the inner electrodes is measured with a high-impedance voltmeter. The measured resistance V/I does not depend significantly on the contact resistances between the electrodes and the sample. It does depend, however, on the current distribution in the sample. The actual sample conductivity can be calculated taking into account the sample geometry and the electrode configuration. For a semi-infinite volume $\rho[\Omega.cm] = 2\pi dV/I$. For wafers and films $\rho[\Omega.cm] = 4.532Vt/I$ and $R_s[\Omega$ per square]$= 4.532V/I$. Here, t is the wafer thickness and d is the electrode spacing [5].

2.4. THE HALL EFFECT

This well known phenomenon in semiconductors was discovered by Hall [6] in 1879. Consider charge carriers drifting through a semiconductor in the X-direction, due to a suitably applied electric field (e.g. a voltage applied between two ohmic contacts). An additionally applied magnetic field, perpendicular to the direction of current flow (e.g. in the Z-direction) will cause the charge carriers to experience a Lorentz force, deflecting them from the original current line. They will thus obtain a drift component in the Y-direction, and a potential difference (the Hall-voltage, V_H) will appear across the two sides of the sample. By measuring V_H, one can determine the density, n, or sheet density, n_s, of charge carriers in semiconductors. The Hall mobility $\mu_H = (qn_sR_s)^{-1}$, where q is the elementary charge, can be obtained [7] by combining a Hall-effect measurement of n_s with a 4-point resistivity measurement of R_s.

2.5. THERMALLY STIMULATED CURRENT (TSC)

In this technique [8], a semiconductor sample with two ohmic contacts is mounted into a vacuum cryostat and cooled in the dark to a starting temperature (e.g. 77 K). The sample is then illuminated, to create excess charge carriers and fill all traps. The illumination is maintained until a steady state condition is established. The temperature is then raised at a constant rate in the dark. The variation of the current, I_{TSC}, associated with the emptying of the traps is recorded as function of temperature. The dark current of the sample, I_d, due to the normal dark conductivity, can be subtracted by repeating the experiment without the illumination step. The energy position of the traps within the energy gap of the semiconductor can be determined from the initial rise slope of I_{TSC} vs. T. The density of traps of given depth, or their capture cross sections, can be determined from the total charge associated with the respective TSC peaks. Alternatively, the thermally stimulated capacitance of a semiconductor in a diode configuration can be measured.

2.6. DEEP LEVEL TRANSIENT SPECTROSCOPY (DLTS)

DLTS allows the determination of virtually all parameters associated with traps in semiconductors. A semiconductor sample is prepared in a diode configuration (p-n or Schottky). The deep levels are filled (charged) via a short pulse of forward-biased voltage (or light). After switching to reverse bias, the capacitance or current transients associated with the deep levels emitting their trapped carriers are recorded.

The amplitude of the measured signal is proportional to the density of the traps. The emission rate, measured as capacitance or current decay, depends on the temperature and the trap depth. The measurement is performed periodically with the filling of the traps with charge (filling pulse) and the emission of the trapped charge (emission pulse). Thus, synchronised measurements and averaging can be used for better signal recovery from noise. Increased sensitivity is the major advantage of DLTS over the thermally stimulated current or capacitance methods. New approaches to the signal processing of DLTS measurements, using digital averaging [9], Laplace [10] and wavelet [11] techniques have recently been developed.

2.7. THE SCANNING KELVIN PROBE

The Kelvin probe is a non-contact and non-destructive method for measurement of the contact potential difference (CPD) between two materials. A flat vibrating metal electrode (the probe), mounted plane-parallel to the sample, is used to form a vibrating capacitor with the sample under test. If the work functions of probe and sample are different, the varying capacitance of the probe-sample assembly will cause charging and discharging capacitive currents to flow through the external circuit. These currents can be set to zero by means of a bias voltage, called a "backing potential", applied between the sample and probe, and equal to the CPD. The work function difference is measured for metals, and the surface potential difference for non-metals. The method was developed by Lord Kelvin in 1898 and essentially improved by Zisman in 1932.

The work function is extremely sensitive to surface contamination, surface structural defects, the existence of adsorption or oxide layers, etc. This makes the determination of the absolute value of the sample work function very difficult. If the vibrating probe electrode is made small and is scanned over the sample, valuable information on the X-Y spatial distribution of the CPD can be obtained. The applications of a scanning experiment include 3D topography measurements of: work function, surface charge, surface stress, different kinds of surface contamination and inhomogeneity, adsorption and oxide layers, surface photovoltage spectroscopy etc [12].

2.8. ELECTRON PARAMAGNETIC SPECTROSCOPY (EPS)

This spectroscopy technique is often called electron spin resonance (ESR) or electron paramagnetic resonance (EPR). Due to its spin, each electron in a crystal lattice possesses a small magnetic moment. Normally, this is cancelled out by a paired electron with reverse spin. Unpaired electrons have permanent magnetic moments and appear as paramagnetic centres.

An external magnetic field ($\sim$0.3 T) is applied to split the initial energy level for the electron's magnetic moment into two discrete energy levels. Their energy difference depends on the magnetic field strength. Electron transfers from the lower to the higher energy level can be induced by selective absorption of low-amplitude microwave electromagnetic radiation in the 10 GHz range. The ESR spectrum is obtained by monitoring the microwave absorption as the magnetic field strength is changed. Usually the first derivative of the adsorption is plotted against the magnetic field strength.

The electron spin resonance effect is used to detect and investigate paramagnetic centres in crystals, e.g. for Si:
- the defect density in hydrogenated amorphous Si,
- the dangling Si-bonds at the $Si-SiO_2$ interface,

and unpaired electrons in molecules, e.g. radicals, molecules in a triplet state and d-metal complexes.

2.9. LIGHT REFLECTANCE MEASUREMENTS

Light reflection reduces the amount of light reaching the solar cell's volume, where photogeneration can occur. Antireflection coatings and/or surface texturing can be used to reduce the reflection of light from the semiconductor surface. The spectral dependence of the light reflection can be measured by means of a spectrophotometer with an integrating sphere. The function of an integrating sphere to spatially integrate the radiant flux is based on its diffuse reflecting surface. The sample under test is placed either in the sphere or at a port opening opposite to the light-entrance-port of the integrating sphere. The light reflected by the sample (diffuse and/or specular) is collected by the integrating sphere, and is measured using a coupled photodetector.

2.10. ELECTROREFLECTANCE SPECTROSCOPY (ERS)

Electroreflectance spectroscopy is a potential-modulated UV-visible reflectance spectroscopy technique. A modulated electric field is applied to the sample by means of a transparent front contact (e.g. immersion in an electrolyte), or in a contactless manner using a capacitor-like configuration. The change in the reflectance of monochromatic light at an electrode surface, in response to the potential modulation, is recorded over a range of wavelengths. ERS is sensitive to optical transitions in the band structure. Sample properties (and electrode processes in the case of a semiconductor-electrolyte interface) that influence the optical properties at the interface may be investigated by electroreflectance spectroscopy. The observed spectra yield information about the band gap and any built-in electric fields.

2.11. PHOTOREFLECTANCE SPECTROSCOPY

Photoreflectance spectroscopy is a contactless variation of electroreflectance spectroscopy. It can be successfully applied to investigate the surface Fermi level and the internal electric fields of semiconductors. Information about material parameters such as the band gap, the doping level, the surface state density, and the direction of band bending can be obtained.

2.12. MEASUREMENTS OF LIGHT ABSORPTION

The semiconductor absorption behaviour determines significantly the rate of electron-hole pair photogeneration, which is the most important process in photovoltaic conversion [13,14]. Upon illumination, a fraction of the incident light is lost due to reflection. The remaining light can be absorbed or be transmitted without absorption through the sample. In absorption spectroscopy, the absorption of nearly monochromatic incident light is measured, over a range of wavelengths.

The absorption coefficient depends strongly on the wavelength of the incident light. The physical processes which cause light absorption also vary with wavelength. Photons with energy higher then the semiconductor band gap (hv > E_g) are almost completely absorbed, due to band-to-band electron excitation. For some semiconductor materials, excitons (bound electron-hole pairs), which do not contribute to photoconductivity can also be generated near the absorption edge. Direct and indirect band gap semiconductors show different absorption behaviour for photon energies near the band gap energy. In some cases, light absorption can also occur below E_g. Near the absorption edge, the light absorption depends strongly on the dopant and defect concentrations.

Heavily doped semiconductors, e.g. the emitter region of *p-n* junction solar cells, show band-gap narrowing. Very high dopant concentrations can lead to the so called "Burstein-Moss" shift and an increased optical band gap.

Optical transitions between tail states below the band gap energy become possible for heavily doped partly compensated semiconductors. The band structure of semiconductors, and thus the light absorption, depend also on the concentration of lattice defects, e.g. in polycrystalline thin films with a high density of grain boundaries. Amorphous silicon is also a typical example. The distribution of the density of states in the band gap of a-Si:H can be determined from the spectral dependence of the optical absorption. High electrical fields, as in *p-i-n* thin film solar cells, can also affect the absorption of light (Franz-Keldysh effect).

The infrared and Raman spectra, due to specific vibrational excitations of molecular functional groups, are used in identifying the components in solids, liquids and gases.

2.12.1. *IR-VIS-UV Spectrophotometry*

A suitable light source, lenses, a monochromator and light detector can be used to configure a spectrophotometer for the measurement of light absorption in a semiconductor sample. The choice of light sources and detectors depends on the desired wavelength range. Usually, a deuterium discharge lamp is used in the UV range, and a halogen lamp is used in the visible and near-infra-red ranges. Photodiodes, photomultipliers and cooled PbS photodetectors are employed as transmitted light detectors. In some cases, care should be taken to correct the measured transmittance data for reflection losses. Some spectrophotometers have the necessary accessories, e.g. an integrating sphere. Double-beam spectrophotometers can be used to automatically subtract the substrate absorption in measurements on thin films, as well as to eliminate common background effects.

The light absorption in semiconductors can vary over many decades. As mentioned above, light is almost completely absorbed for photons of high energy ($hv > E_g$). On the other hand, the absorption is very low for photons with energies near the band gap. Other methods (see below) are more suitable for absorption measurements on almost transparent thin films, e.g. in the sub-band gap region.

2.12.2. *Photocurrent Methods*

Photocurrent methods are applicable only to photoconductive semiconductor materials. Two coplanar Ohmic contacts are used to apply a constant voltage. The sample is illuminated with light of given wavelength and intensity, and the photocurrent is measured. It is assumed that the light absorption is due to the excitation of photocarriers. Different experimental approaches are available:

- Measurement of the photocurrent and normalisation of it to the incident light intensity.
- Photocurrent measurement at a constant photon flux.
- The constant photocurrent method (CPM) - in this case the measured value is the number of incident photons necessary to maintain a constant photocurrent. The occupation of an energy level does not depend on the energy of the incident light, if the photocurrent is kept constant. Thus a CPM measurement directly yields information about the spectral dependence of the photoionisation cross-section of the defects present.
- Photocurrent measurements at different illumination levels.

In all these cases, a precision voltage supply and a current measurement device, as well as a suitable light source, a monochromator and a light detector, are used to configure the measurement system. Calibrated light detectors, or pyrodetectors with flat spectral characteristics, should be used to permit the determination of the number of photons in the incident light. Chopped light illumination and a lock-in technique can additionally improve the signal-to-noise ratio. Photocurrent methods (especially in CPM) can be applied for light absorption measurements in the low absorption region (10^{-1}–10^3 cm^{-1}). This permits the investigation of defect states in a-Si:H thin films.

2.12.3. *Cavity Ring Down (CRD) Absorption Spectroscopy*

The Cavity Ring Down technique [15] is based upon the measurement of the rate of decay of a light pulse confined in a closed optical cavity with a high Q-factor. A short light pulse is coupled into an optical cavity, formed by means of two plano-concave mirrors. The light pulse which entered the cavity is reflected many times by the two mirrors. Its intensity inside the cavity is monitored by measuring the small fraction of light leaking out through one of the mirrors. Due to losses at the mirrors, the light intensity inside the cavity decays exponentially with time, with a given time constant τ. The sample under test (solid, liquid or gas), when mounted in the optical cavity, will add absorption and/or scattering losses, and will thus increase the rate of decay of the light intensity inside the cavity. With typical mirror reflectivity losses of 10^{-4}, absorption coefficients below 10^{-9} cm^{-1} can be measured.

2.12.4. *Photo-thermal deflection spectroscopy (PDS)*

If charge carriers excited by absorption of light photons are allowed to recombine (e.g. there is no electric field to separate the generated electrons and holes) the absorbed light energy will be transferred into a heating of the crystal lattice, (in case of nonradiative recombination).

Photo-thermal deflection spectroscopy [16] is a method for absorption spectra investigation, based upon the heating of a sample by the absorbed light energy. A chopped (3 – 30 Hz) monochromatic light beam is focused on the sample under test. Light absorption causes the periodic heating, and thus periodic temperature changes and temperature gradients, in the sample and in the surrounding medium. Due to the temperature dependence of the refraction index of the sample and of the surrounding medium, periodic changes and gradients of the index of refraction arise due to the periodic illumination. A probe laser beam is directed parallel to and just above the sample surface, traversing the region with a high refraction index gradient. A position-sensitive photodetector can be used to measure the laser light deflection. The resulting deflection data and measurement set-up parameters can be used to calculate the absorption coefficient of the sample.

PDS can be applied over a wide wavelength range, and is several orders of magnitude more sensitive than standard transmission spectroscopy. It is ideally suited to measurements of weak absorption in highly transparent films, e.g. in the sub-band gap region of semiconductors.

2.12.5. *FT-IR Spectroscopy*

The absorption of infrared radiation results in vibrational excitation of molecule functional groups and polar bonds. A molecular asymmetry, or change of the dipole moment, is required for photon absorption to be allowed. Since chemical functional groups are known to absorb at specific wavelengths, the infrared spectroscopy of semiconductors permits the study [17] of:

- the type and concentration of incorporated impurities and defects,
- the concentration, effective mass and mobility of free carriers,
- the oscillator strength, frequency and damping of optical phonons,
- the geometry and optical functions of layered structures.

At low temperatures, before their thermal ionisation, dopant atoms can absorb IR-radiation at characteristic wavelengths. Impurities can be detected through their local vibrational mode, and extended defects can be detected if decorated with impurities. Free carrier concentrations can be obtained from the IR-reflection spectrum.

Modern spectrometers in the infrared range mostly use the Fourier transform (FT) technique, and are based on the principle of the Michelson interferometer. FT-IR spectrometers have higher sensitivity and are much faster than conventional devices. In contrast to a spectrophotometer using a monochromator, which scans single wavelengths in a given range, the FT-IR spectrometer yields IR-spectra 'at once'. Thus, complete IR-spectra can be monitored almost continuously in time-resolved measurements.

2.12.6. *Raman Spectroscopy*

Upon illumination with light of a given wavelength λ_o, reflected light and Rayleigh scattered light at λ_o are re-emitted from the sample. In addition, Raman scattering at different wavelengths ($\lambda \neq \lambda_o$), due to inelastic interactions of light with the sample molecules, is observed.

In Raman spectroscopy, the energy levels of molecules are explored by examining the frequencies present in the scattered light with respect to the frequency of the incident light – the so-called Raman shift. The Raman spectral lines arise as a result of the specific vibrations in the molecules scattering the light. Thus, the Raman spectrum appears to be a vibrational signature of a molecule or complex system. The information obtained is thus similar to that obtained from IR-spectroscopy. On the other hand, both spectroscopies (IR and Raman) complement each other. Raman spectra are more intense for more polarizable molecular vibrating groups. IR-spectra depend on the change in the dipole moment of the vibrating molecular group. Thus, there are functional groups which are well visible in Raman and are not visible in IR spectroscopy.

An advantage of Raman over IR-spectroscopy is that the incident radiation and scattered light can be in the visible range. However, Raman shifts are very small, so that purely monochromatic incident light is needed. Raman scattering has a very low intensity, so the incident light source must have a very high intensity to produce measurable scattering. Therefore, focused lasers are used as light sources and high quality monochromators are required to reliably separate the Raman lines from the much more intense Rayleigh line.

2.13. PHOTOLUMINESCENCE

Photoluminescence is the spontaneously emitted light radiation (luminescence) of a sample, caused by absorption of light of higher energy (shorter wavelength). Photoluminescence with decay times less than 100 µs is generally described as fluorescence, and with longer decay times as phosphorescence. In quantum-mechanical terms, phosphorescence is due to spin-forbidden and fluorescence to spin-allowed spontaneous electronic transitions. Photoluminescence spectroscopy investigates the spectral content of emitted light.

Excitation of photoluminescence can be used to probe the electronic band structure of a material. For compound semiconductors, photoluminescence offers rapid assessment of the uniformity, thickness, and composition of epitaxially grown wafers. Often photoluminescence studies are performed in a scanning mode. A scanning laser beam is used to excite the light emission. A monochromator, coupled to a suitable photodetector analyses the wavelength, intensity and spectral width of the emitted light. Maps of the photoluminescence intensity in a chosen wavelength window as a function of position, or maps showing the wavelength at maximum intensity, are recorded.

Applications include investigations of:

- radiative electron-hole recombination and measurements of minority carrier lifetime,
- optical transitions through point defects and impurity centres,
- photoluminescence exhibited by light-emitting porous silicon and nanoscale crystalline silicon (*nc*-Si) clusters, etc.

2.14. ELLIPSOMETRY

Ellipsometry is an optical technique for the analysis of very thin films. Linearly polarised light is directed onto the probe. The reflected light generally has an elliptic polarisation, with parameters depending on the wavelength of the incident light and on the Fresnel reflection coefficients of the probe material. The state of polarisation of the light reflected from the sample is measured. The complex dielectric constant and refractive index of the probe can be calculated. However, the sample may constitute a complex optical system, build up of various layers of different optical properties. In order to obtain a higher number of independent measurable quantities, measurements at different wavelengths or different angles of light incidence have to be performed. Provided a proper optical model exists, information about layer thickness, morphology, or the chemical composition of layers even thinner than the wavelength of the incident light can be obtained.

Nulling ellipsometers make use of polarisers and retarders in the optical path of the reflected light, to determine its polarisation change. The polariser and the retarder are rotated manually or under computer control, until the light intensity reaching a photodetector approaches null. New ellipsometers use phase modulation instead of 'nulling' to achieve a short response time, a higher sensitivity and a lower noise level. The thickness detection limit can be 1% of a SiO_2 monolayer. Ellipsometry measurements can be performed in a scanning mode, so that 3D images of the respective optical, morphological or chemical properties can be obtained.

2.15. SURFACE PHOTO-VOLTAGE (SPV) [18,19]

The contact potential difference of a system (see Section 2.7) containing a semiconductor sample depends in part on band bending in the space charge region (SCR) near the semiconductor surface. Illumination of the semiconductor, leading to changes of the SCR build-in potential and/or capture of carriers in surface states will result in a measurable surface photo-voltage.

SPV spectroscopy measurements with sub-band gap illumination can be used to study the energy distribution of surface states. The flat band potential of the semiconductor, or the open-circuit photovoltage of a photovoltaic system, can be determined upon illumination at a high light intensity and a sufficient photon energy ($h\nu > E_g$). Illumination with light of different wavelengths can yield information about the diffusion length of minority carriers. For low light intensities, the SPV is proportional to the SCR width, and thus depends on the doping concentration.

Various techniques can be used to measure the SPV. A vibrating Kelvin probe, using a transparent (conducting) probe electrode, can be used in a scanning mode for SPV surface imaging and diffusion length mapping. Alternatively, the SPV can be measured with chopped light and a non-vibrating capacitively-coupled transparent or semitransparent probe electrode.

2.16. PHOTOCONDUCTIVITY DECAY (PCD)

Study of the photoconductivity decay is a technique used for the determination of the minority carrier lifetime. A light pulse is used to create excess carriers in the semiconductor sample. The time constant, τ, of the photoconductivity decay is measured. The value of τ corresponds to the recombination lifetime of the excess carriers. The methods for measurement of PCD differ in the way the excess carrier concentration is detected:

- photoconductivity measurements with Ohmic contacts,
- measurement of the impedance of a coil inductively coupled with the sample, using e.g. a balanced impedance bridge,
- measurement of the microwave reflection, which depends on the electrical (conductivity), magnetic and geometrical parameters of the sample.

The PCD is sensitive to any imperfections acting as recombination centres. It allows a separation of surface and bulk recombination properties by using excitation light of appropriately different wavelengths. If a focused modulated laser beam is used to create the excess carriers, a scanning mode of operation can be utilised and minority carrier lifetime maps can be obtained.

2.17. QUASI-STEADY-STATE PHOTO CONDUCTANCE (QSSPC)

In contrast to the PCD, in the QSSPC-technique the information regarding the minority carriers lifetime is taken from the rising part of the transient photoconductivity, in response to a square pulse of light [20].

2.18. MODULATED FREE CARRIER ABSORPTION (MFCA)

In contrast to the PCD and QSSPC techniques, the MFCA [21] is based on the detection of the excess carrier concentration optically, by absorption of infra-red (IR) light. Therefore, this method is contactless and permits the measurement of minority carrier lifetime in a scanning configuration, with high spatial resolution. A sinusoidally modulated light source is used to excite excess charge carriers by homogeneous illumination of the whole sample. The IR-absorption through of the sample is measured using a focused laser beam. The phase shift between the incident light and the concentration of excess charges (i.e. IR-absorption) is used to calculate the lifetime. A map of the lifetime can be obtained by moving the sample with an X-Y positioner.

2.19. ELECTROLYTICAL METAL ANALYSIS TOOL (ELYMAT)

The ELYMAT [22] is a measurement technique for mapping the minority carrier diffusion length in Si wafers. A Si-wafer is mounted in a double electrochemical cell; i.e. both sides of the wafer are contacted with an electrolyte. The edge of the wafer is contacted by tungsten carbide tips. Both half-cells are filled with a HF-containing electrolyte. This results in the build-up of Schottky contacts on both sides of the wafer. Platinum wires immersed in the electrolytes on both sides of the cell are used as electrodes. A focused laser beam can be scanned onto the front side of the wafer. The resulting photocurrent can be collected either from the front or the back side of the wafer, or from both sides simultaneously. For (clean) Si in HF, surface recombination can be neglected. Thus, the measured photocurrents depend mainly on the local diffusion length. Using lasers with different wavelengths provides information on the diffusion length depth profile, and allows one to distinguish between bulk and surface recombination processes.

ELYMAT applications include:
- the mapping of defects, such as metal contamination or oxygen precipitates, in crystalline Si wafers,
- mapping the diffusion length in multicrystalline and crystalline Si wafers, for solar cell applications.

2.20. X-RAY ANALYSIS

X-rays have wavelengths in the Å-range ($\sim 10^{-8}$ cm), and thus can be scattered by the electron clouds of atoms. The diffraction can be used for structure analysis – the X-ray diffraction (XRD) technique. An X-ray photon of sufficient energy can eject an electron from an inner shell of the absorbing atom. Hence, if a sample is subjected to X-ray illumination, the incident photons can be absorbed - an effect used in the well-known technique of radiography, which will not be discussed here.

The kinetic energy of the emitted electron can be measured and used to determine the associated binding energy in the X-ray photoelectron spectroscopy (XPS) technique. The excited atom is unstable, since the electron vacancy which is created is short lived. An electron from a higher level in the atom can 'jump down' and fill the vacancy. In most cases, the innermost K and L shells are involved. This process of atomic relaxation is accompanied by emission of a characteristic X-ray photon. The effect forms the basis of X-ray fluorescence (XRF) spectroscopy, and finds extensive use for compositional analysis. A competing process to the fluorescence is the ejection of one of the outer-shell electrons – the Auger process.

2.20.1. *X-Ray Diffraction (XRD)*
Crystalline solids feature long range structural order, and behave like diffraction gratings for incident light of an appropriate wavelength, ~ 1Å (X-rays, but also electron and neutron beams). Measurement of the XRD can provide information about the crystalline structure, phase content or lack of crystallinity of a sample.

A highly collimated monochromatic X-ray beam is directed onto the sample, which is mounted on a goniometer. Diffracted X-rays arise from the constructive (in-phase) interference between two or more scattered X-ray waves. The intensities of the diffracted X-rays are recorded using a single photon counter, an integrating CCD camera, an image-plate detector, or a high-resolution photographic plate. The recorded diffraction pattern allows the reconstruction of the periodic structure of molecules or atoms in a crystal, or the electron density at an atomic scale. Perfect crystals yield perfectly sharp scattering, and the XRD pattern consists of a set of delta functions. For imperfect crystals, the peaks are broadened. Amorphous materials do not yield characteristic lines.

XRD can be performed using single crystal or powder samples. Measurements on single crystals permit the determination of the intensity of each reflection in three dimensions. They are thus used for precise structure determinations. Powder diffraction measurements are used mainly with polycrystalline materials. The powder samples should be finely ground to a uniform particle size, to avoid the appearance of preferred orientations. The sample appears as a collection of many randomly orientated small crystallites, and yields all possible reflections.

2.20.2. *X-ray Photoelectron Spectroscopy (XPS)*

As mentioned above, X-ray photoelectron spectroscopy is based on the photoelectric effect caused by X-ray photons. Monochromatic X-rays photons knock out core electrons from the absorbing atoms. If these photoelectrons are ejected from atoms near the top 10 nm of the surface, they can escape from the sample. Their kinetic energy can be measured and used to calculate the binding energy for electrons in the core shells of the respective atoms. A series of photoelectron peaks, characteristic of each element present close to the surface of the sample, can be obtained. The area under the measured peaks can be used to determine the relative concentrations of the atoms present, and hence the composition of the sample.

The X-ray photoelectric effect cannot be observed with hydrogen or helium, but with all other elements present at concentrations greater than 0.1 atomic %. XPS can also provide information about the molecular environment of the absorbing atoms - their oxidation state, bonding atoms, etc. This is because the electron binding energy is altered by the chemical state of the absorbing atom. Thus, XPS is also called Electron Spectroscopy for Chemical Analysis (ESCA). Angular-dependent XPS studies yield non-destructive elemental depth profiles up to 10 nm into the sample. Scanning XPS measurements with a spatial resolution down to 1 μm can be performed for surface composition mapping.

2.20.3. *X-ray Fluorescence (XRF)*

X-ray fluorescence spectroscopy is a fast, non-destructive technique for compositional analysis. As described briefly above, an atom which is excited by the absorption of an X-ray photon can re-emit another X-ray photon with lower, but characteristic, energy upon relaxing. This effect, called X-ray fluorescence, can be used to identify the elements in a sample by measuring the wavelengths of the emitted X-rays, i.e. by fluorescence spectroscopy. The intensity of the fluorescent radiation depends on the concentration of the respective elements in the sample. Therefore, XRF is simultaneously a qualitative and a quantitative technique.

An XRF-spectrometer consists of a high intensity X-ray source, a probe holder and an X-ray spectral analyser, including a sensitive X-ray detector. Instead of X-rays, irradiation with high energy electrons, alpha-particles or protons can be used for some applications. An XRF measurement needs no vacuum. The most critical element of the XRF system is the analyser crystal, which must have an exactly known and properly chosen (for the wavelength range of interest) atomic spacing. Fluorescence X-rays are collimated and directed onto the surface of the analyser crystal. After reflection according to the Bragg condition, X-rays of a given wavelength pass through an auxiliary collimator and enter the detector for intensity measurements. The analyser crystal and the detector can be rotated, so that sequential wavelength-dispersion measurements, i.e. XRF-spectroscopy, can be performed.

The XRF spectrum plots X-ray intensity against binding energy. The peaks which are obtained are characterised in terms of their energies and heights. The peak energy identifies which element is present in the sample. The peak height yields information about how much of this element is present (in relative units). Scanning XRF measurements with a spatial resolution approaching the micron level can be performed for surface heterogeneity assessment.

2.21. ELECTRON BEAM IMAGING AND ANALYSIS

A beam of electrons represents a type of "ionising radiation". Electrons can be scattered within a sample, through collisions with atoms. This:

(i) affects their trajectories, but not their kinetic energy – elastic scattering, or

(ii) involves a process in which energy is first transferred to the atoms – inelastic scattering, leading to a subsequent generation of phonons, plasmons, electron-hole pairs, secondary electrons, Auger electrons and X-rays.

In the first case, at high scattering angles, backscattered electrons from the top part of the electron beam / sample interaction volume can be detected. The backscattering increases for lower tilt angles of the incident electron beam. At higher incident energies, the electrons penetrate deeper. The interaction volume increases, but the probability for elastic scattering decreases.

In the second case, absorption of the electron beam energy may be accompanied by a mobilisation of conduction band electrons, the promotion of valence band electrons to the conduction band (generation of electron-hole pairs, conductivity increase, photocurrent) and by the ejection of secondary electrons from the sample. X-rays can be emitted, due to bremsstrahlung and to electron transfers in the atoms; the latter leading to characteristic X-ray emission.

Electron beams can be used for imaging and analysis. The electron beam penetration volume depends on the electron energy, the sample structure and the atomic number of the absorbing atoms. It is usually of the order of 1 μm. Scanning electron microscopy (SEM) and transmission electron microscopy (TEM) use detection of transmitted scattered, backscattered and emitted secondary electrons. X-ray images can be used for elemental (compositional) mapping. The detection of the emitted light is used in (scanning) cathodoluminescence. The spectral content of the emitted X-rays is characteristic for the absorbing atoms.

X-ray spectroscopy is employed in electron microprobe analysis (EMPA) using energy dispersive (EDX) or wavelength dispersive (WDX) techniques. The kinetic energy of the emitted Auger electrons yields information about the elemental composition. The current due to the generation of electron-hole pairs in the sample is used in electron beam induced current (EBIC) measurements. A comparison of the different analytical methods can be found in the literature [23].

2.21.1. *Scanning Electron Microscopy (SEM)*

Optical microscopes are limited to a maximum resolution of about 200 nm (magnification < 1000), due to diffraction restrictions for visible light. If electron beams are used instead, a magnification in the order of 10^6 and up to a nanometer resolution can be obtained. SEM also features a large depth of field – e.g. structural details up to 2 μm in height appear well focused at 10 nm resolution.

According to the principle of SEM operation, a focused electron beam (0.5 – 50 keV) is scanned across the surface of the sample. Detection of low energy (< 50 eV) secondary electrons permits the imaging of the sample surface topography. To avoid surface charging effects, dielectric samples must be pre-coated with a thin C or Au layer. Higher energy backscattered primary electrons give information on the spatial distribution of atoms within the top micron of the sample – atomic number (or orientation) contrast mode.

The applications of SEM include the imaging of grain size, roughness, porosity, material homogeneity, mechanical damage, micro-crack location, film and coating thickness determination, dimension verification, etc.

2.21.2. *Transmission Electron Microscopy (TEM)*

Transmission electron microscopy is used for the microstructural analysis of solid materials, down to a sub-nanometer scale. TEM is analogous to optical microscopy, but the light is replaced by a high-energy (100 keV – 3 MeV) electron beam. Glass lenses are replaced with electromagnetic lenses. As in optical microscopy, the beam needs no scanning. Due to the very short wavelength of high-energy electrons, a very high (atomic) resolution (0.1 nm) and a large depth of field are achievable. Magnifications of 350,000 times can easily be obtained. In some cases, atoms can be imaged at magnifications up to 15 million times.

The sample should be stable in vacuum, and resistant to the electron beam. It should be (or be made) thin enough (< 0.5 μm) to transmit the incident electrons. Dependent on the lens configuration, TEM can yield structural information either in a topographic view or as a diffraction pattern. In addition to high resolution topographic imaging, the TEM apparatus can be configured to record scanning

TEM images, energy dispersive spectroscopy (EDS) for x-ray elemental analysis, parallel electron energy loss spectroscopy (PEELS) for low atomic number element analysis, and electron holography for the determination of thickness, surface topography and electric field distribution.

2.21.3. *Energy Dispersive X-ray Spectroscopy (EDX, EDS)*
Energy dispersive X-ray spectroscopy can provide a rapid non-destructive qualitative analysis of the elemental composition of the top 1-2 μm of a sample. Usually this technique is implemented in a SEM or TEM for electron microprobe chemical analysis (EMPA).

A highly energetic (> 10 keV) focused electron beam is directed onto an electrically conducting solid sample, in vacuum. X-rays are emitted from the target. White X-rays with a continuous spectrum are due to the bremsstrahlung. Characteristic X-rays are emitted via excitation and subsequent electron transfer in atoms which have absorbed sufficient energy from the bombarding electrons. The emitted X-rays are quantified as number of photons vs. energy. The measurement is performed using a solid-state *p-i-n* Si(Li) detector, cooled with liquid nitrogen.

2.21.4. *Wavelength Dispersed X-ray spectroscopy (WDX, WDS)*
Wavelength dispersive X-ray spectroscopy can also provide a non-destructive quantitative analysis of the elemental composition of the top 1-2 μm of a sample. WDX differs from EDX only in the way the X-rays emitted from the sample surface are detected. An X-ray monochromator based on a diffraction crystal and a detector (gas ionisation chamber) are used to disperse and quantify the characteristic X-ray emission spectrum.

2.21.5. *Auger Electron Spectroscopy (AES)*
An inner shell electron may be ejected from an atom as a result of the inelastic scattering of electrons bombarding a sample. In a subsequent relaxation of the atom (electron from an upper shell filling the vacancy) a characteristic X-ray photon can be emitted, or an electron from the outer shell can be ejected with characteristic kinetic energy. The process depends on the chemical element involved, and in particular on the energy levels of the first ejected electron, of the electron filling the vacancy and the level from which the Auger electron leaves the atom.

AES is an analytical technique, based on the measurement of the kinetic energy of the Auger electrons, ejected upon electron beam bombardment of conducting surfaces. AES yields information about:

- elemental composition (except H and He) close to the surface (~ 10nm), because electrons can only escape from the solid if they originate from atoms near the surface,
- the relative quantity of each element,
- the bonding (chemical) condition of the elements present, provided the Auger kinetic energy is measured with high resolution,
- the spatial elemental distribution on solid surfaces, if combined with SEM,
- depth profiling, if combined with ion sputtering.

2.21.6. *Electron Energy Loss Spectroscopy (EELS)*

This technique utilises the inelastic scattering of low energy electrons in order to determine the vibrational spectra of surface species. Electrons impinging on the sample surface may lose their energy through a variety of mechanisms. A substantial number of electrons undergo elastic scattering, leading to a strong elastic peak in the spectrum. Additional weak peaks, corresponding to lower kinetic energies are superimposed. If a beam of mono-energetic incident electrons is used, measurement of the kinetic energy of the reflected or transmitted electrons can give information about the composition of the sample, plus the bulk and surface plasmon and phonon excitations. EELS is particularly sensitive for detecting trace quantities of transition elements and low-atomic-number elements. Plasmons are collective excitations of the electron gas, with energies of several eV. Therefore, measurements at high resolution, as in high resolution EELS (HREELS) permit the investigation of the vibrational spectra of chemisorption bonds, i.e. of adsorbed species on the surface.

The experimental configuration uses an electron monochromator to filter out a well-defined beam of electrons with a given incident energy, and an analyser to measure the energy distribution of the scattered electrons. The resulting spectrum consists of a strong peak due to the elastically scattered electrons, plus a series of characteristic weak peaks, due to electrons which have undergone discrete energy losses during inelastic scattering. EELS is used also in association with TEM.

2.21.7. *Reflection High Energy Electron Diffraction (RHEED)*

RHEED is a technique for the determination of surface structure. It can be applied for the investigation of interatomic spacings in the crystal surface, and for determination of its lattice unit cell sizes, crystal orientation at the surface and surface reconstruction. RHEED is commonly used in molecular beam epitaxy (MBE) to monitor the layer growth and to count the individual layers.

A high energy electron beam (3-100 keV) strikes the sample surface at a grazing angle. After diffraction from the crystal surface structure, the reflected electrons produce a diffraction pattern on a phosphor screen mounted opposite the electron gun. An atomically flat surface produces sharp RHEED patterns, while a partially completed or rougher surface leads to more diffuse patterns.

2.22. SECONDARY ION MASS SPECTROSCOPY

SIMS is an ion beam surface analytical technique. It is used for obtaining elemental and molecular chemical information about surfaces, and for detecting traces of impurities in semiconductors and metals. All elements are detectable. SIMS provides low detection limits and an excellent depth resolution.

A primary ion beam is directed onto the sample surface. Ions, atoms, molecules, charged and uncharged molecular fragments are sputtered from the sample surface. Re-sputtered primary ions, electrons and photons are also emitted from the surface. The emitted secondary ions are collected and analysed by mass spectrometry. In some cases, neutral atoms, molecules and fragments can also be ionised before being collected for mass spectroscopy. A time-of-flight or a quadrupole mass-spectrometer can be used, depending on the requirements. The elemental composition of the uppermost 10 to 20 Å of analysed surface can be identified.

The primary ion can be focused to a spot less than 1 μm in diameter. Thus, the lateral distribution of elements and molecules on the surface can be mapped. During the SIMS analysis, the surface is slowly sputtered away, so that depth profiling with a resolution of a few Å is possible. Usually, Cs^+, O^{2+}, Ar^+ or Ga^+ are used as primary ions. As a side effect, primary ions are implanted and mix with sample atoms to a depth of up to 10 nm.

2.23. ELECTROCHEMICAL MEASUREMENT METHODS

The ELYMAT technique, as presented above, is one example of how electrochemical methods can be used for the characterisation of a silicon wafer before it undergoes further processing in a solar cell production line.

- Electrochemical measurement methods are well known in electrochemistry, but find less attention in semiconductor characterisation. As Tomkiewicz [24] wrote: "One reason is an apparent reluctance of the solid-state community to get involved with 'wet chemistry'...". On the other hand, electrochemistry offers [25]:
- controlled electrochemical and photoelectrochemical etching, and recently methods for growing nano- and micropore Si and A_3B_5 materials [26,27],
- liquid-junctions to semiconductors which behave as Schottky barriers and can be used in photoelectrochemical devices (e.g. for solar energy conversion) and for 'contactless' investigations (as in ELYMAT).

Here, we will only mention some of the electrochemical techniques which can be employed for the investigation of semiconductor properties.

- cyclic voltammetry in the dark and upon illumination can yield information about semiconductor electrode reactions and their kinetics, adsorption processes, photoconductivity, flat-band potential etc.,
- spectral response of a semiconductor-electrolyte junction,
- impedance spectroscopy in a wide frequency range and with different electrode potentials can give information about reaction kinetics, adsorption and surface states, doping levels, the flat-band potential (Mott-Schottky plots), etc.

3. Measurement Methods for Solar Cells

The main purpose of measuring ready-to-use solar cells or modules is to obtain information about their energy-conversion efficiency. Related parameters are the open-circuit voltage, the short-circuit current, the point of maximum power, the fill-factor. These can be obtained by current-voltage (I-V) measurements in the dark and under illumination. Also of great interest are data obtained by spectral response measurements. Imaging techniques (IR-thermography, LBIC, EBIC, CELLO) provide useful information about the spatial distribution of the solar cell local parameters, e.g. the diffusion length of minority carriers, the series and shunt resistances, plus identification of defects and inhomogeneities. The information obtained can help to improve the technology for production of efficient and reproducible solar cells. Diagnostic analysis of solar cells and modules is performed on just-produced devices, during their operation (field-tests) and after prolonged operation, to test ageing effects.

3.1. THE I-V CHARACTERISTIC OF A SOLAR CELL

Measurements of the I-V characteristic of a solar cell are performed either in the dark or under illumination. I-V curves obtained in the dark can yield useful information regarding important solar cell parameters, i.e.:
- the series resistance (including the contact resistance between emitter and grid-fingers),
- the shunt resistance,
- the diode factor(s) and diode saturation current(s),
- the long-time stability of the cell.

Recently, the dark current I-V technique was extended for applications to photovoltaic modules [28].

I-V curves measured under illumination are routinely used to characterise a solar cell under test, and permit the estimation of:
- the open-circuit voltage,
- the short-circuit current (and short-circuit current density),
- the maximum power and working point of maximum power,
- the cell efficiency, provided the illumination conditions are known,
- the fill-factor,
- the series resistance,
- the shunt resistance.

The evaluation of gathered I-V data needs a proper theoretical model [29] in order to obtain a good fit to measured data and to extract the model parameters. The simplest model is based on a one-diode equivalent circuit, consisting of a current source, one diode, a shunt resistance and a series resistance. More elaborate investigations use a two-diode model for a more precise fit to the experimental results. Large area solar cells may need models which take into account the distributed series resistance [30] of the cell.

Precise I-V measurements need:
- a solar simulator for controlled illumination,
- cell temperature control,
- four-point electrical connections to the cell - to eliminate external series resistances,
- precise electronics for voltage supply and current measurement.

Computer control for data gathering and fit-parameter evaluation is a usual requirement. For example, a program for evaluation of solar cell IV-characteristics is available at http://www.uni-konstanz.de/FuF/Physik/Bucher/ivcc.html.

A critical element of the measurement equipment is the illumination source. The photocurrent and the photovoltage of a solar cell depend on light intensity. Solar cells have different spectral sensitivities. The sun radiates an essentially continuous spectrum of light, closely approximating the emission from a 6000 K black-body. However, the solar radiation arriving at the Earth is selectively attenuated, due to absorption and scattering in the atmosphere. In order to permit a meaningful comparison of solar cells performance tests obtained from different laboratories, and to rely on practically relevant results, standardised light sources have to be used for illumination. Solar light simulators are commercially available, or can be assembled in the laboratory. In most cases they consist of a high-pressure xenon-lamp equipped with optics and a set of filters. Important parameters are the integral intensity, and the spectral distribution and homogeneity of the light flux. Flash-light solar simulators are used for fast I-V measurements, or as a 'low-cost' alternative for measurements of large area solar modules.

3.2. FLASH-LAMP TECHNIQUES FOR I-V CHARACTERISATION

In some cases, solar cell characterisation can be performed under flash illumination, combined with fast data acquisition electronics. Fast I-V measurements have the advantage of eliminating the risk of solar cell heating during the measurement.

- I-V curves can be obtained under flash-illumination in a few milliseconds [31]. The measurement system can consist of a digital-to-analog converter programmed to produce a sweep bias voltage and an analog-to-digital converter to measure the current. In fast measurements, however, care should be taken to avoid or take into account the capacitive effect of the p-n junction of the cell.

- In a new version of the flash-illumination technique [32], he bias voltage is held constant and the light intensity is varied. A family of I vs. light intensity curves are measured at different bias voltages. A standard (disco-strobe) flash lamp can be used. No capacitive effects have to be considered, because the potential does not change during the measurement. A drawback of the method is that due to the rapid light intensity variation (i.e. non-steady-state condition), significant errors in current measurement may arise.

- A quasi-steady-state open-circuit voltage method for solar cell characterisation has been developed [33]. The open-circuit voltage of a cell is measured simultaneously with the light intensity change during a pulse from a flash-lamp. The data can be used to construct the implied I-V curve of the cell, neglecting the effect of the series resistance. The open-circuit voltage vs. light intensity data can be used to determine the saturation current densities and ideality factors. Because no current is collected, only a crude contact to the doped region of the cell is required. Thus, cells can be tested in the early stages of production, e.g. after junction formation.

3.3. MEASUREMENT OF SOLAR CELL SPECTRAL RESPONSE

Data from photocurrent spectral response measurements yield information about quantum efficiency, excess carrier generation, recombination and diffusion. Measurements are performed with bias wide-band light illumination superimposed on modulated monochromatic light. Grating monochromators or systems equipped with interference filters are usually used to deliver light of a given wavelength [34,35]. A lock-in voltmeter is used to measure the modulated photocurrent synchronously. An essential part of the equipment is the reference detector, used to measure the incident light intensity at any given wavelength. Calibrated photodetectors (photodiodes with known spectral and power characteristics) or calibrated pyroelectric detectors are usually employed.

3.4. INFRARED THERMOGRAPHY IMAGING

A typical photovoltaic module at 50 °C emits IR-radiation in the range 3-20 μm with a peak at around 9 μm [36]. An infrared imaging camera can be used for measuring the absolute and relative temperature distribution. Such measurements provide information about localised shunts, poor solder bonds, module bypass diode functionality, reverse bias heating in solar modules, and the temperature distribution in flat-plate and concentrator cells. Even storage batteries and the power processing electronics can be tested for defects. Localised heating due to currents through shunts can be stimulated by operating the solar cell upon reverse bias conditions in the dark. Bypass diodes in modules can also be localised under reverse bias. Poor solder bonds can be IR-imaged under forward bias conditions in the dark. The current will predominantly flow through the well-functioning bonds. Thus, the bad ones can be determined.

A lock-in thermographic technique applying periodic heat generation has been developed to improve substantially the sensitivity of the IR-imaging [37].

3.5. LIGHT BEAM INDUCED CURRENT (LBIC)

LBIC is a well-known technique for mapping the spatial distribution of the photocurrent in solar cells [38-40]. It is usually employed under short-circuit current conditions, and allows determination of the local diffusion length of the cell material from local photocurrent data, if surface reflectivity is taken into account.

In a typical LBIC measurement, an intensity-modulated focused laser beam of suitable wavelength is directed at the solar cell under test. The cell is mounted on a precise X-Y positioning table, and can be additionally illuminated from a bias-light source. The short circuit current is measured, and from this the locally excited short-circuit current component can be determined using a lock-in voltmeter. The local reflection of the laser light can be detected using an integrating sphere with a photodetector and a second lock-in voltmeter. It is important to note that even with low-power lasers, a laser beam focused to an illumination spot of diameter less than a micrometer can easily result in local overheating and a non-linear collection efficiency, due to very intense local photogeneration. Therefore, the laser beam intensity must be controlled, and very sensitive current measurement devices should be used.

A commercially available mapping analyser - PVScan 5000 by NREL [41] - can be used to map defects and grain boundaries, using reflectivity data and special surface etching, as well as being suitable for LBIC.

3.6. AN ADVANCED LBIC MEASUREMENT - CELLO

An advanced LBIC technique for solar cell local characterisation has been recently developed [42]. A solar cell is illuminated with near AM1 light intensity, and IS additionally subjected to an intensity modulated, scanning, focused IR-laser beam. The response (a.c. and d.c. current or potential) of the solar cell is measured for various fixed global conditions (different pre-set voltage or current values) during scanning. Applying an advanced fitting procedure to the measured data (current and voltage response) yields the spatial distribution of the photocurrent, the series and shunt resistance, the lateral diffusion of minority carriers, the quality of the back surface field and even allows the calculation of local I/V curves.

3.7. ELECTRON BEAM INDUCED CURRENT (EBIC)

EBIC is an imaging technique [43,44] which provides information about the sub-surface electrical and physical properties of a semiconductor device (e.g. solar cell), with a resolution typical for SEM. Thus, EBIC is an extremely sensitive technique to locate failure sites, such as junction defects, surface and subsurface damages, recombination centres and other defects affecting the short-circuit current of a solar cell.

EBIC is usually build-up around a SEM. The electron beam of the SEM generates electron-hole pairs in the semiconductor. Due to the relatively high energy of the SEM electron beam, a single primary electron can produce up to 10,000 free electron-hole pairs. Due to the built-in field of the solar cell junction, an external current, typically 1000x higher than the beam current, is generated. This multiplication effect is an important advantage of the EBIC technique. The EBIC current is measured with an external current amplifier. Its output value can be mapped synchronously with the regular secondary electron signal of the SEM, to produce the EBIC image.

3.8. TEST OF SOLAR CELLS FOR SPACE APPLICATIONS

Efficiency and reliability are the main qualities desired for solar cells designed for space applications. Temperature variations, electron and proton irradiation, the impact of meteorites and space debris, materials ageing, etc. can cause significant degradation of solar cell performance during prolonged operation in the space environment. Therefore, it is not surprising that solar cells developed for use in space are subjected to elaborate tests.

Here on Earth, space simulators are used to test spacecraft equipment and solar cells under conditions simulating the space environment: AM0 intensity and spectral content of illumination, temperature, vacuum, electron, proton and neutron radiation [45,46]. Solar cells are being used in space, and some are tested after years of use in orbit, as for example a solar array panel from the Mir station [47]. Tests are also performed on solar cells during space flights [48-49]. Measurements on cells subjected to intensive irradiation with flux of electrons have shown that the degradation of their efficiency is due to an increased concentration of recombination and compensating centres [50].

3.9. DIAGNOSTIC ANALYSIS OF SOLAR CELLS AND MODULES AFTER PROLONGED USED

An interesting report [51] presents the results of performance and materials diagnostic tests on a solar cell array which was in use for 20 years for power production. Field observations and laboratory tests were performed. Dark I-V tests and IR-thermal scans have shown some degradation due to reflection loss, and increased resistance in module wiring and terminations. Cross-sectioning and microanalysis have been applied to study the solder-joint metallurgy. The glass surface and the encapsulant were examined. Adhesional strength tests, SEM, XPS and AES were employed to study the encapsulant and its interfaces.

4. Conclusions

In this paper, we have described only *some* of the methods and techniques used to investigate and test solar cells, cell modules and the materials used in their production. Most of these measurement techniques are of general application (in other fields of science), but some have been specially developed for investigation or parameter determination of solar cells. Some of the techniques are already standard, and are commercially available, while others are only available in individual laboratories as custom equipment.

5. References

1. Schroder, D.K. (1998) *Semiconductor Materials and Device Characterization*, John Wiley & Sons, Inc., New York.
2. Sardin, G., Andreu, J., Delgado, J.C. and Morenza, J.L. (1988) Characterization of intrinsic and doped amorphous silicon through thermal hydrogen effusion, *Solar Energy Materials* **17**, 227 – 234.
3. Beyer, W., Hapke, P. and Zastrow, U. (1997) Diffusion and effusion of hydrogen in microcrystalline silicon, *Mat. Res. Soc. Symp. Proc.* **467** 343.
4. van der Pauw, L.J. (1958) A method of measuring specific resistivity and Hall effect of discs of arbitrary shapes, *Philips Research Reports* **13**, 1-9.
5. WWW-link: http:/four-point-probes.com/jannotes.html
6. Hall, E.H. (1879) On a new action of the magnet on electrical currents, *Amer. J. Math.* **2** 287.
7. WWW-link: http:/www.eeel.nist.gov/812/effe.htm

8. Milnes, A.G. (1973) *Deep Impurities in Semiconductors*, John Wiley & Sons, N.Y.
9. Kolev P.V., Deen, M.J. and Alberding, N. (1998) Averaging and recording of digital DLTS transient signals, *Rev. Scientific Instruments* **69**, 2464-2474.
10. Dobaczewski, L., Kaczor, P., Hawkins, I.D. and Peaker, A.R. (1994) Laplace transform deep-level transient spectroscopic studies of defects in semiconductors, *J. Applied Physics* **76**, 194.
11. Younan, N.H., Lee, H.S. and Mazzola, M.S. (2001) Estimating the model parameters of deep-level transient spectroscopy data using a combined wavelet/singular value decomposition Prony method, *Rev. Scientific Instruments* **72**, 1800-1805.
12. Baikie, I.D. and Estrup, P.J. (1998) Low cost PC based scanning Kelvin probe, *Rev. Scientific Instruments.* **69**, 3902-3907.
13. Bube, R.H. (1978) *Photoconductivity of Solids*, Robert E. Krieger Publishing Company, Huntington, New York.
14. Green, M.A. (1982) *Solar cells – operating principles, technology and system applications*, Prentice-Hall, Inc., Englewood Cliffs, NJ.
15. WWW-link: http://mlfsilly.sci.kun.nl/~richarde/crd.html#intro
16. Boccara, A.C., Fournier, D. and Badoz, J. (1980) Thermo-optical spectroscopy: detection by the 'mirage effect', *Applied Physics Letters* **36**, 130–132.
17. Krcho, D. (1998) FTIR spectroscopy for thin film silicon solar cell characterisation, *2nd World Conference and Exhibition on Photovoltaic Solar Energy Conversion*, Vienna.
18. Schroder, D.K. (2001) Surface voltage and surface photovoltage: history, theory and applications, *Meas. Sci. Technol.* **12**, R16–R31.
19. Datta, S., Ghosh, S. and Arora, B.M. (2001) Electroreflectance and surface photovoltage spectroscopies of semiconductor structures using indium-tin-oxide-coated glass electrode in soft contact mode, *Rev. Scientific Instruments* **72**, 177-183.
20. Cuevas, A., Stocks, M., Macdonald, D. and Sinton., R. (1988) Applications of the quasi-steady-state photoconductance technique, *2nd World Conference and Exhibition of Photovoltaic Solar Energy Conversion*, Vienna.
21. WWW-link: http://www.ise.fhg.de/Research/SWT/mfca.html
22. Carstensen, J., Lippik, W. and Föll, H. (1994) Mapping of defect related silicon properties with the ELYMAT technique in three dimensions, *Conference proceedings: "Semiconductor Processing and Characterization with Laser-Applications in Photovoltaics"*, Stuttgart, *Mat. Science Research Forum* **173-174**, 159.
23. Jaegerman, W. (1992), Forschungsverbund Sonnenenergie – Photovoltaic
24. Tomkiewicz, M. (1992) Photoelectrochemical characterization, in J. McHardy and F. Ludwig (eds.), *Electrochemistry of semiconductors and electronics*, N.J., Noyes Publications.
25. Pleskov, Yu.V. (1990) Solar energy conversion: a photoelectrochemical approach, Springer-Verlag, Berlin.
26. Föll, H., Carstensen, J., Christophersen, M. and Hasse, G. (2000) A new view of silicon electrochemistry, *Physica Status Solidi (a)* **182**, 7-16.
27. Langa, S., Tiginyanu, I. M., Carstensen, J., Christophersen, M. and Föll, H. (2000) Formation of porous layers with different morphologies during anodic etching of n-InP, *J. Electrochemical Society Letters* **3**, 514.
28. King, D.L., Hansen, B.R., Kratochvil, J.A. and Quintana, M.A. (1997) Dark current-voltage measurements on photovoltaic modules as a diagnostic or manufacturing tool, *Proc. 26th IEEE Photovoltaic Specialists Conference*, Anaheim, California.
29. Goetzberger, A., Voß, B. and Knobloch, J. (1997) *Sonnenenergie: Photovoltaik*, B. G. Teubner, Stuttgart.
30. Fischer, B., Fath, P. and Bucher, E. (2000) Evaluation of solar cell I(V) measurements with a distributed series resistance model, *Proc. 16th EPVSEC*, Glasgow, Paper VA1.40. (File D340.pdf on CD version)
31. Lipps, F., Zastrow, A. and Bucher, K. (1995) I-V characteristics of PV modules with a msec flash light generator and a 2MHz data acquisition system, *Proc. 13th European Photovoltaic Solar Energy Conference and Exhibition*, Nice, France.

32. Keogh, W. (2001) *Accurate performance measurement of silicon solar cells*, PhD-thesis, Australian National University.

33. Sinton, R.A. and Cuevas,A. (2000) A quasi-steady-state open-circuit voltage method for solar cell characterisation, *Proc. 16th European Photovoltaic Solar Energy Conversion*, Glasgow, Paper OB3.2 (File D285.pdf on CD version)

34. Field, H. (1998) UV-VIS-IR Spectral responsivity measurement system for solar cells, *NREL - Document CP-520-25654*.

35. Emery, K., Dunlavy, D., Field, H. and Moriarty, T. (1998) Photovoltaic spectral responsivity measurements, *NREL - Document: CP-530-23878*.

36. King, D.L., Kratochvil, J.A., Quintana M.A. and McMahon, T.J. (2000) Applications for infrared imaging equipment in photovoltaic cell, modules and system testing, WWW-Link: http://www.sandia.gov/pv/ieee2000/kingquin.pdf

37. Langenkamp, M., Breitenstein, O., Nell, M.E., Wagemann, H.-G. and Estner, L. (2000) Microscopic localisation and analysis of leakage currents in thin film silicon solar cells, *Proc. 16th European Photovoltaic Solar Energy Conversion*, Glasgow, (File D411.pdf on CD version) WWW-Link: http://www.mpi-halle.mpg.de/~solar/

38. Hiltner, J.F. and Sites, J.R. (2000) High resolution laser stepping measurements on polycrystalline solar cells, *Proc. 16th European Photovoltaic Solar Energy Conversion*, Glasgow, VB1.57, (File B154.pdf on CD version)

39. Litvinenko, S., Ilchenko, L. and Skryshevsky, V. (2000) Investigation of the solar cell emitter quality by LBIC-like image techniques, *Materials Science and Engineering* **71**, 238.

40. Kress, A., Pernau, T., Fath, P. and Bucher, E. (2000) LBIC measurements on low cost contact solar cells, *Proc. 16th European Photovoltaic Solar Energy Conversion*, Glasgow, VA1.39, (File D339.pdf on CD version)

41. NREL – Technology brief (Document: NREL/MK-336-21116, 8/96)

42. Carstensen, J., Popkirov, G., Bahr, J. and Föll, H. (2000) CELLO: An advanced LBIC measurement for solar cell local characterization, *Proc. 16th European Photovoltaic Solar Energy Conversion*, Glasgow, VD3.35, (File D407.pdf on CD version)

43. Corkish, R., Altermatt, P.P. and Heiser, G. (2001) Numerical simulation of electron-beam-induced current near a silicon grain boundary and impact of a p-n junction space charge region, *Solar Energy Materials & Solar Cells* **65**, 63-69.

44. Boudaden, J., Riviere, A., Ballutaud, D., Muller, J.-C. and Monna, R. (2000) *Proc. 16th European Photovoltaic Solar Energy Conversion*, Glasgow, VD3.62, (File D432.pdf on CD version)

45. WWW-Link: http://www.boeing.com/assocproducts/radiationlab/equip/cretc.htm

46. Severns, J.G., Hobbs, R.M., Elliott, N.P., TowsleyR.H., and Conway R.W. (1989) LIPS III-A Solar Cell Test Bed In Space, *Proc. 20th IEEE Photovoltaic Specialists Conference*, Sept 1988, *IEEE Aerospace and Electronics Systems Magazine*, Dec.1989, WWW-Link: http://stromboli.nsstc.uah.edu/

47. Visentine, J. (1998) Mir solar array return experiment, *NASA's Space Environments and Effects Program*, Fall 1998 Issue, p.2-3.

48. La Roche, G., Hoesselbarth, B. and Bogus, K. (2000) Evaluation of the flight data of the equator-s mini modules, *Proc. 16th European Photovoltaic Solar Energy Conversion*, Glasgow, PC1.4

49. Landis, G.A. and Bailey, S.G. (1998) Photovoltaic engineering testbed on the international space station, *presented at 2nd World Conf. on PV Solar Energy Conversion, Vienna*. WWW-Link: http://powerweb.grc.nasa.gov/pvsee/publications/wcpec2/PET.html

50. Bourgoin, J.C. and de Angelis, N. (2001) Radiation-induced defects in solar cell materials, *Solar Energy Materials & Solar Cells* **66**, 467-477.

51. Quintana, M.A., King, D.L., Hosking, F.M., Kratochvil, J.A., Johnson, R.W., Hansen, B.R., Dhere, N.G. and Pandit, M.B. (2000) Diagnostic analysis of silicon photovoltaic modules after 20-year field exposure, *presented at 28th IEEE PV Specialists Conf., Anchorage, Alaska*, WWW-Link: www.sandia.gov/pv/ieee2000/quinking.pdf.

STRUCTURAL AND OPTICAL PROPERTIES OF MICROCRYSTALLINE SILICON FOR SOLAR CELL APPLICATIONS

R. CARIUS

Institut für Photovoltaik Forschungszentrum Jülich GmbH,
52425 Jülich, Germany

1. Introduction

Recently, microcrystalline silicon (μc-Si:H) has become a very attractive material for thin film solar cells because it is seen as a materials system that is fully compatible with the thin film technology presently used for amorphous silicon. In addition to the advantages of this technology, i.e. low temperature processes on large areas with little material consumption on foreign substrates, μc-Si:H provides the extended spectral response of crystalline silicon to the near infrared region. Due to the indirect band gap of this material, the thickness of μc-Si:H absorber layer has to be in the micrometer range for sufficient carrier generation, even when using light trapping schemes. However, light induced metastability, which is a critical issue for amorphous silicon solar cells, seems not to be a problem in this material.

In contrast to (mono-)crystalline Si, μc-Si:H can not be considered as a homogeneous material. The microstructure is a complicated phase mixture of crystalline grains, grain boundaries, amorphous regions and voids. This paper is devoted to the techniques commonly used to determine the structural and optical properties of μc-Si:H and to the interpretation of the data obtained on material prepared by glow discharge techniques.

The first studies already noted that hydrogen dilution is a very important parameter in changing the gross microstructure of μc-Si:H from the growth of amorphous to microcrystalline material [1]. In particular, μc-Si:H can be obtained in a similar manner to a-Si:H, in a PECVD process at the standard (RF) frequency of 13.56 MHz by admixture of large amounts of hydrogen (H) to the process gas silane (SiH$_4$), denoted here as silane concentration in hydrogen, **SC**. However, under standard low pressure conditions, this leads to very low deposition rates, of the order of 0.1 Å/s. Studies by the Neuchatel group (e.g. [2]) and of the Jülich group [3] have demonstrated, that by using **V**ery **H**igh **F**requency PECVD, the deposition rate of μc-Si:H can be significantly improved. Meanwhile, high deposition rates have also been achieved using RF-PECVD at high pressures [4], in addition to preparation by the so called 'hot wire' or Catalytic CVD process [5].

The results presented in the following figures have been obtained on samples prepared by VHF-PECVD at 95 MHz, a substrate temperature of ~ 200 °C, a

93

J.M. Marshall and D. Dimova-Malinovska (eds.),

Photovoltaic and Photoactive Materials - Properties, Technology and Applications, 93–108.

pressure of 200 mTorr and a power of typically 5 W for a substrate size of 10x10 cm². As will be noted where appropriate in the text, similar results are obtained for material prepared by the other previously mentioned techniques.

In the following sections, three methods are presented which in combination provide a good insight into the microstructure of microcrystalline silicon, namely Raman spectroscopy, x-ray diffraction (XRD) and transmission electron microscopy (TEM).

2. Experimental Techniques and Results

2.1 RAMAN SPECTROSCOPY

Raman spectroscopy is the most frequently used technique for characterising the structural properties of μc-Si:H films. It is based on the inelastic scattering of photons by interaction with phonons, i.e. lattice vibrations. For microcrystalline silicon, the following considerations are of importance:

i)   In crystalline silicon only the excitation (or annihilation) of phonons which obey certain selection rules is allowed
$\Rightarrow$ a narrow Raman line at 520 cm^{-1} is observed, attributed to optical phonons

ii) In amorphous silicon, there is no translation symmetry due to the structural disorder, and therefore no selection rules apply
$\Rightarrow$ a broad spectrum at about 480 cm^{-1} reflecting the phonon density-of-states

iii) Changes of the inter-atomic distance of the lattice by compressive (tensile) strain lead to a shift of the Raman lines towards higher (lower) frequencies. An inhomogeneous strain distribution leads to a shift and a broadening of the Raman line

iv) Breaking the translation symmetry by structural defects (stacking faults, point defects, or even a high impurity level) leads to a 'softening' of the selection rule. Therefore, forbidden transitions become partially allowed, leading to an asymmetric tail of the Raman spectrum at low phonon frequencies

v)   Localisation of phonons, e.g. due to grain boundaries (size effects), leads to a broadening and a shift of the Raman spectrum

Raman spectroscopy on μc-Si:H is usually performed in the (quasi-) back-scattering geometry, i.e. at an angle of incidence almost perpendicular to the sample surface. For the excitation, laser lines of an Argon laser (λ_o = 488 or = 514 nm) or a HeNe laser (λ_o = 633 nm) are often used as the excitation source. The choice of the laser wavelength is of importance as it determines (i) the penetration depth of the light and (ii) due to resonance effects also its sensitivity to the amorphous and crystalline phases. Because the absorption coefficient depends on the microstructure of the film (see the section on optical properties) the probe depth for Raman spectroscopy may vary from ~50 nm (λ_o = 488 nm, a-Si:H) to ~500 nm (λ_o = 488 nm, c-Si).

Figure 1 shows typical Raman spectra for c-Si, a-Si:H and µc-Si:H, at λ_o = 488 nm. c-Si exhibits a symmetric spectrum centred at 520 cm^{-1}, with a width (full width at half maximum; FWHM) of about 3.5 cm^{-1} which is limited here to ~ 6 cm^{-1} by the chosen resolution of the system. The spectrum of the a-Si:H film is centred at about 480 cm^{-1}, with a FWHM of about 60 cm^{-1}. The µc-Si:H film shown is one which is 'fully microcrystalline' according to XRD, and exhibits an asymmetric spectrum located at 518 cm^{-1} with a FWHM of 12 cm^{-1}

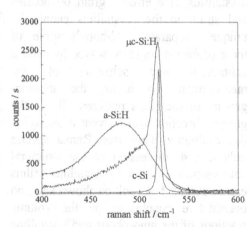

Figure 1. Typical Raman spectra of crystalline (c-Si), microcrystalline (µc-Si:H) and amorphous (a-Si:H) silicon.

and a tail at low phonon frequencies. This tail is sometimes interpreted as indicative of an amorphous phase. We will discuss this important aspect below.

Figure 2 shows the Raman spectra of thin (~ 0.5 µm) films prepared on glass substrates under the conditions described above, with the silane concentration in the gas phase as the parameter [6]. These spectra were taken in the standard configuration, i.e. from the sample surface. It is obvious that with increasing silane concentration, the spectrum changes from a 'fully crystalline' one at SC = 2% to an almost 'fully amorphous' one at SC = 6.2% (compare with Figure 1). At intermediate SC, the spectra are composed of a superposition of crystalline and amorphous ones, with varying contributions. In particular, they can be deconvoluted into three parts: a broad band centred at 480 cm^{-1} caused by the disordered phase, a peak at about 518 cm^{-1} attributed to the crystalline phase in the 'regular' (diamond-like) structure and a weak peak around 495 cm^{-1} which can be attributed to the presence of a hexagonal structure [7]. This structure is a consequence of a high density of stacking faults (twins) [8].

To obtain a semi-quantitative measure for the crystalline volume fraction from Raman spectra, the ratio of the scattering intensity of the crystalline phase to the total scattering intensity can be used, i.e.

$$I_{crs} = I_c / (I_a + I_c) \qquad (1)$$

where I_c denotes the area of the peaks at 520 cm^{-1} and 495 cm^{-1} and I_a the area of the peak at 480 cm^{-1}, determined from Gaussian fits to the spectra. This procedure has to be used with caution, as for several reasons it cannot provide a unique measure of the structure of the films. To name just a few problems that arise: The Raman cross sections and optical absorption coefficients of the amorphous and crystalline phases are generally different, and may depend on the size of the

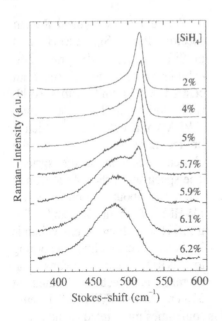

Figure 2. Raman spectra of μc-Si:H for different silane concentrations

crystallites, size effects, grain boundaries and strain in the crystallites cannot be uniquely separated. Although some of these problems could be solved by careful calibration and selection of the measurement conditions, the inhomogeneity of the films preferentially in the growth direction will prevent a satisfying data evaluation. Thus, from Raman spectra only a first impression of the structural composition of 'microcrystalline' films can be obtained. As will be shown later, no quantitative conclusion on the volume fractions of the amorphous and crystalline phases can be derived (see also [9]).

When the Raman spectra are excited and monitored through the glass substrate, the signal of the amorphous phase is much more pronounced for the samples with SC = 4 - 5%, indicating a higher amorphous volume fraction at the glass/film interface, as compared to the sample surface in this films. Such a 'structural development' from an amorphous growth at the glass/film interface to an almost completely crystalline growth is often observed. It has to be taken into account for the interpretation of the structural properties of the complete film, based on Raman spectra.

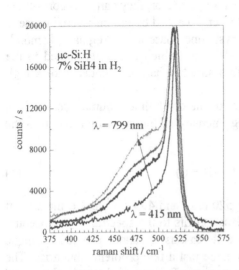

Figure 3. Raman spectra for different excitation wavelengths (penetration depths)

In many cases, the substrate or the device structure does not allow the investigation of films from the substrate side. In such circumstances, different laser wavelengths can be used to change the probe depth from the near surface region into the bulk of the film. An example is shown in Figure 3, where Raman spectra of a μc-Si:H film are shown for different excitation wavelengths, i.e. probe depths. The spectra provide a good survey of the structural development of the film, which seems highly crystalline in the near surface region and appears to become increasingly amorphous with increasing probe depth. Whereas this information can be taken as a rough fingerprint, a detailed interpretation in terms of the amorphous or crystalline volume fraction as a function of film thickness cannot be given. On the one hand, the penetration depth (probe depth) cannot be easily calculated, as it depends on the microstructure. On the other hand, the Raman cross section will vary, and the spectra are integrals composed of contributions from different depth profiles, which makes a more quantitative evaluation of the data very difficult. However, the fingerprint that can be taken from Figure 3 is consistent with the spectra taken through the glass substrate, i.e. it indicates an increasing amorphous volume fraction from the sample surface towards the substrate film/interface, and thus an inhomogeneous structural composition in the growth direction.

A quantitative evaluation of the fractions of the amorphous and crystalline volumes in the film is possible from XRD measurements. In addition, a comparison of XRD and Raman data may allow a better interpretation of the Raman data. In the next section, experimental aspects of XRD measurements on thin films will be briefly described, and data for a series of μc-Si:H films will be discussed with reference to their Raman spectra.

2.2   X-RAY DIFFRACTION (XRD)

In X-ray diffraction, photons of wavelengths similar to the atomic distances (here $\lambda \approx 0.154$ nm for the K_α-line) are diffracted by interactions with the atoms of the solid. By this, the inter-atomic distances can be determined. For the application of XRD measurements to the characterisation of microcrystalline silicon films, the following are of importance:

i)   Grazing incidence of the incident X-rays is used to enhance the sensitivity and to suppress the substrate contribution, as the penetration depth of the K_α-line is ~ 70 μm at normal incidence and ~ 1 μm at 1° relative to the sample surface

ii)   In this configuration, the reflecting lattice planes are **not** parallel to the sample surface

iii)   The diffraction patterns of crystalline silicon exhibit sharp peaks reflecting the <111>, <220> and <311> planes for diffraction angles up to 70° at $2\Theta_{hkl} = 28.4°$, $47.3°$ and $56.1°$; respectively.

iv)   Amorphous silicon is characterised by broad peaks: the so-called 'first sharp diffraction peak' at about 27° and a second peak at about 52°. The width is determined by variations of the atomic distances and the dihedral angles.

v)   Variations of the interatomic distances, e.g. by a high density of structural defects or grain boundaries, will lead to a shift and a broadening of the peaks.

vi)   The crystalline volume fraction in films exhibiting powder spectra (no texture) can be determined from the integrated intensities of the amorphous ($I_{a,XRD}$) and crystalline ($I_{c,XRD}$) peaks, after calibration by standards of crystalline and amorphous powder spectra:

$$X_{c,XRD} = 1/[1 - c\,(I_{a,XRD}/I_{c,XRD})] \qquad (2)$$

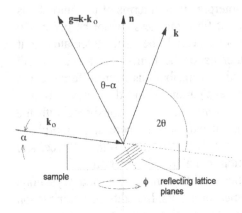

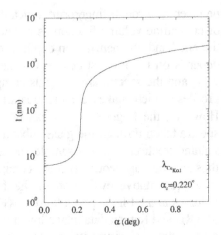

Figure 4. Scattering geometry of the asymmetric grazing incidence XRD.

Figure 5. Penetration depth of the amplitude of the electromagnetic wave of the Cu-K$_\alpha$ line, as a function of the angle of incidence.

In Figure 4, the scattering geometry of the XRD experiment is shown for the asymmetric grazing incidence case. Note that the reflecting lattice planes are not parallel, but are tilted with respect to the sample surface in contrast to the commonly used Θ-2Θ configuration.

As noted in (i), XRD at grazing incidence is important for obtaining signals from the thin film and for suppressing contributions from the substrate. The penetration depth, $l(\alpha)$, for the amplitude of the electric field of the K$_\alpha$-line in silicon is plotted in Figure 5 (on a logarithmic scale!), as a function of the angle of incidence α. It is obvious that above the so called critical angle $\alpha_c = 0.220°$, the penetration depth increases by about 2 orders of magnitude from about 10 nm below α_c to about 1 µm in a very narrow angular range. To avoid the difficulties that can arise at the critical angle, and still maintain a good suppression of the substrate, $0.5° < \alpha < 1°$ was chosen for our experiments.

In figure 6, the X-ray diffractograms for the same sample series as shown in figure 2 are presented, for diffraction angles up to 62° [9]. The spectrum for the sample with SC = 2% consists of sharp peaks from the crystalline phase, in particular the reflecting <111>, <220> and <311> lattice planes and a broad background of the substrate (Corning glass 7059) at about 25°. The substrate is visible only because of the small thickness of this sample (~200 nm), whereas the other samples are much thicker (> 500 nm).

With increasing SC, the crystalline peaks are broadened and the spectra exhibit an increasing contribution from the first and second sharp diffraction peaks of the amorphous phase, which dominates the spectrum for SC = 6.2%. Whereas the contribution of the amorphous phase is clearly visible and easily interpreted, the broadening of the crystalline peaks might be due to disorder, strain effects and/or a decreasing domain size. It is important to note that the domain size is not necessarily the size of a crystallite in a disordered matrix, but should be considered

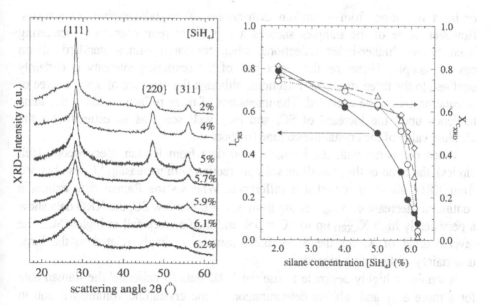

Figure 6. Grazing-incidence X-ray diffractograms for different silane concentrations [SiH₄]. Compare to figure 2.

Figure 7. Crystalline volume fraction of µc-Si:H films as function of SC; Raman: •, XRD: ◇[111], △[220], o[311].

as the size (extension) of coherently scattering domains perpendicular to the reflecting lattice planes [9].

In order to obtain volume fractions for the amorphous and the crystalline phases, the scattering intensities of the individual phases are calculated. This is valid since the weak scattering allows the assumption of an incoherent superposition of the signals from the different phases, and thus the signals are proportional to their scattering volume. The crystalline diffraction peaks can be fitted with Lorentzian functions. For the amorphous phase and the glass substrate, reference spectra have to be used. The crystalline volume fraction for the different diffraction peaks, *hkl*, is then calculated from equation (2). The constant c, which corrects for the proportionality between the scattering volume and the intensity, was experimentally obtained from a thermally crystallised microcrystalline film. Please note that in order to obtain a valid calibration, the a-Si:H and the crystallised reference samples were measured under identical experimental conditions, including film thickness and specimen shape.

In Figure 7, the volume fractions for the three strongest reflections, $X_{c,XRD}(111)$ $X_{c,XRD}(220)$ and $X_{c,XRD}(311)$ are shown on the right hand side, together with the results obtained from the Raman spectra on the left hand side [9]. There is a small discrepancy between the different $X_{c,XRD}(hkl)$, which is most prominent at higher silane concentrations. This difference is related to a slight deviation from a single-crystal powder pattern. A careful inspection of Figure 6 reveals an unsystematic variation of the ratio of the peak power of the reflections, upon variation of the silane concentration. This may be caused by differences in the structural properties or by a deviation from a random distribution of crystallographic orientations. However, none of the samples showed a significant redistribution of scattering intensity into higher-index reflections, when compared with a standard silicon powder sample. Therefore, the major part of the scattering intensity is certainly confined to the three strongest reflections, although the presence of some degree of orientation cannot be excluded. The intensities of these reflections show the same tendency upon the increase of SC, and may well serve as an estimate for the absolute value of the crystalline volume fraction.

When compared with the intensity-ratio data from Raman spectroscopy (full circles), the trend of the crystalline volume fraction with increasing SC, as obtained from XRD, shows a remarkable difference. Whereas the Raman data exhibit a continuous decrease of $X_{c,rs}$, starting from $X_{c,rs} \approx 0.8$ at low SC, the XRD data show a persistently high $X_{c,XRD}$ up to SC = 5% and then a steep fall at higher SC. The reverse order of X_c for the SC = 6.2% sample is simply related to the large uncertainty of the fitting procedure.

It would be highly desirable to use the XRD data to 'calibrate' the Raman data for a more easy and reliable determination of the crystalline volume fraction in μc-Si:H films by optical means. Unfortunately, the data presented here are not suitable to serve even for a rough calibration as:

- the penetration or probe depths of the two experiments were different

- the influence of the domain size on the Raman spectra is not yet clear.

While the penetration depth could in principle be adjusted by a proper choice of the angle of incidence (XRD) or excitation wavelength (Raman), the second point would still inhibit an accurate data evaluation. The choice of the penetration depth is of particular importance for samples prepared in the region of the transition from crystalline to amorphous growth. This will become more clear from investigations by transmission electron microscopy, which are the subject of the next section.

2.3 TRANSMISSION ELECTRON MICROSCOPY (TEM)

TEM allows the investigation of the structure of solids, with atomic resolution, by using the diffraction of high energy electrons. The strong interaction of the electrons with matter on the one hand reveals a much better contrast as compared to XRD, allowing a larger variety of investigations. On the other hand, the interpretation of the images (even though intuitively more easy) is often difficult. In addition, the sample preparation is difficult, and the samples may deteriorate during preparation. This prevents TEM from being routinely applied, and requires careful checks of the samples at different stages, to assure reliable results. In the present case, the cross-sectional specimens were prepared by argon ion milling. The bright- and dark-field images were recorded with a JEOL 2000EX microscope operated at 200 kV. For high resolution lattice fringes, a JEOL 4000EX operated at 400 kV was used [9, 10].

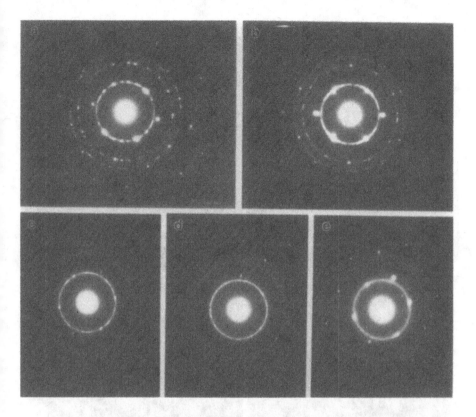

Figure 8. Selected-area diffraction patterns obtained from a cross-sectional TEM microcrystalline silicon specimen. The samples are from the same series as those in figures 2, 5 and 6. They were deposited at SC = 2% (a), 4% (b), 5.7% (c), 5.9% (d) and 6.2% (e).

For the application of TEM measurements in the characterisation of microcrystalline silicon films, the following are of importance:

i)   In the bright field images, the image is formed only by the 'non-diffracted' electrons, i.e. crystallites appear dark.

ii) In the dark field images, the image is formed by those electrons which have been diffracted into a certain angle, determined by the position of an aperture. Crystallites that fulfil the 'Bragg condition' appear bright.

iii) Individual lattice planes are shown in high resolution TEM

iv) By imaging the diffracted electrons, diffraction patterns are obtained which give additional information on the crystallinity, on the distance between the lattice planes, and on the preferential orientation.

In figure 8, selected area diffraction patterns are shown. These were obtained from the same sample series as in previous sections (figures 2, 6 and 7). The area contributing to the pattern has a diameter of approximately 500 nm. It encloses the

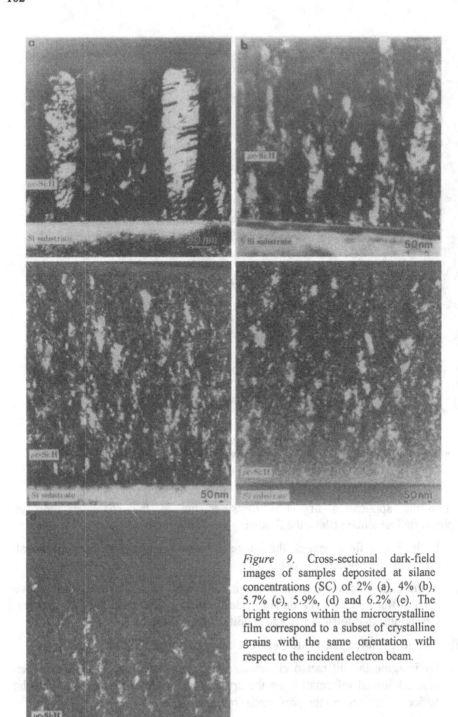

Figure 9. Cross-sectional dark-field images of samples deposited at silane concentrations (SC) of 2% (a), 4% (b), 5.7% (c), 5.9%, (d) and 6.2% (e). The bright regions within the microcrystalline film correspond to a subset of crystalline grains with the same orientation with respect to the incident electron beam.

film-substrate interface, and therefore bright spots related to the zero-order Laue zone of the Si(001) substrate are present. The diffraction patterns exhibit a picture typical of μc-Si:H, i.e. for the films with high crystalline volume fraction diffraction spots originating from coherently scattering domains arrange to yield spotty Debye-Scherer rings. Most prominent are the inner three rings, with radii equal to the reciprocal distances of the <111>,<220> and <311> lattice planes of silicon. They are most clearly seen in figures 8a and b. In figures 8c and d, the outer rings are hardly visible because of the poor contrast. The <111> ring still exhibits a spotty character, but in addition a diffuse halo pattern emerges. This is most prominent in figure 8e, and corresponds to amorphous silicon. The distribution of the spots along the ring lacks any regularity in all samples, i.e. the related distribution of the orientation of the crystallites appears to be random.

In figure 9, dark field images of the same series of samples shown in the previous figures are presented [9]. The bright regions within the film correspond to crystalline regions with nearly identical crystallographic orientations with respect to the electron beam. In figure 9a, the columnar structure of the μc-Si:H film with a high crystalline volume fraction is obvious. The crystalline columns are aligned perpendicular to the substrate, i.e. parallel to the growth direction. They can be subdivided into two classes, one of which contains columns that extend over the whole film thickness. For the 2% sample, these columns have a width of approximately 50 nm. The second class consists of columns with restricted height, which in the case of the 2% sample are located close to the interface between the substrate and the film. Within the columns, the contrast behaviour reveals an inner structure. On the right hand side of figure 9a, an irregular fringe contrast is found in the bright column, which can be attributed to twinning. The column on the left hand side exhibits a fragmented contrast which can be ascribed to crystalline regions with different crystallographic orientation, or to disordered regions. In this case, it is not clear whether these columns represent crystalline grains with inner defects such as twins or whether they are agglomerates of interconnected individual grains with irregular shapes but similar orientations. A similar structure has been observed in previous studies [11], for highly crystalline samples deposited on glass substrates.

The columnar structure is also visible in the film prepared with SC = 4% (figure 8b), but most of the columns do not extend over the whole film thickness and their width is smaller. Although it is not so obvious, the twinning and the fragmented contrast is also found in these columns. For higher SC, i.e. 5.7 and 5.9 %, the columnar growth gradually disappears and the microstructure can be characterised by agglomerates of crystallites (see figures 9c and 9d). These agglomerates exhibit a preferential alignment parallel to the growth direction. Simultaneously, the number density of crystallites decreases, the size of the coherent regions is greatly reduced and an amorphous phase appears. This amorphous phase dominates the structure close to the interface between the film and the substrate. In particular, the film with SC = 5.9% shown in figure 9d exhibits an initial layer which is predominantly amorphous up to 30 nm in height. However, nucleation of crystallites occurs at a low density within this 'incubation' layer; these being the nuclei for the development of the crystalline structure. The film with SC = 6.2% contains a number of randomly dispersed small and irregularly shaped crystallites embedded in an amorphous matrix. Thus, the indication of a small crystalline volume in the Raman spectrum of this sample is validated, and the TEM appears to be more sensitive to the Raman spectroscopy in this case.

From the structural investigations, and in particular from TEM results on a large variety of samples, a schematic diagram of the prominent structural development of microcrystalline films has been derived [12], as shown in figure 10. Here, we have chosen the more general parameter of crystalline volume fraction instead of SC, since the transition of the microstructure from highly crystalline to fully amorphous can be achieved by variation of deposition conditions. The details, i.e. the width and position of this transition, are sensitive to the deposition parameters, e.g. SC, the applied plasma power or the plasma excitation frequency.

The general trend of the microstructure of microcrystalline silicon can be described in the following way [12]: At high SC, crystalline growth starts at nucleation centres near the interface between the film and the substrate. During the competitive growth between the nucleation centres, the diameter of the remaining crystallites increases, resulting in a conical shape of the crystallites near the substrate surface. The space between these crystalline grains can be filled with amorphous silicon and/or voids, strongly dependent on the deposition conditions and the substrate [13]. Since no grain coalescence is found, a columnar structure evolves due to equally fast growing grains which form stable grain boundaries. Characteristics like the grain sizes and phase composition are then independent of the film thickness. A disordered network can only be found in the form of grain

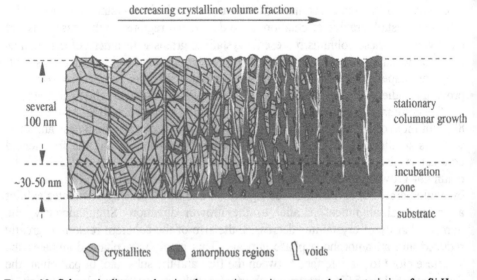

Figure 10. Schematic diagram showing the prominent microstructural characteristics of μcSi:H. From the left to the right, the film composition changes from highly crystalline to predominantly amorphous.

boundaries between the crystalline columns. Thus the amorphous volume fraction is low. The diameters of the crystalline columns depend on the deposition conditions and increase with increasing plasma excitation frequency. However, the material within the columns is not perfectly single crystalline, but exhibits a large number of twin defects. Thus, the size of the perfect single crystalline domains is smaller. In the transition from highly crystalline to predominantly amorphous growth, the column size decreases while more amorphous phase is incorporated between the crystalline grains. This interrupts the columnar crystalline structure in the lateral direction. Note that the material on the right hand side of the figure is *not* typical for the amorphous silicon used in a-Si:H solar cells, but shows the transition region between µc-Si:H and a-Si:H. This material is characterised by columnar growth of amorphous silicon with very small crystalline domains embedded in the amorphous matrix. It is obvious that the electrical and optical properties of the material will be affected by such structural changes.

A word of caution should be given here: It is well known that, besides the parameters mentioned above, the surface structure of the substrate has a very strong influence on the nucleation of the microcrystalline film. Therefore, the results presented here, although supported by many investigations of other groups, should not be considered as unique.

In conclusion, we have shown that microcrystalline silicon has a very complex microstructure which can be altered by variation of the deposition parameters. The complexity of the microstructure requires the application of elaborated diagnostic tools such as TEM, XRD and Raman spectroscopy to obtain useful results. Up to now, a unique link between the 'more easy' and thus more frequently applied Raman spectroscopy and (in terms of crystalline volume fraction and domain size) the more quantitative XRD method is not available. Instead, TEM provides images which seems to 'image the reality'. However the interpretation of these images is not always straightforward, and great care is needed in order to avoid misinterpretation.

3. Optical Properties

The most direct, and perhaps the simplest method for probing the band structure of a semiconductor is optical absorption. In the absorption process, a photon of known energy excites an electron from a lower- to a higher-energy state. The rapid rise of the absorption as a function of photon energy, which is a signature of the fundamental absorption, can be used to determine the energy gap of the semiconductor. The energy dependence of the absorption coefficient gives evidence for the direct or indirect nature of the gap.

Usually, the absorption coefficient, α, is determined by a straightforward measurement of the intensity of the incident light, I, the transmitted light, T, and the reflected light, R:

$$T = \frac{(1-R)^2 \exp(-\alpha d)}{1 - R \exp(-\alpha d)}, \qquad (3)$$

with d being the sample thickness. For large αd this can be approximated as

$$T \approx (1 - R)^2 \exp(-\alpha\, d). \qquad (4)$$

For thin films (d $\approx$ 1 µm), an accurate determination of $\alpha < 10^3$ by transmission/reflection measurements is not possible.

3.1. PHOTOTHERMAL DEFLECTION SPECTROSCOPY (PDS)

Photothermal spectroscopy is based on the measurement of the heat generated by the absorbed photons. The signal I^{PDS} is proportional to the absorbed light power I^{abs}. It is orders of magnitude more sensitive than transmission/reflection measurements, and is insensitive to film roughness.

Here, a conventional transverse PDS experimental set up is used, with some additional features, i.e. the synchronous measurement of the PDS signal and its phase shift with respect to the excitation, and the measurement of the amplitudes of both the incident and the transmitted light. This offers a convenient and reliable way to determine the absorption coefficient, to distinguish between film and substrate absorption and to obtain further important parameters.

Below, a brief description of the basic idea of the experiment is given. The sample is illuminated by a monochromatic light source. Dependent on the incident power and the absorption, the sample temperature changes. This results in a change of the temperature gradient in the deflection medium (usually CCl_4) which is in contact to the film surface. The temperature gradient induces a refractive index gradient in which a laser beam, which is guided parallel to the film surface, is deflected. At low power, the deflection of the beam, measured by a position sensitive detector, is proportional to the absorbed power in the film. To achieve the required signal-to-noise ratio, the incident monochromatic light is chopped and the PDS-signal is measured by a Lock-In technique.

A temperature-stabilised semiconductor laser with a high pointing stability is used. In our system, the spectral resolution was 5-10 nm, corresponding to about 4-8 meV at a photon energy of 1 eV.

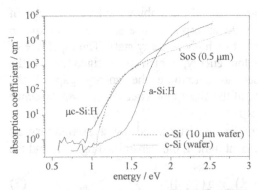

Figure 11. Comparison of the absorption coefficient of c-Si and device quality µc-Si:H and a-Si:H measured by PDS.

In figure 11, a comparison between the energy dependence of the absorption coefficients of µc-Si:H, c-Si and a-Si:H is shown. µc-Si:H exhibits significant absorption well below the band gap of crystalline Si ($E_g \approx 1.12$ eV) and also a higher absorption coefficient above 1.8 eV. However, the main features are still closer to c-Si than to a-Si:H. The absorption below the silicon band gap can tentatively be attributed to

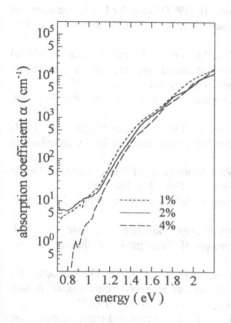

Figure 12. absorption spectra of a μc-Si:H film for different SC, as indicated.

transitions involving defects and/or band tails. As can be seen in figure 12, the absorption coefficient in this energy region decreases with increasing SC which is in agreement with a decreasing defect concentration and an improved device performance but a quantitative link has not yet been established.

The enhanced absorption at higher energies has frequently been attributed to enhanced light scattering, in addition to a contribution from the amorphous phase at high amorphous volume fractions. Via a detailed investigation, we found that the contribution of the amorphous phase is not sufficient to explain these effects when a simple effective medium approach is used. Instead, a contribution from disorder due to strain should be considered in order to understand them.

From the solar cell point of view, the enhanced absorption can be considered favourable, as long as the transport and recombination properties are not seriously affected. For our VHF-PECVD material, the best solar cells are prepared when material close to the transition to amorphous growth is used. According to absorption measurements, it has a reasonably low subgap absorption, similar to that for the SC = 4% material shown in figure 12.

4. Acknowledgements

The author gratefully acknowledges the indispensable support and fruitful collaboration with his colleagues, in particular F. Finger, M. Luysberg, L. Houben, P. Hapke and O. Vetterl.

5. References

1. Matsuda, A. (1983) Formation kinetics and control of microcrystallites in μc-Si:H from glow discharge plasma, *Journal of Non-Crystalline Solids* **59/60**, 767 - 747.
2. Prasad, K., Finger, F., Curtins, H., Shah, A., and Baumann, J. (1989) Preparation and characterization of highly conductive (100S/cm) phosphorus doped μc-Si:H films deposited using the VHF-GD technique, *Proc. Mat. Res. Soc. Symp.* **164**, 27-32.
3. Finger, F., Hapke, P., Luysberg, M., Carius, R., Wagner, H. and Scheib, M. (1994) Improvement of grain size and deposition rate of microcrystalline silicon by use of very high frequency glow discharge, *Appl. Phys. Letters* **65**, 2588 - 2590.
4. Rech, B., Roschek, T., Müller, J., Wieder, S. and Wagner, H. (2000) Amorphous and microcrystalline solar cells prepared at high deposition rates using RF (13.56 MHz) plasma excitation frequency, *Solar Energy Materials and Solar Cells* **66**, 267.

5. Heya, A., Nakata, K., Izumi, A. and Matsumura, H. (1998) Guide for low temperature and high-rate deposition of device quality poly-silicon films by CAT-CVD method, *Mat. Res. Soc. Symp. Proc.* **507**, 435 - 439.

6 Hapke, P., Luysberg, M., Carius, R., Tzolov, M., Finger, F. and Wagner, H. (1996) Structural investigation and growth of <n>-type microcrystalline silicon prepared at different plasma excitation frequencies, *J. Non-Crystalline Solids* **200**, 198-927.

7. Kobliska, R.J. and Solin, S.A. (1973) Raman spectra of wurtzite silicon, *Phys. Rev. B* **8**, 3799 - 3802.

8. Houben, L., Luysberg, M., Hapke, P., Carius, R., Finger, F. and Wagner, H. (1998) Structural properties of microcrystalline silicon in the transition from highly crystalline to amorphous Growth, *Phil. Mag. A* **77**, 1447.

9. Ossadnik, A., Veprek, S. and Gregora, I. (1999) Applicability of Raman scattering for the characterisation of microcrystalline silicon, *Thin Solid Films* **337**, 148 - 151.

10. Houben, L. (1999), Ph.D. Thesis, Berichte des Forschungszentrums Jülich; 3753; ISSN 0944-2952.

11. Luysberg, M., Hapke, P., Carius, R. and Finger, F. (1997) Structure and growth of µc-Si:H: Investigation by TEM and Raman spectroscopy of films grown at different plasma excitation frequencies, *Phil. Mag. A* **75**, 31 - 47.

12. Vetterl, O., Finger, F., Carius, R., Hapke, P., Houben, L., Kluth, O., Lambertz, A., Mück, A., Rech, B. and Wagner, H. (2000) Intrinsic microcrystalline silicon: A new material for photovoltaics, *Solar Energy Materials & Solar Cells* **62**, 97 - 108.

13. Tzolov, M., Finger, F., Carius, R. and Hapke, P. (1997) Optical and transport studies on thin microcrystalline silicon films prepared by very high frequency glow discharge for solar cell applications, *J. Appl. Phys.* **81**, 7376 - 7385.

CRYSTALLINE SILICON P-N JUNCTION SOLAR CELLS - EFFICIENCY LIMITS AND LOW-COST FABRICATION TECHNOLOGY

J. SZLUFCIK

Interuniversity Micro-Electronics Centre (IMEC)
Kapeldreef 75, 3001 Leuven, Belgium

1. Introduction

1.1. THE NEED FOR RENEWABLE ENERGY

Economic growth depends on energy use. Worldwide, energy accounts for 25 to 30% of the present investments in development and economic growth. The highest future energy needs are envisaged for developing countries, where 90% of world's population growth will take place. In poorer economies, an average person annually uses only 2.5 to 10 percent of the commercial fuels used in Europe, Japan or the USA [1,2], and around 2 billion people are still not connected to an electric grid [3]. As the Third World countries become increasingly industrialised, the growth in energy demand is increasing rapidly.

By 2025, the demand for fuel is projected to increase by 30%, and that for electricity by 265% [4]. As a result of such a high demand for energy, fossil fuels are currently being depleted 100,000 times faster than they are being formed. The remaining stocks of recoverable fossil fuels are estimated to last less than 170 years [1]. Additionally, burning fossil fuels for energy generation has a negative environmental impact, due to the emission of oxides of carbon and sulphur. These cause acid rain and the greenhouse effect, and also lead to many health problems.

Most of the environmental problems can be abated substantially. However, the costs related to massive retrofitting or replacement of existing facilities can amount to 30% of current US costs of fossil fuels and the electricity generated from them [5]. Such costs are far too high for those countries in which capital is scarce. Nuclear power plants have, in principle, no major climatological and ecological impacts. However, expanded use of nuclear energy is unlikely, due to a fear of radioactive wastes and accidents.

It is therefore clear that even with more efficient energy use and conservation, a transition to new sustainable energy sources is required.

109

J.M. Marshall and D. Dimova-Malinovska (eds.),
Photovoltaic and Photoactive Materials - Properties, Technology and Applications, 109–130.
© 2002 *Kluwer Academic Publishers.*

1.2 SOLAR RADIATION

Solar radiation is one of the biggest energy resources [4,6]. Solar radiative energy originates from a nuclear fusion reaction in the sun. The net mass loss of the sun is converted through the Einstein relation ($E = mc^2$) to energy at a rate of 4×10^{26} J/sec. This energy is emitted mainly as electromagnetic radiation, in the spectral region from 0.2 to 3 µm. Calculations show that the sun can generate radiative energy at a nearly constant rate, for a period of 10 billion (10^{10}) years.

The intensity of solar radiation in free space at the average distance of the earth from the sun is called the solar constant, and has a value of 1353 W/m^2. The sunlight is absorbed and scattered when it passes through the atmosphere on its way to the earth's surface. The degree of the attenuation is defined as the "air mass" (AM). This measures the atmospheric path length relative to the minimum path length when the sun is directly overhead, and is equal to $1/\cos(\Theta)$, with Θ being the angle between the sun and the zenith. The term AM0 describes the solar spectrum outside the earth's atmosphere. AM1 is the equivalent spectrum at the earth's surface when the sun is at zenith, and AM1.5 is that for which the sun is positioned at 45° above the horizon. The AM1.5 spectrum is used to characterise the solar energy-weighted average for terrestrial conditions. Although the total incident power for AM1.5 is 844 W/m^2, a standardised value of 1000 W/m^2 is used for simplicity in terrestrial solar device characterisation. Figure 1 shows the solar spectrum for different conditions [7].

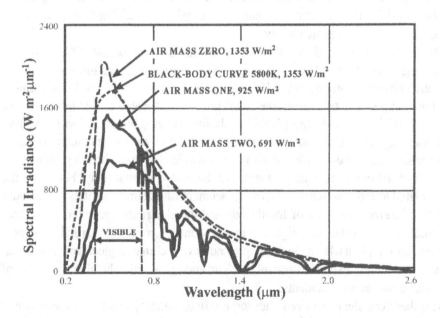

Figure 1. Solar spectral irradiance curves [7].

The average solar radiation over a year depends on the location. For example, in Western Europe the annual energy on a horizontal plane coming from the sun is around 1000 kWh/m^2. For more ideal locations, this energy amounts to 2000 to 2500 kWh/m^2. The total solar radiative energy delivered over the whole earth surface gives 10,000 times the amount of primary energy currently needed worldwide [4]. It is evident that converting even a small percentage of the solar energy, in an efficient way, into a useful form of energy can largely solve the problem of energy demands.

1.3. ATTRACTIVENESS OF PV SYSTEMS

The advantages of solar cells utilising the photovoltaic (PV) effect as an energy conversion process are [8]:

- they are environmentally benign, with no emission during operation and no unexpected fuel costs,
- they convert solar energy directly into a high level energy, i.e. electricity,
- they are static (no moving parts except for concentrator applications), and therefore require little maintenance and are reliable,
- they are available everywhere, but most abundantly in areas of future energy demand (Africa, Asia, South America and Australia),
- they can be integrated into buildings and other constructions,
- they are modular, and therefore can serve small as well as very large power demands in a centralised or decentralised way,
- PV systems have a short lead time, reducing the problem of installing under-used capacity.

2. Efficiency Limits of p-n Junction Crystalline Silicon Solar Cells

2.1. PRINCIPLES OF p-n JUNCTION CRYSTALLINE SOLAR CELLS

Although the photovoltaic effect was first observed by Becquerel in 1839, it took almost 120 years before, in 1954, researchers at the Bell Laboratories [9] developed the first diffused silicon solar cell, with a reasonable efficiency (4.5%). Since then, the rapid development of silicon solar cell technology has been driven mainly by interest in photovoltaics as sources of power for satellites. With the help of the planar device technology developed by the microelectronics industry, the solar cell structure evolved in the 1960s to that shown in Figure 2.

The main features of this structure; a phosphorus doped front junction, a grided front contact, an antireflection coating, and a rear surface totally covered by a metal contact, are still major features of current solar cell structures. Since the mid-1970s, terrestrial applications have increasingly become a driving force in the development of the photovoltaics industry.

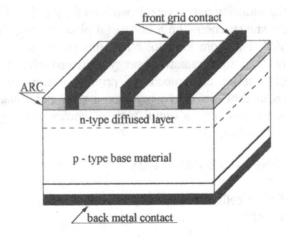

Figure 2. Basic p-n junction solar cell structure.

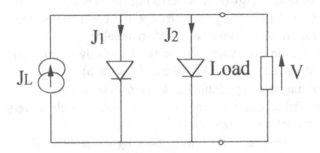

Figure 3. Equivalent circuit of an ideal p-n junction solar cell.

Solar cell operation can be modelled by considering the ideal equivalent scheme shown in Figure 3. This comprises a current source and two diodes in parallel. The current source models the light generated (J_L), or short-circuit current density (J_{sc}). The first diode models the diffusion current from the base and emitter regions, and the second diode represents generation-recombination current in the junction space-charge region.

The illuminated current voltage characteristics are described by equation (1).

$$J = J_L - J_1 - J_2 = J_L - J_{01}[\exp(qV/kT) - 1] - J_{02}[\exp(qV/2kT) - 1] \quad (1)$$

J_{01} represents the "first diode saturation current density" which results from the thermal generation of minority carries within the emitter and base regions. J_{02} is the "second diode saturation current density", which arises from the generation-recombination current in the space charge region. Very often, a simpler model with only one diode in parallel with the current source is used. The J-V curve under illumination is shown in Figure 4.

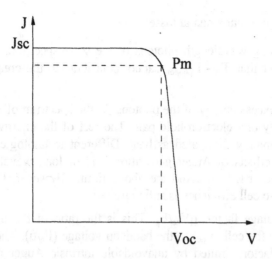

Figure 4. The J-V characteristic of an illuminated solar cell. J_{sc} is the short-circuit current density, V_{oc} is the open–circuit voltage, and P_M the maximum power point. The fill factor (FF) is defined as $P_M/(J_{sc}V_{oc})$.

In the ideal case, the efficiency η is defined as the ratio of the maximum output power P_M to the power of the incident light, P_L:

$$\eta = P_M/P_L \tag{2}$$

Equation (2) can be rewritten as:

$$\eta = (J_{sc}V_{oc}FF)/\int_0^\infty P(\lambda)d\lambda \tag{3}$$

where $P(\lambda)$ is the solar power density at wavelength λ. The solar cell efficiency can be expressed by taking into account different loss factors, i.e.

$$\eta = \frac{\int_0^{\lambda g} P(\lambda)d\lambda}{\int_0^\infty P(\lambda)d\lambda} \cdot \frac{E_g \int_0^{\lambda g} N(\lambda)d\lambda}{\int_0^\infty P(\lambda)d\lambda} \cdot \frac{qV_{oc}}{E_g} \cdot FF \cdot (1-R) \cdot \frac{A_f}{A_t} \cdot \eta_d \cdot \eta_{col} \tag{4}$$

$$\quad\text{(i)}\qquad\qquad\text{(ii)}\qquad\qquad\text{(iii)}\quad\text{(iv)}\quad\text{(v)}\quad\text{(vi)}\ \text{(vii)}\ \text{(viii)}$$

Some of these have fundamental limits and are unavoidable. The others come from technological limitations, and can sometimes be avoided [10].

The following are fundamental losses:

- (i) Loss by long wavelength photons of the solar spectrum. Photons with an energy smaller than $E_g - E_{phonon}$ cannot contribute to the creation of electron-hole pairs.

- (ii) Loss by excess energy of the photons. In the spectrum of interest, a photon generates only one electron-hole pair. The rest of the energy, larger than the bandgap, is mainly dissipated as heat. Different technological approaches to increase the effects of Auger generation [11] or long wavelength absorption [12] have been tried, to avoid the above limits. However, their practical net contribution to cell efficiency is still marginal.

- (iii) The voltage factor, qV_{oc}/E_g. This is the ratio of the maximum voltage developed by the cell (V_{oc}) to the bandgap voltage (E_g/q). The upper bound of the voltage factor, limited by unavoidable intrinsic Auger recombination, is 0.65 for thick silicon cells. It rises to 0.72 for a 20 μm thick cell [13].

- (iv) The fill factor (FF). In an ideal solar cell, where the J-V characteristic is represented by the Boltzmann expression, the ideal fill factor is 0.89.

The voltage and fill factors can be much lower in a practical cell, due to carrier recombination, series resistance and shunt resistance losses.

The most common technological factors influencing cell efficiency are:

- (v) Loss by reflection. Part of the incident energy is reflected at the non-metallised surface of the cell. The reflective losses are to be considered as a technological problem that can be solved by use of special surface treatments and anti-reflective coatings.

- (vi) Loss by metal coverage: A_f/A_t. The parameter A_f is the area of the front surface not covered by the metal contacts, and A_t is the total area. This is a technological limitation, caused by the coverage factor $1-A_f/A_t$. The coverage factor is a compromise between power losses coming from shadowing and FF losses caused by series resistance. The compromise results in a grating structure of the front contact. In some solar cell structures (back contact cells), both contacts are on the rear surface, and $A_f/A_t = 1$;

- (vii) Loss by incomplete absorption due to the limited cell thickness. Special light trapping techniques can significantly increase the absorption, even in very thin solar cells.

- (viii) Collection efficiency. Not all generated carriers reach the junctions and are collected. Some recombine in the bulk or at the surfaces. Recombination processes such as Auger, SRH (Shockley-Read-Hall) and radiative recombination are unavoidable. However, appropriate silicon and cell processing techniques can reduce their contributions to a fundamental minimum.

More detailed analyses of fundamental efficiency limits have been presented by many authors [14 -16]. Table 1 summarises the highlights of these.

TABLE 1. Fundamental efficiency limits of silicon solar cells, according to different models.

Author	Model assumptions	Efficiency
Shockley-Queisser [14]	- "black body" cell - only radiative recombination - hv > Eg - global AM 1.5 solar spectrum	32.9 %
Tiedje *et al.* [15]	- includes Auger recombination - measured absorption and Auger coefficients in silicon - practical cell geometry	29.8%
Campbell-Green [16]	- effective light trapping - concentrated light	37.0%

A more accurate description of a real solar cell is obtained by inclusion of:

- series resistance, which accounts for power losses mainly due to current flow through highly resistive emitter and contacts (R_s),
- shunt conductance representing the leakage current across the junction (R_{sh}),
- additional current components which are included in the second diode ideality factor, with n ~ 2 or greater.

The equivalent circuit from Figure 3 and the related J-V characteristic (equation (1)) are now modified to those represented by Figure 5 and equation (5).

$$J = J_L - J_{01}[\exp(q(V+JR_s)/kT)-1] - J_{0n}[\exp(q(V+JR_s)/nkT)-1] - (V+JR_s)/R_{sh} \quad (5)$$

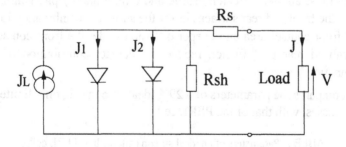

Figure 5. Equivalent circuit of a real p-n junction solar cell.

2.2. HIGH EFFICIENCY LABORATORY CELLS

By high efficiency laboratory solar cells, we mean here cells for which the processing employs all available high quality materials and techniques, in order to reduce energy conversion losses and to approach the fundamental efficiency limit. These cells usually have a small area (1-4 cm^2) and are processed on high quality and expensive float zone (FZ) silicon. Efficiency is a primary goal, and processing costs are not of major importance. During the last decade, the efficiency of crystalline laboratory solar cells has increased considerably. This improvement is important for all industrial solar cells, since it shows that higher efficiencies are possible and because it can be used as a guide in finding improved processes for future commercial manufacturing. The problems related to high efficiency solar cells are thoroughly treated in the literature [17]. The most important achievements from a practical point of view will be briefly treated below. Today, the most efficient crystalline silicon solar cell is a PERL ("passivated emitter and rear locally-diffused") structure (figure 6), yielding efficiencies of about 24% on FZ material [18].

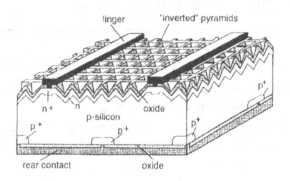

Figure 6. The PERL cell [18].

The essential features of this cell are:

- front surface reflection losses reduced by inverted pyramids,
- a well tailored emitter diffusion profile and a high quality passivating thermal oxide on the front and rear surfaces lower the surface recombination losses,
- a small front contact area and deeper diffusion under the front contact fingers and localised rear p$^+$ diffusion reduce the contact contribution to the total recombination losses.

Table 2 compares the parameters of a 29% ideal silicon cell, only limited by unavoidable losses, with that of the PERL cell.

TABLE 2. Parameters of an ideal silicon cell, the best PERL cell, and achievable parameters of a PERL cell (AM1.5).

	Jsc (mA/cm^2)	Voc (mV)	FF (%)	Eff. (%)
Ideal cell [15]	43.0	769	89.0	29
PERL [18]	40.8	708	83.1	24
Achievable cell [19]	42.5	730	84.0	>26

The limiting efficiency of the ideal cell has been calculated [15]. It includes both Auger and radiative bulk recombination, but assumes zero surface recombination and an ideal light trapping scheme, yielding the maximum optical path length enhancement in the cell. It is generally agreed that by fully optimising all design parameters, the PERL structure is able to demonstrate 26% efficiency under one sun illumination, as outlined in [19]. It includes improvement of the short circuit current, the fill-factor and the open circuit voltage. This can be achieved by a better optical and light trapping scheme, and by improving the interface quality. Slight modifications in the cell design, such as the use of floating junctions at the rear side, offer interesting possibilities for commercial solar cells.

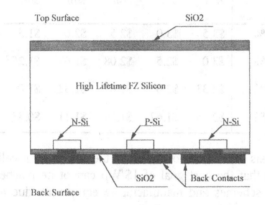

Figure 7. Cross section of a point-contact solar cell [20].

Another example of high efficiency design is the point contact solar cell [20]. This (Figure 7) is derived from the Interdigitated Back Contact Cell [21] with an interdigitated back metallization, avoiding front metal coverage and giving a low series resistance. It is much thinner than the PERL solar cell, allowing good carrier collection at the back. The top surface is textured, and very well passivated. Bulk recombination is further minimised by using high lifetime very high resistivity silicon for the base. The one-sun version of the point contact cell has achieved an efficiency near 23% [22]. The true potential of the point contact solar cell is demonstrated under concentrated sunlight, as efficiencies over 28% at a concentration factor of 100X have been achieved [23].

3. Industrial Multicrystalline Silicon Solar Cells

For industrial solar cells, there is still an efficiency objective, but low cost is the highest objective. Processing techniques and materials are selected to give the maximum cost reduction, while maintaining a relatively good efficiency.

In the last decade, world PV module shipments have grown at an annual rate of around 25%, with the total world production reaching nearly 300 MWp in the year 2000 [24]. Crystalline silicon solar cells constitute more than 85% of the world PV market, with a tendency to increase their market share. Therefore, the development of fast and cost effective crystalline silicon solar cell processing technologies plays a key role in the large scale penetration of PV in the total energy system. The European and US PV industry roadmaps have defined the mid-term cost goal for a bulk crystalline silicon solar cell as 1$/Wp. Table 3 below shows how cell efficiency and direct manufacturing costs influence the final PV module cost [25].

TABLE 3. Impact of efficiency and direct manufacturing cost for crystalline silicon technology cost [$/Wp].

Cost Efficiency	350 $/m^2	300 $/m^2	250 $/m^2	200 $/m^2	150 $/m^2
10%	$3.5	$3.0	$2.5	$2.0	$1.5
12%	$3.0	$2.5	$2.08	$1.67	$1.25
15%	$2.33	$2.0	$1.67	$1.33	$1.0
18%	$2.05	$1.67	$1.39	$1.11	$0.83

A detailed analysis of the-state-of-the-art in the silicon solar cell manufacturing technologies shows that the cost goal of 1$/Wp cannot be reached based on the existing processing schemes and manufacturing equipment, due to the following limitations:

- shortage and high price of silicon feedstock,
- expensive and thick substrate material,
- high cost of diversified and proprietary processes and equipment with numerous wafer handling operations,
- modest efficiencies of industrially produced solar cells,
- large cost and low automation of PV module fabrication.

Large efforts are being undertaken world-wide, towards improving the efficiency of industrial-type multicrystalline silicon solar cells. This is done in parallel with production process simplification and throughput increase. The main novelties with reference to existing production multi-Si cell processes are:

- high throughput isotropic surface texturing,
- a simple low-cost selective emitter process,
- novel processing equipment for very high throughput diffusion and PECVD SiN deposition,
- back-contacted cell structures and modules.

Wherever possible, the fabrication processes are executed by means of screen printing. This reduces the equipment diversity, increases automation and limits the wafer handling operations.

A close analysis of Table 3 leads to the conclusion that the low-cost goal can only be achieved if a cell efficiency above 18% can be obtained in parallel with a drastic reduction of direct manufacturing cost, to 150 \$/m². Translating this into technological requirements means that one should manufacture high efficiency multicrystalline solar cells ($\eta \geq$ 18%) on thin ($\leq$ 100 µm), low-cost substrate material (wafers or ribbons) by means of low-cost, high throughput manufacturing techniques. Several cell structures and processes, essential for cost reduction, have already been proposed and will be reviewed in the chapters below.

3.1. DEVELOPMENT OF HIGH EFFICIENCY PROCESSING STEPS

3.1.1. *High-Throughput Texturing Techniques*
Effective surface texturing plays an important role in efficiency enhancement. It not only minimises reflection losses, but also enhances light trapping and increases carrier collection. Although alkaline texturization effectively reduces the reflectance from monocrystalline silicon to around 10%, it does not work as effectively on multicrystalline substrates. This is a direct consequence of the anisotropic nature of the etching process, which depends on the crystallographic orientations of the different grains. The importance of developing a texturing technique specific to multicrystalline substrates triggered the development of several promising methods, as listed in Table 4.

Three texturing techniques: mechanical grooving, reactive ion etching and isotropic acid etching have already been developed for high-throughput production techniques. Mechanical grooving is the most mature one, and is well documented in the technical literature [26]. Isotropic acid etching and RIE etching are briefly described below.

TABLE 4. Overview of the texturing techniques applicable to multicrystalline substrates.

Texturing technique	Properties
Mechanical grooving	+ applicable to all types of multi-Si substrate, very low reflectance, high-throughput production systems available - requires thick wafers and a new metallization technique
Reactive ion etching	+ applicable to all types of multi-Si substrate, low reflectance - difficult to control, expensive vacuum processing equipment
Porous silicon	+ applicable to all types of multi-Si substrate, combines texture and ARC - lack of effective surface and bulk passivation - high absorption for short wavelength photons
Low cost masking technique and wet or dry etching.	+ applicable to all types of multi-Si substrate - many-step process
Isotropic acid etching	+ combines saw damage etching and texturing + high throughput etching system under construction - applicable to wire saw cut wafers

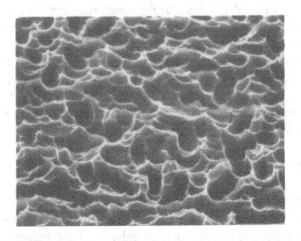

Figure 8. SEM photograph of a multi-Si surface after isotropic acid etching.

Isotropic Acid Etching. A simple and low cost technique based on isotropic etching with acidic solutions was introduced a few years ago, and has lately been upgraded to an industrial process [27]. The acidic aqueous solution consists of a mixture of nitric and hydrofluoric acids, plus some additives. The texturing process starts with "as-cut" multi-Si wafers. The removal of saw damage and the surface texturing are done simultaneously in one short single chemical step. The resulting etch pits, 1-10 μm in diameter, are uniformly distributed over the wafer surface, independently of crystallographic orientation, giving a homogenous reflectance over the surface of the wafer and an absence of steps between grains (see Figure 8).

The obvious advantage of an isotropically textured surface is the reflectance reduction relative to anisotropically etched multi-Si wafers. There are, however, several additional process benefits originating from such a structured surface:

(1) no steps between grains, so narrower and uninterrupted front contact lines can be printed,

(2) increased contact area of the metal-silicon interface decreases the contact resistance,

(3) only a very thin silicon layer (4-5 μm) is etched away from the "as-cut" silicon wafer,

(4) multicrystalline wafers have a nice uniform appearance, as required for implementation in BIPV modules.

The direct profit from (1) and (2) is that cells with reduced shadowing losses and fill factors up to 79% can be processed on multi-Si cells. The increased contact interface area allows re-optimisation of front contact pastes, and consequently an increase in the emitter sheet resistance to above 65 Ohm/square, significantly improving the cell's "blue response". Summing up all benefits, a short circuit current increase of up to 1.5 mA/cm^2 could be obtained.

An automatic wet-bench with temperature control of the etching solution and automatic replenishment of chemicals has already been designed for industrial implementation of the acid isotropic texturing technique in multi-Si cell production lines. The cost calculations revealed that isotropic chemical etching contributes only 0.02-0.03 $/Wp to the total cell fabrication cost.

Figure 9. SEM picture of a RIE etched multi-Si wafer [29].

Isotropic texturing by Reactive Ion Etching.

RIE texturization in a chlorine plasma has been implemented by Kyocera [28]. It is a dry, contactless and stress-free isotropic etching process, perfectly applicable to large area, thin and fragile silicon substrates. A chlorine plasma is used to create a surface which is less rough than "silicon grass", comprising a very high density of etch pits (Figure 9). The shape of these depends on the crystallographic orientation of the etched surface, but the average texture on every different surface orientation can have the same reduced reflectivity [29]. Therefore, this texturing method is suitable for uniform light trapping on multicrystalline solar cells. The increased roughness of texture produces a reduced reflectivity, but also causes increased carrier recombination in the emitter [29,30].

During texturing, 3-10 μm of silicon is removed from the surface, depending on the process parameters. An optimized RIE surface texturing process brings about an increase of up to 1.4 mA/cm^2, relative to an anisotropic KOH texture.

3.1.2. *Optimisation of Industrial Solar Cell Emitters for Maximum Cell Efficiency*

An emitter optimisation study for cells with screen-printed contacts [31] pointed out that for the cells with a non-selective emitter, the phosphorus surface concentration should be at least 10^{20}/cm^2. Due to the high contact resistivity, the fill factor of cells decreases drastically for lower surface concentrations. Moreover, given the characteristics of screen-printed metal contacts, the junction depth should be at least 0.3-0.4 μm, to avoid shunting of the emitter. These constraints leave only a small region in the surface–concentration/junction depth plane where a near-maximum cell efficiency can be obtained. Small deviations from the optimum emitter profile result in drastic changes in cell efficiency.

For cells with a selective emitter, the best efficiency is not much higher, but is less sensitive to small variations in emitter surface concentration or junction depth. This lower sensitivity for deviations from the optimum processing conditions can be considered as the main advantage of using a selective emitter structure for industrial cells with screen-printed contacts. Below, we give a review of the homogenous and selective emitter processes developed for screen-printed metallization.

Homogeneous Emitter

After many years of investigations seeking process simplification and efficiency improvement, an industrially transferable high efficiency processing sequence, based on screen-printed contacts and a homogenous emitter, has been established. It is currently possible to completely avoid the emitter "dead layer" always present in the old types of industrial screen printed solar cell. New types of silver paste, when fired through the PECVD SiN_x anti-reflection coating layer, can contact the shallow emitter with a sheet resistance above 60 Ohm/square, and produce cells with a good "blue response". Fill factors above 77 % are routinely obtained. The simplicity of the cell process is preserved, while yielding a considerable improvement in multi-Si cell efficiency.

Selective Emitter

Several low-cost selective emitter processes applicable to the industrial environment have recently been introduced [32-34]. All of the proposed processes require only one diffusion step to obtain a selective emitter structure. The selective emitter process proposed by IMEC [32] is shown in Figure 10. A phosphorus paste is selectively applied to the front surface of the Si substrate, by screen-printing. The printed pattern is similar to the metallization pattern that has to be printed at the end of the cell processing sequence. Deep and shallow emitter regions are formed simultaneously in the same high temperature process. Deeply diffused regions with a high surface concentration are present in those regions onto which screen printed metallization is applied later in the processing sequence. At the same time, shallow emitter regions with a low surface concentration are formed elsewhere, via gas phase diffusion.

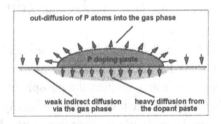

Figure 10. Principle of a single diffusion step selective emitter process [32].

IQE measurements prove that an excellent cell "blue response" can be achieved by this process. Large area multicrystalline silicon solar cells with a record efficiency of 16.8% for a screen printed process have been obtained and independently verified [32].

A very elegant process of selective emitter formation has recently been introduced by Ebara Solar and Georgia Tech [33]. The principle is based on "self-doping" from silver metallization pastes, as presented in Figure 11.

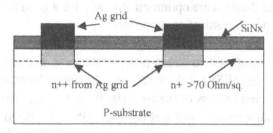

Figure 11. Principle of a self-aligned selective emitter process.

First, a shallow homogeneous emitter (> 70 Ohm/square), optimised for a good "blue response", is diffused in a standard way. Next a PECVD SiN_x layer is deposited, for perfect surface and bulk passivation. This is followed by screen printing the phosphorus-doped silver paste and firing at a temperature above 835°C. The process is absolutely self-aligned, since the phosphorus present in the metallization paste takes care of heavy n^{++} diffusion under the metal grid. Multicrystalline solar cells with an efficiency of 15.5% have been obtained by this very simple process.

3.1.3. *Surface and Bulk Passivation*
Passivation from PECVD SiN_x Layers
A low surface recombination velocity and significant improvements in bulk quality are key issues for efficiency improvements of solar cells based on a large variety of multicrystalline silicon materials. It has been proven that PECVD silicon nitride layers, being perfect antireflection coatings, provide excellent surface and bulk passivation. Furthermore, their deposition processes can be executed with a high throughput, as required by the PV industry [35,36]. There are two basically different methods of PECVD SiN_x:H deposition: direct and remote PECVD. High quality SiN_x layers and a damage free surface are formed in a remote plasma deposition process. Consequently, record low surface recombination velocities, below 100 cm/s, are obtained [36]. The heavy ion bombardment present in a low frequency direct plasma produces surface damage. Therefore, a lower quality silicon surface passivation is achieved than in the case of high or remote plasma deposited SiN_x:H layers. On the other hand, it is believed that some surface damage enhances the in-diffusion of hydrogen into silicon, and helps in bulk passivation.
Emitter Passivation
The direct plasma silicon nitride effectively passivates the emitter of a silicon solar cell. If the nitride coating is not thermally treated, damage resulting from the ion bombardment during plasma deposition degrades the surface quality, resulting in a doubling of J_{0e} (1.3×10^{-12} A/cm^2 compared to 6.2×10^{-13} A/cm^2). With thermal treatment at temperatures above 700°C, this damage is annealed. In practice, the

annealing process is executed in a "firing-through" front contact metallization process step. The front silver screen printed contacts are fired through the nitride layer, to make contact with the underlying emitter. The paste composition, temperature profile and duration are optimised, not only for a good contact but also for the best surface and bulk passivation properties.

Al-BSF Formation

On the back surface of most modern screen-printed solar cells, aluminium is used to create a Back Surface Field (BSF). A thick (20 μm) aluminium layer is alloyed with silicon during the sintering process of the contacts. If high enough temperatures (> 800°C) are used, the aluminium concentration in the BSF can rise to 5×10^{18} cm^{-3}, while the base doping level for material with a typical resistivity of 1 Ωcm is only 1.5×10^{16} cm^{-3}. This high/low junction effectively repels minority carriers from the back surface. An effective surface recombination velocity at the p/p$^+$ junction, as low as 200 cm/s, has been measured [37,38]. Furthermore aluminium alloying is known to be an effective gettering process of metallic impurities.

As mentioned above, high temperature treatment improves the surface and bulk passivation properties of PECVD SiN$_x$ layers. This creates the opportunity for merging these two separate processing steps into a single one.

This in-situ BSF-formation using the firing-through process is a complementary advantage to the effect of hydrogen bulk passivation. While materials of lower quality (low base diffusion length) will be more dependent on bulk passivation, those of higher quality (high base diffusion length) that are less dependent on bulk passivation will benefit more from the created back-surface field. The most important effect of simultaneous Al-alloying and high temperature treatment of the SiN$_x$ layer is the strong positive synergy in SiN$_x$ passivation and Al gettering. The lifetime enhancement achieved by simultaneous processing is much greater than the additive effects from the separate steps. This has been confirmed on cast multicrystalline wafers [39] as well as on silicon ribbons [40].

3.2. A GENERIC PROCESS FOR MULTICRYSTALLINE SOLAR CELLS

Based on the latest developments in the low-cost advanced processing steps described above, one can propose a generic low-cost solar cell process applicable to all available multicrystalline materials [41-44].

The cell process flow consists of only a few processing steps:

- saw damage removal and isotropic texturing,
- shallow emitter or selective emitter diffusion,
- parasitic edge junction isolation,
- PECVD SiN$_x$ deposition,
- Screen-printing of front and back contacts,
- Co-firing using a firing-through PECVD SiN$_x$ process.

TABLE 5. The best reported large area multi-Si solar cells
fabricated by means of a screen printing firing-through process

	Area cm^2	Jsc mA/cm^2	Voc mV	FF %	Eff. %
Mitsubishi Electric Corp. [19]	225	34.27	626	78.2	16.8
Kyocera [20]	225	34.16	621	78.5	16.1
IMEC [18]	156	34.5	634	76	16.6 SARC
	156	35.3	635	76	17.0 DARC

The pilot line solar cell process established in IMEC using 5-inch multicrystalline silicon substrates has reached an average efficiency level above 16%, with the best cell near 17% [41]. The best reported results obtained on multi-c Si with industrially-compatible substrate sizes (>156 cm^2) are presented in Table 5.

A test PV module of 36 cells, using 5-inch multicrystalline wafers with an average efficiency of 16%, reached a record maximum power of 90 Wp, as confirmed by the calibration laboratory of the Joint Research Centre in Ispra, Italy [41].

3.3. TOWARDS THE COST OBJECTIVE OF CRYSTALLINE SOLAR CELLS

Improving the efficiency of production solar cells from 13% to the level of 17% will not lead to the cost goal of 1$/Wp imposed by the European and US PV industry roadmap. As can be concluded from Table 3, a drastic reduction of the direct manufacturing cost, to 150$/Wp, must be followed by a further efficiency improvement above 18%. The most obvious way to achieve the cost reduction and simultaneous efficiency improvement is to use much thinner substrate material.

It is well known that the cost of the silicon material itself is a very large part of the final module cost figure. The cost of silicon crystal growth and the subsequent wafering into multicrystalline substrates has been estimated to amount to 55% of that of the finished module [45]. Therefore, the use of thin wafers (thickness < 200 μm) can result in substantial savings compared to state-of-the-art cells. All manufacturers of commercial silicon ribbon materials are running the pilot line development of ribbon production, with thicknesses of 100 μm [46]. Photowatt has mastered a high yield wafer cutting process, with thicknesses down to 150 μm [47].

The question arises as to whether high efficiency thin solar cells can be manufactured with current processing techniques and cell structures. Figure 12 presents simulated efficiencies for conventional screen-printed solar cells, as a function of cell thickness and minority carrier diffusion length [48].

The most important conclusion from the simulations is the increased demand on rear surface passivation quality with decreasing cell thickness. A rear surface recombination velocity of about 100 cm/s and a diffusion length above 200 μm are

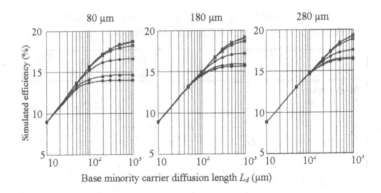

Figure 12. Simulated efficiency of conventional screen-printed solar cells, as a function of cell thickness and minority carrier diffusion length. The parameter in the graphs is the rear surface recombination velocity: ✱-1cm/s ☐-10 cm/s ✖-10^2 cm/s ◆-10^3 cm/s ▲-10^4 cm/s ●-10^5 cm/s. Internal reflectance front and back: 70%, front outer reflection: 10%.

required for an 80 μm thick cell to reach an efficiency close to 18%. Higher efficiencies can be reached if a more effective rear surface reflector can be applied. Diffusion lengths of the order of a few hundred microns and back surface recombination velocities around 200 cm/s have already been demonstrated, using a firing-through SiN$_x$ process.

However a problem arises when transferring the process based on the simultaneous processing of SiN$_x$ and Al for these thin cells: the expansion coefficients of Al (23.9×10^{-6} K^{-1}) and Si (7.6×10^{-6} K^{-1}) are significantly different. As a result, a wafer can be severely bent after alloying a 20 μm thick Al-layer onto a 200 μm thick cell. This makes the module manufacturing of the cells much more difficult, and can result in a sharp and unacceptable decrease in mechanical yield. Therefore, an alternative method has to be developed. Originally it was thought that the Al-layer was only important for the creation of the BSF, and therefore another back surface passivation method (Boron BSF, oxide or nitride passivation) would yield the same or even better results. However several studies [39,40] clearly indicate that the simultaneous high temperature processing of the hydrogen rich PECVD-SiN$_x$ coating and the printed Al-layer is paramount for obtaining good bulk passivation.

In order to solve this problem, several routes can be followed:

• Removal of the full Al back contact in a hot HCl solution, after the firing of the SiN$_x$ and Al-layers (the BSF is not removed). When the Al-layer is etched off, the warping of the wafer is also removed. However, this process is complicated and environmentally unfriendly, and also necessitates a second firing of the nitride layer, the front contact and the back contact (for instance a Ag/Al grid).

• Finding another method or material to obtain the same effect as that of SiN$_x$/Al co-firing (e.g. a material that has an expansion coefficient better suited to silicon, but also alloys in a similar manner). Another possibility is the deposition of an amorphous silicon layer on the back surface, with a wide bandgap (ideal for back surface passivation). This layer could also be a source of vacancies.

- Optimisation of an Al back contact pattern that diminishes the stress level and bending problem.
- For very thin (~ 100 μm) cells, the firing of silicon nitride on the front surface (without the presence of Al on the back) might be sufficient for effective bulk passivation. In this case, other techniques such as PECVD-SiN$_x$ deposition can be used for back surface passivation.

It is clear that this topic is becoming one of the most important research tasks at the level of cell processing, since the lower costs associated with thinner cells can only be benefited from if the cell efficiency can be maintained (or preferably increased). The ideal processing method for the back surfaces of thin cells should have the following features:

- Excellent back surface passivation (comparable to or better than a state-of-the-art Al-BSF).
- Suitable for thin cells, from the viewpoint of handling (high mechanical yield).
- High internal back surface reflection, to reduce the loss of long wavelength photons.
- If possible, enhancement of the in-diffusion of hydrogen into the cells from the silicon nitride layer on the front.

Recently, several low-cost processes for back surface passivation and light trapping have been proposed. A multicrystalline silicon solar cell, with front and back surfaces passivated only by silicon nitride layers [49], reached an efficiency of 18.1%. This excellent value can be attributed to the superior quality of the rear passivation scheme, consisting of a SiN$_x$ film combined with an aluminium local BSF.

A very elegant way to fabricate the LBSF rear cell structure by a low-cost process has been proposed by FhG-ISE, Freiburg, Germany [50]. The process involves three steps: (1) deposition of a SiN$_x$ passivating layer, (2) evaporation of an aluminium layer, (3) firing of the contact points (on 1% of the rear surface) through the passivating layer.

A screen printing back surface reflector process has recently been introduced. Low cost wafer materials are often also of a lower quality. Even for thin wafers, the diffusion length can be smaller than the cell thickness. Cell structures with collecting junctions on both surfaces (emitter wrap through (EWT) or metallization wrap through (MWT)) can increase the carrier collection volume and make the cell efficiency less dependent on the diffusion length. The other important advantage of these cell structures is that the collecting electrodes are placed on the cell's rear side. Removing or reducing the area of the front surface metal grid results in a higher effective semiconductor area, and thus the potential for increased cell efficiency. Bringing the external contacts to a single surface allows significant cost reductions in module assembly. Furthermore, the architectural requirements for uniform appearance is fulfilled, as the visual disturbance caused by the presence of the bus bars and the tabbing is displaced to the rear of the module. The availability of these kinds of high performance module, featuring a uniform dark visual aspect, will be an important step in the dissemination of the potential of photovoltaics to the building community. A few low cost fabrication processes have already been introduced [51].

4. Summary

Crystalline silicon solar cells almost completely dominate worldwide solar cell production. Excellent stability and reliability, plus continuous development in cell structure and processing, make it very likely that crystalline silicon cells will remain in this position for several years. Laboratory solar cells, processed by means of sophisticated techniques originating from the microelectronics industry and using high quality FZ-Si substrates, have yielded energy conversion efficiencies of over 24%.

An overview of the latest achievements in low-cost processing technologies of industrial-type large area multicrystalline silicon solar cells has been given. A generic cell process based on isotropic texturing, a shallow emitter or single diffusion selective emitter, combined with screen printed metallization fired through a PECVD SiN_x anti-reflection coating has been commonly accepted as a universal low-cost process. A significant efficiency enhancement, relative to existing production processes, has been measured in the cases of all available substrate materials, including silicon ribbons. Novel dedicated fabrication equipment for isotropic texturing, emitter diffusion and PECVD SiN_x deposition has been developed and implemented, thereby removing the processing bottlenecks. As a result, the efficiency and throughput of industrial screen printed multi-Si solar cells can be increased far beyond the state-of-the-art production cells.

The technological requirements linked to cell design and low-cost processing techniques needed for drastic cost reduction of multicrystalline silicon solar cells down to 1\$/Wp have been briefly described.

To keep bulk crystalline silicon solar cells competitive with thin film technologies, low cost processing techniques for rear surface passivation and light trapping must be developed and applied to thin substrates.

5. References

1. Davis, G.R. (1990) Energy for planet earth, *Scientific American* **263**, 21-27.
2. Amulya, K.N. and Goldemberg, J. (1990) Energy for the developing world, *Scientific American* **263**, no. 3, 63-71.
3. Palz, W. (1994) Power for the world: a global photovoltaic action plan, *Proc. 12th. European PV Solar Energy Conf.*, 2086-2088.
4. Hoagland, W. (1995) Solar energy, *Scientific American* **273**, 170-173.
5. Holdren, J.P. (1990) Energy in transition, *Scientific American* **263**, 109-115.
6. Weinberg, C.J. and Williams, R.H. (1990) Energy from the sun, *Scientific American* **263**, 99-106.
7. Thekaekra, M.P. (1974) Data on incident solar energy, *Suppl. Proc. 20th Annual Meeting Inst. Environ. Sci.* 21.
8. Nijs, J. (1994) Photovoltaic cells and modules: technical and economic outlook towards the year 2000, *Int. J. Solar Energy* **15**, 91-122.
9. Chapin, D.M., Fuller, C.S. and Pearson, G.L. (1954) A new silicon p-n junction photocell for converting solar radiation into electrical power", *J. Appl. Phys.* **25**, 676-677.
10. Backus, C.E. (1984) Principles of photovoltaic conversion, in G. Furlan, N.A. Mancini and A.A.M. Sayigh (eds.) *Nonconventional Energy* Plenum Publishing, pp. 297-348.

11. Kolodinski, S., Werner, J.H. and Queisser, H.J. (1994) Quantum efficiency exceeding unity in silicon leading to novel selection principles for solar cell materials, *Sol. Energy Mat. Sol. Cells* **33**, 275-285.

12. Keevers, M.J. and Green, M.A. (1993) Efficiency improvements of silicon solar cells by the impurity photovoltaic effect, *Proc. 23rd IEEE PV Specialist Conference*, 140-146.

13. Green, M.A. (1984) Limits on the open-circuit voltage and efficiency of silicon solar cells imposed by intrinsic Auger process, *IEEE Trans. Electron Dev.* **ED-31**, 671-678.

14. Shockley, W. and Queisser, H.J. (1961) Detailed balance limit on efficiency of p-n junction solar cells *J. Appl. Phys.* **32**, 510-519.

15. Tiedje, T., Yablonovitch, E., Cody, G. and Brooks, B.G. (1984) Limiting efficiency of silicon solar cells *IEEE Trans. Electron Dev.* **ED-31**, 711-716.

16. Campbell, P. and Green, M.A. (1987) The limiting efficiency of silicon solar cells under concentrated sunlight, *IEEE Trans. Electron Dev.* **ED-33**, 234-239.

17. Green, M.A. (1987) High efficiency solar cell, *Trans Tech Publications*.

18. Zhao, J., Wang, A., Altermatt, P.P., Wenham, S.R. and Green, M.A. (1996) 24% efficient PERL silicon solar cell: recent improvement in high efficiency silicon cell research, *Solar Energy Materials and Solar Cells* **41/42**, 87-99.

19. Green, M.A., Wenham, S.R. and Zhao, J. (1992) High efficiency silicon solar cells, *Proc. 11th European PV Solar Energy Conference*, 41-44.

20. Swanson, R., Verlinden, P., Crane, R. and Tilford, C. (1992) High efficiency silicon solar cells, *Proc. 11th European PV Solar Energy Conference*, 35-40.

21. Schwarz, R.J. (1982) Review of silicon solar cells for high concentration, *Solar Cells* **6**, 17-38.

22. King, R.R. (1990) Studies of oxide-passivated emitters in silicon and applications to solar cells, *Ph.D. Thesis*, Stanford University.

23. Sinton, R.A. and Swanson, R.M. (1987) An optimization study of Si point-contact concentrator solar cells, *Proc. 19th IEEE PV Specialist Conference*, 1201-1208.

24. Maycock, P. (2001) *PV-News*, February 2001.

25. King, R.J. (1998) Opening remarks, *Proc. 8th Workshop on Crystalline Silicon Solar Cell Materials and Processes*, Colorado, 1-6.

26. Willeke, G., Nussbaumer, H., Bender, H. and Bucher, E. (1992) A simple and effective light trapping technique for polycrystalline silicon solar cells, *Solar Energy Materials and Solar Cells* **26**, 345-356.

27. De Wolf, S., Choulat, P., Vazsonyi, E., Einhaus, R., Van Kerschaver, E., Declercq, K., and Szlufcik, J. (2000) Towards industrial application of isotropic texturing for multi-crystalline silicon solar cells *Proc. 16th European Photovoltaic Solar Energy Conf.*, Glasgow, 1521-1524.

28. Inomata, Y., Fukui, K. and Shirasawa, K. (1997) Surface texturing of large area multicrystalline Si solar cells using reactive ion etching method, *Solar Energy Materials and Solar Cells* **48**, 237-242.

29. Dekkers, H., Duerinckx, F., Szlufcik, J. and Nijs, J. (2000) Silicon surface texturing in a chlorine plasma *Proc. 16th EC PVSEC*, 1532-1535.

30. Ruby, D., Zaidi, S., Narayanan, S., Damani, B. and Rogathi, A. (2001) RIE-texturing of multicrystalline silicon solar cells, *Tech. Digest Intern. PVSEC-12*, Korea, 273-274.

31. Demesmaeker, E. (1993) *PhD Thesis, Kath. Univ. Leuven*, Belgium.

32. Horzel, J., Szlufcik, J. and Nijs, J. (2000) High efficiency industrial screen printed selective emitter solar cells. *Proc. 16th European Photovoltaic Solar Energy Conf.*, Glasgow, 1113-1115.

33. Rohatgi, A., Hilali, M., Meier, D., Ebong, A., Honsberg, C., Carroll, A. and Hacke, P. (2001) Self-align self-doping selective emitter for screen-printed silicon solar cells *Proc. 17th European Photovoltaic Solar Energy Conference and Exhibition*, Munich, in press.

34. Kinderman, R., Bultman, J., Hoornstra, J., Koppes, M. and Weeber, A. (2001) First xSi cell results using selective emitters formed with diffusion barriers in one step *Tech. Digest Intern. PVSEC-12*, Korea, 229-230.

35. Szlufcik, J. and Duerinckx, F. (2001) Defect passivation of industrial multicrystalline solar cells based on PECVD silicon nitride *Symposium E, E-MRS Spring Meeting, 2001: Crystalline Silicon for Solar Cells*, Strasbourg, France; *Solar Energy Materials and Solar Cells*, in press.

36. Aberle, A. (2001) Overview on SiN surface passivation of crystalline silicon solar cells, *Sol. Energy Materials and Solar Cells* **65**, 239.

37. Lolgen, P., Leguit, C., Eikelboom, J.A., Steeman, R.A., Sinke, W.C., Verhoef, L.A., Alkemande, P.F.A. and Algra, E. (1993) Aluminium back surface field doping profile with surface recombination velocities below 200 cm/sec *Proc. 23rd IEEE PVSC*, 236-242.

38. Narasimha, S. and Rogathi, A. (1997) Optimised aluminium back surface field techniques for silicon solar cells, *Proc. 26th IEEE PVSC*, 63-66.

39. Duerinckx, F., Allebé, C. and Szlufcik, J. (2000) Enhanced passivation for multicrystalline silicon solar cells by co-sintering of PECVD-SiNx and aluminium. *Proc. 10th Workshop on Crystalline Silicon Solar Cell Materials and Processes*, Colorado, 190.

40. Rohatgi, A., Yelundur, V. and Jeong, J. (2001) Lifetime enhancement and low-cost technology development for high efficiency manufacturable silicon solar cells *Proc. 11th Workshop on crystalline silicon solar cells materials and processes*, Colorado, 80-84.

41. Duerinckx, F., Frisson, L., Michiels, P.P., Choulat, P. and Szlufcik, J. (2001) Towards highly efficient cells and modules from multicrystalline silicon, *Proc. 17th European Photovoltaic Solar Energy Conference and Exhibition* Munich, Germany, in press.

42. Arimoto, S., Nakatani, M., Nishimoto, Y., Morikawa, H., Hayashi, M., Namizaki, H. and Namba, K. (2000) Simplified mass-production process for 16% efficiency multicrystalline Si solar cells. *Proc. 28th IEEE PVSC*, Alaska, USA, 188-193.

43. Fujii, S., Fukawa, Y., Takahashi, H., Inomata, Y., Okada, K., Fukui, K. and Shirasawa, K. (2001) Production technology of large area multicrystalline silicon solar cells. *Solar Energy Materials and Solar Cells* **65**, 269-275.

44. Final publishable report of the EC funded HIT project, to be published.

45. Szlufcik, J., Sivoththaman, S., Nijs, J., Mertens, R. and Van Overstraeten, R. (1997) Low-cost industrial technologies of crystalline silicon solar cells, *Proc. IEEE* **85** (5), 709-730.

46. Hanoka, J. (2001) An overview of silicon ribbon growth technology, *Solar Energy Materials and Solar Cells* **65**, 231-237.

47. Sarti, D., private communication.

48. Van Kerschaver, E. (2001) *PhD Thesis Kath. Univ. Leuven*, Belgium.

49. Mittelstadt, L., Dauwe, S., Metz, A., Hezel, R. *et al.*, *Proc. 17th European Photovoltaic Solar Energy Conference and Exhibition*, Munich, Germany, to be published

50. Glunz, S., Dicker, J., Kray, D., Lee, J., Preu, R., Rein, S., Schneiderlochner, E., Solter, J., Warta, W. and Willeke, G. (2001) High-efficiency cell structures for medium-quality silicon, *Proc. 17th European Photovoltaic Solar Energy Conference and Exhibition*, Munich, Germany, to be published

51. Van Kerschaver, E., DeWolf, S. and Szlufcik, J. (2000) *Proc. 16th European Photovoltaic Solar Energy Conf.*, Glasgow, 1517.

APPLICATION OF III-V COMPOUNDS IN SOLAR CELLS

V.M. ANDREEV
Ioffe Physico-Technical Institute
Polytechnicheskaya 26, St.Petersburg, 194021, Russia

1. Introduction

An obvious disadvantage of sunlight as a power source is its low energy density. To generate appreciable electrical power in space and on the earth, it is necessary to collect sunlight from large areas, having covered them with expensive semiconductor solar cells. The cost of electricity obtained in such a way exceeds substantially that generated by conventional methods, and this is the main reason retarding the development of a large-scale solar power industry.

One way to solve these problems is by reducing the cost of semiconductor materials and solar cells. Research in this direction has lately been very extensive. For example, due to the development of progressive technologies for the production of solar cells based on single crystal and multi-crystalline silicon, their cost has been reduced below \$4/W of installed peak power. The highest efficiency of silicon crystalline cells achievable is 23-24% (AM1.5). Solar cells with 17% efficiency have been fabricated on the basis of cheaper multi-crystalline silicon. Fairly high efficiency values have been achieved in thin-film solar cells based on amorphous silicon, CdTe and $CuInGaSe_2$, which rely on a very low cell cost per unit area. However, to introduce thin-film solar cells into large-scale energy production, it is necessary to solve a number of problems, among which is a reproducible technology for cheap solar cell fabrication and high parameter stability.

This paper considers another way for decreasing the cost of solar energy conversion - photovoltaic conversion of concentrated sunlight on the basis of III-V heterostructure solar cells. In this case, the required cell area and cost can be reduced in proportion to the sunlight concentration ratio, using cheap mirrors or Fresnel lenses.

Some problems arise in the practical realization of techniques for converting concentrated sunlight. The first of these is that as the sunlight density increases, the photocurrent generated by a solar cell grows proportionally, which complicates the cell design required to decrease resistive losses. Secondly, the temperature of the solar cells increases, requiring an efficient heat removal system. Thirdly, it is necessary to design highly efficient and cheap sunlight concentrators. Fourthly, we

131

J.M. Marshall and D. Dimova-Malinovska (eds.),
Photovoltaic and Photoactive Materials - Properties, Technology and Applications, 131–156.
© 2002 *Kluwer Academic Publishers.*

must have precise tracking of the sun, which adds complexity to the installation design and its performance.

Via the use of concentrators, it is possible to employ expensive semiconductor materials; for example, heterostructures based on gallium arsenide, for the fabrication of high-efficiency, thermally stable, and high-power solar cells.

Concentrator cells are more *efficient* than one-sun cells, and also permit the use of the combined thermal, photon and injection annealing effects to remove radiation defects arising during cell operation in space. Since comparatively small-area cells are used in this case, a better protection from adverse environmental factors can be provided, owing in particular to the screening action of concentrators and heat removal systems.

The method of photovoltaic conversion of concentrated sunlight has evolved, both theoretically and practically, from two basic ideas. One is that the cost of the electrical power generated can be made lower by reducing the area of expensive solar cells, in proportion to the sunlight concentration, using relatively cheap mirrors and lenses. The other idea is that the photovoltaic conversion efficiency of p-n junction solar cells can be made higher, since the operating voltage can be increased due to a higher photocurrent density provided by light concentrators. Moreover, because the cost of solar cells is not too critical at high concentration ratios, they can be fabricated from more suitable semiconductor materials and contain more complex structures - multilayer heterostructures and cascade cells providing the highest efficiency. As a result, concentrating photovoltaics today is a rapidly developing area of science and technology, the results of which largely determine trends in many other areas. Further progress in concentrated sunlight conversion primarily relies on whether we can achieve maximum efficiencies for the light concentrators, and hence for the solar cells.

The monographs [1-7] and review papers [8-11] on photovoltaic solar energy conversion have dealt predominantly with research on solar cells operating under conditions of natural non-concentrated sunlight. In recent years, however, extensive experimental and theoretical work has been done on concentrator solar cells. Some of these results have been discussed in the literature [12-16].

The basic materials used in the fabrication of concentrator cells are Si, GaAs and related III-V compounds. An optimal concentration ratio (K_s) for cells on AlGaAs/GaAs and GaInP/GaAs heterostructures is about an order of magnitude higher than for Si concentrator cells, due to lower internal ohmic losses and a greater temperature stability of conversion efficiency. This permits the use, in GaAs concentrator modules, of rather simple, low-cost cooling systems for high electrical power yields per cell. The maximum efficiency obtainable with such cells can exceed 30% for $K_s = 100$-1000, and varies insignificantly in the operating temperature range 20-70°C. An additional advantage of GaAs-based cells is their high resistance to radiation, making them viable under space conditions.

2. Solar Cells on AlGaAs/GaAs Heterostructures

The use of III-V heterostructures in the fabrication of solar cells can further increase the sunlight conversion efficiency. Among the large number of heterojunctions investigated with regard to their applicability in solar cells, AlGaAs/GaAs heterojunctions, together with GaInP/GaAs, have found the widest application. This is due to the good lattice matching of semiconductors in these heterostructures, and because gallium arsenide has an optimal band gap for the effective conversion of sunlight.

The first solar cells based on AlGaAs/GaAs heterojunctions were produced at the Ioffe Physico-Technical Institute, Russia [17-20]. The basic low bandgap material was GaAs, and a wide-gap window was made of $Al_xGa_{1-x}As$. Such a heterostructure is illuminated through the window, and light of photon energy exceeding the band gap value of GaAs is absorbed in it, while the p-n junction field separates the minority carriers. Because of the close lattice parameters of the contacting materials, the interface in $Al_xGa_{1-x}As/GaAs$ heterojunctions has a low density of surface states, providing a highly effective accumulation of carriers. Thin layers of $Al_xGa_{1-x}As$ solid solutions close in composition to aluminium arsenide ($x = 0.8-0.9$) are almost completely transparent to sunlight, making solar cells very sensitive in the spectral range $\lambda = 0.4-0.9$ µm. Extensive investigations have resulted in the fabrication of Al-Ga-As heterocells exhibiting high performance characteristics for both un-concentrated and concentrated ($K_s = 100-2500$) sunlight [17-28]. Before analyzing concentrator solar cells, let us consider the properties of heterojunctions in the Al-Ga-As system.

2.1. PROPERTIES OF AlGaAs/GaAs HETEROJUNCTIONS

The lattice parameters of gallium arsenide (a' = 0.5653 nm) and aluminium arsenide (a' = 0.566 nm) differ by 0.12% at room temperature. Owing to the different thermal expansion coefficients of GaAs ($\alpha_T = 6.5 \times 10^{-6}$ deg^{-1}) and AlAs ($\alpha_T = 5.2 \times 10^{-6}$ deg^{-1}) at the epitaxial growth temperatures of solid solutions (T= 600-900°C), the lattice parameters of the substrate and layers become still closer. This facilitates the growth of high-quality $Al_xGa_{1-x}As/GaAs$ solid solutions, free from misfit dislocations, on a GaAs substrate. However, as the temperature decreases to ambient, elastic strain arises because of the different α_T values. Decreasing the thickness of the $Al_xGa_{1-x}As/GaAs$ layer can reduce the strain in the active region of an AlGaAs/GaAs cell.

Aluminium arsenide is a material sensitive to corrosion. Yet, when doped with a stabilizing component (gallium) to form a solid solution with AlAs, its stability becomes much higher in a humid atmosphere. Addition of 10-20 atomic percent of gallium makes $Al_xGa_{1-x}As/GaAs$ solid solutions with $x = 0.8-0.9$ applicable for the production of solar cells and other devices possessing stable characteristics.

Aluminum arsenide is an indirect-gap semiconductor with E_g = 2.17 eV. The energy gap between the "direct" minimum of the conduction band and the top of the valence band in it is about E_g = 3.0 eV. The relationship between the band gap and the composition of $Al_xGa_{1-x}As/GaAs$ solid solutions can be obtained by interpolating the direct and indirect energy band gaps of AlAs and GaAs. In the range of $Al_xGa_{1-x}As/GaAs$ compositions from x = 0 (GaAs) to x < 0.4 and of band gap energy E_g = 1.40-1.95 eV (300K), this material has a direct energy band structure, but it is indirect in the range x > 0.4.

The $Al_xGa_{1-x}As/GaAs$ layer functioning as a window in AlGaAs/GaAs solar cells should be transparent to sunlight. The absorption edge in "indirect" solid solutions is largely determined by the direct energy gap. The conclusion important for the fabrication of solar cells from these materials is that almost complete transparency of $Al_xGa_{1-x}As/GaAs$ layers to short-wavelength light may be attained with fairly thin layers (0.1 μm) and x values in the range 0.8 to 0.9.

Another parameter important for designing antireflection coatings is the refractive index n of the front layer. In $Al_xGa_{1-x}As/GaAs$ solid solutions, the refractive index varies nearly linearly from n = 3.6 for GaAs to n = 3.0 for AlAs at λ = 0.9 μm. Hence, the refractive index at x = 0.8-0.9 and λ = 0.9 μm is n = 3.1-3.05 and increases with decreasing λ. So, with λ varying from 0.9 to 0.71 μm, the refractive index of $Al_xGa_{1-x}As/GaAs$ rises from 3.34 to 3.5.

The band diagram of an abrupt heterojunction differs from that of a homo p-n junction in the appearance of offsets in the valence and conduction bands (ΔE_v and ΔE_c, respectively). Considering the small difference in the lattice parameters of the contacting materials, $Al_xGa_{1-x}As/GaAs$ heterojunctions may be placed in the category of structures which exhibit a low recombination rate at the interface [17,18,29,30]. The dependence of ΔE_v on the parameter x for $Al_xGa_{1-x}As/GaAs$ heterojunctions [30] is ΔE_v = 0.55x. Using these data, the conduction band offset for direct solid solutions is ΔE_c = 0.9x, and the contributions to the net offset from ΔE_c and ΔE_v, amount to 60% and 40% respectively. In the range of indirect compositions, the value of ΔE_c drops with increasing x. For example [30], ΔE_c = 0.2 eV and ΔE_v = 0.5 eV, for $Al_{0.9}Ga_{0.1}As/GaAs$; i.e. most of the offset will be due to the valence band.

The appearance of band offsets in abrupt N-p and P-n heterojunctions (the first symbol indicates the type of conductivity in the wide-gap material) gives rise to additional potential barriers (kinks), which present an obstacle to minority carrier separation. A barrier in the conduction band of an N-p heterojunction impedes the separation of electrons, and a barrier in the valence band of a P-n heterojunction does the same to holes generated in the low band gap material. High-energy carriers can overcome the barrier or tunnel through it. The efficiency of these processes increases with the doping level of the wide band gap material, due to the decreasing barrier width. It also decreases in heterojunctions operating in reverse bias, when the probability for a carrier to overcome the barrier increases.

One way of eliminating these barriers is to incorporate, in the heterojunction, intermediate layers with a composition gradually varying with thickness. In these graded heterojunctions, provided the buffer layers are of sufficient thickness, there are no barriers to minority carriers. This is the case not only in reverse or zero bias, but also in forward bias, corresponding to the operating regime of the solar cell. An important problem here is optimization of the thickness of the graded layer necessary for complete elimination of the potential barrier to minority carriers. The required thickness of the graded layer increases with increasing width of the space-charge region in the *p-n* junction, i.e., with decreasing doping level in the contacting materials. For majority carrier concentrations typical of $Al_xGa_{1-x}As/GaAs$ solar cells, the graded layer thickness must be about 5 nm. Buffer layers of such a thickness can be obtained in $Al_xGa_{1-x}As/GaAs$ heterostructures grown by liquid-phase epitaxy, using isothermal mixing of melts containing different amounts of aluminium. This can also be done by sub-melting the substrate and subsequent crystallization of a solid solution of graded composition, as well as by high-temperature diffusion of the heterojunction components - Al and Ga.

A simple and common way of eliminating the potential barriers to minority carriers in solar cells based on AlGaAs/GaAs heterojunctions involves displacing the *p-n* junction from the interface into the gallium arsenide (Figure 1), by a distance smaller than the electron diffusion length in *p*-GaAs. In such a structure, electron-hole pairs generated near the *p-n* junction in both *n*-GaAs and *p*-GaAs meet no obstacles on their way to the space-charge region of the junction, while the potential barrier in the isotype heterojunction *p*-AlGaAs/*p*-GaAs does not hinder the movement of majority carriers (holes).

2.2. FABRICATION OF SOLAR CELLS BASED ON AlGaAs HETEROSTRUCTURES

Figures 1 and 2 show some major types of AlGaAs heterostructure for solar cells. The band diagrams relate to heterostructures with graded buffer layers incorporated into AlGaAs/GaAs heterojunctions. Much research effort has been made to obtain a wide spectral range of sensitivity and low sheet resistance, to provide an efficient conversion of concentrated sunlight.

The most common type of heterostructure developed in the early 1970s was that with the composition *n*-GaAs/*p*-GaAs/*p*-AlGaAs. The basic material for the fabrication is *n*-GaAs with a majority carrier concentration of $1-5 \times 10^{17}$ cm^{-3}, grown mostly by liquid-phase epitaxy. The *p*-GaAs layer ($p = 10^{18}-10^{19}$ cm^{-3}) of 0.5-3.0 μm thickness is either grown epitaxially or by zinc or beryllium diffusion during the growth of an $Al_xGa_{1-x}As/GaAs$ solid solution doped with one of these impurities. The diffusion produces a quasi-electric field arising from the acceptor concentration gradient (Figure 1), which enhances the effective diffusion length of electrons generated by light in the *p*-GaAs layer, if the acceptor concentration in the *p*-GaAs is sufficiently high (up to 10^{19} cm^{-3}) near the interface to reduce the sheet resistance in the front layers of the concentrator cells. Figure 1 shows the first structures of this type [18-28], in which a low sheet resistance of the front p-region is mainly due to a thick p-GaAs layer (2).

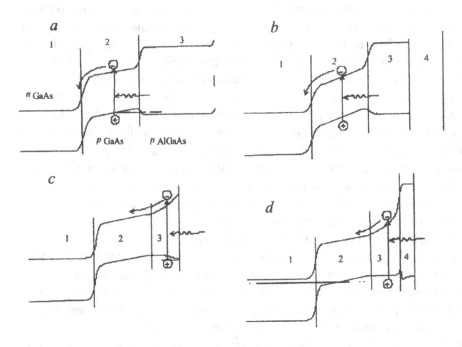

Figure 1. Band diagrams of p-AlGaAs/p-GaAs/n-GaAs heterostructures for concentrator solar cells.

Higher conversion efficiencies have been obtained in structures with the wide band gap layer having a smaller thickness [24-28]. A fairly low sheet resistance can be obtained by increasing the doping level of the p-GaAs layer. To reduce the contact resistance, an additional p^+-GaAs layer (4) is sometimes grown onto the solid solution layer, and subsequently etched off in places free from metallization (Figure 1b).

Another way of enhancing the short-wavelength photosensitivity is to use, for the front layer (3), a solid solution of graded composition with the band gap increasing towards the illuminated surface. The field due to the E_g gradient significantly enhances the effective electron diffusion length, and suppresses the surface recombination of electron-hole pairs generated near the surface by short-wavelength light (Figure 1b,c).

The surface recombination in structures with an E_g gradient can also be quenched by introducing an additional thin layer (4) of a wide band gap material (Figure 1d) transparent to short-wavelength light. Introduction of this layer makes it possible to lower the E_g gradient in layer (3), while ensuring complete collection of light-generated electrons. In this way, the sheet resistance in the front layer is reduced, ensuring operation at high light concentration ratios.

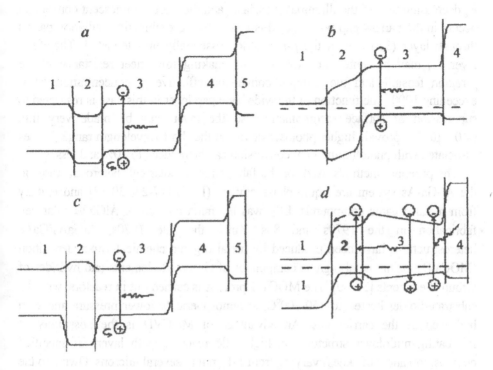

Figure 2. Band diagrams of AlGaAs heterostructures for concentrator solar cells with a back-surface field (*a,c*), with a thin wide band gap layer in the space-charge region (*b*), and with a structure giving intermediate conversion of sunlight into luminescence (*d*).

As noted above, the introduction of a potential barrier at the back of the photoactive region improves the collection efficiency of minority carriers generated by low-energy photons in the base. This barrier is made either by growing a buffer layer 2 (Figure 2*a*) of *n*-GaAs, doped to a level exceeding that in active layer (3), or of *n*-Al$_x$Ga$_{1-x}$As (Figure 2*c*). The thin wide band gap layer (2) (*n*-Al$_x$Ga$_{1-x}$As) was introduced in the structure in Figure 2*b*, to separate the *p*- and *n*-layers [31]. This buffer ensures that electron-hole pairs generated in layer (3) should be effectively separated by the *p-n* junction. The band gap in the space-charge region of the *p-n* junction is significantly increased, and the recombination component of the reverse saturation current is lowered by several orders of magnitude. Thus, the current is entirely due to the injection component under one-sun illumination (j = 3x10^3 A/cm^2).

The high (nearly 100%) internal quantum yield of radiative recombination in GaAs and Al$_x$Ga$_{1-x}$As solid solutions of compositions close to GaAs provides a highly efficient conversion of light to recombination radiation. This effect is exploited in solar cells with intermediate conversion of sunlight to luminescence (Figure 2*d*), which is then utilized for the generation of electron-hole pairs in the *p-n* junction [26]. Such a solar cell contains an additional layer (3) with

E_g decreasing toward the illuminated surface, and the photoluminescent conversion occurs in the narrow-gap portion of this layer. The recombination radiation passes through layer (3) to reach the p-n junction essentially un-attenuated. Therefore, layer (3) may be quite thick (30-50 μm), making low sheet resistance in the p-region feasible and the sunlight conversion effective at concentration ratios exceeding 1000. The function of the wide band gap layer in this case is restricted to suppression of surface recombination, so the layer may be made very thin (< 0.1 μm) to provide higher photosensitivity in the short-wavelength range. Losses associated with photoluminescent conversion can be reduced to 10% and less.

The principal methods used for the fabrication of solar cell heterostructures in the Al-Ga-As system are liquid-phase epitaxy (LPE) [17-27, 30-37] and epitaxy from metal organic compounds. LPE was the main method of AlGaAs solar cell fabrication in the 1970s and 80s. Since the late 1970s, AlGaAs/GaAs heterostructures have been produced by metal organic chemical vapor deposition (MOCVD), using metal organic compounds of Group III elements and hydrides of Group V elements [38-40]. The MOCVD process is carried out in reactors, with the substrate holder heated to 650-700°C, at atmospheric or lower pressure and with hydrogen as the carrier gas. An advantage of MOCVD is the possibility of fabricating multilayer structures in high-yield reactors, with layers of specified composition and a thickness varying from 1-10 nm to several microns. Owing to the fact that MOCVD is capable of producing GaAs single crystal layers on silicon and germanium substrates, it has a potential for the fabrication of low-cost, high efficiency AlGaAs solar cells on these substrates and cascade cells. Of course, a successful application of the MOCVD technique necessitates the use of up-to-date automated equipment and high purity carrier gases and materials. In addition, special precautions should be taken when handling toxic metal organic compounds and hydrides.

An important phase in solar cell fabrication is the formation of stripe contacts to the thin front layer. The metallization grid of heterojunction cells designed for concentrated sunlight conversion is the same as in silicon concentrator cells. It represents, for example, a network of radial stripes and concentric rings. Multilayer or composite coats are deposited for better adhesion and lower contact resistance: Au/Zn, Ag/Mn, Ag/Zn, Cr+Ni, Pd+Ni and Pd+Zn+Au. The Au/Ge eutectic alloy is usually used to make contacts to the back surface of the n-GaAs substrate.

An antireflection coating for solar cells based on AlGaAs structures is made either by anodic oxidation of the $Al_xGa_{1-x}As$ surface, or by deposition of thin films of Si_3N_4, ZnS, Ta_2O_5 and others. With a Si_3N_4 antireflection coating (74 nm), the reflection loss is about 8%. A two-layer coat composed, for example, of Ta_2O_5 (53 nm) and SiO_2 (76 nm) can reduce the reflection loss down to 4%. The same result can be obtained for a two-layer coat of ZnS and MgF_2.

2.3. PERFORMANCE OF SOLAR CELLS ON AlGaAs/GaAs

2.3.1. *Spectral Characteristics and Photocurrent*

The photoresponse spectra of solar cells based on heterostructures of the n-GaAs/p-GaAs/p-AlGaAs type vary with the collection efficiency of carriers generated in the n- and p-GaAs, and with the transmission spectrum of the wide band gap solid solution layer. The first factor determines the shape of the long-wavelength photosensitivity edge and the spectral variation of the photoresponse. The second factor determines the shape and position of the short-wavelength edge of the photosensitivity.

The short-wavelength edge of the photoresponse depends on the solid solution composition and thickness. The AlAs content in the $Al_xGa_{1-x}As$ layer is usually chosen in the range x = 0.8-0.9. As seen from Figure 3, as the $Al_xGa_{1-x}As$ layer thickness decreases from 10 to 0.05 μm, the short-wavelength photoresponse edge shifts from λ = 0.55 to 0.4 μm, which means that a wide band gap layer of thickness 0.05-0.1 μm is almost completely transparent to sunlight. The maximum calculated short-circuit current in these cells at AM0 decreases with increasing thickness (D) of the solid solution layer from i_{sc} = 35mA/cm^2 at D = 0.05-0.1 pm to i_{sc} = 26mA/cm^2 at D = 5 μm. Simultaneously, the 1 sun efficiency decreases from 21.5% to 16.5%. For sunlight at AM1.5 - 2, characterized by a lower intensity of short wavelengths, light absorption in the solid solution decreases, giving a smoother slope to the function i_{sc} = $f(D)$. For example, i_{sc} decreases from 23 to 19.5 mA/cm^2 at AM2, with D increasing from 0.1 to 5 μm. That is, the relative decrease in the efficiency is a factor of 1.5 less than for space sunlight. Maximum efficiency values have been obtained in solar cells with a wide band gap layer thickness, D = 0.02-0.05 μm. These ultra thin layers were grown either by liquid-phase epitaxy or by MOCVD epitaxy.

In addition to the composition and thickness of the $Al_xGa_{1-x}As$ layer, the short-wavelength spectrum is strongly influenced by the antireflection coating, especially around λ = 0.4 μm. The maximum value of external quantum yield has been obtained for solar cells with D = 0.03 μm, coated with a Si_3N_4 layer (Figure 3, curve 1), or with ZnS+MgF layers.

High photosensitivity in the short-wavelength range has also been obtained for structures in which the AlAs content in the wide band gap layer is gradually increased, producing a quasi-electric field (Figure 3, curves 2 and 4). In layers 0.2 and 0.5 μm thick, the field exceeds 10^4 V/cm, and significantly suppresses the recombination of minority carriers generated by light with λ < 0.5 μm, near the surface of the graded band gap layer.

In solar cells based on n-GaAs/p-GaAs/p-AlGaAs heterostructures, in which the p-GaAs layer thickness d is larger than 0.5 μm, the photocurrent is mainly determined by the relation between the layer thickness and the electron diffusion length (L_n). The photocurrent at L_n = 3-6 μm was found to be close to its maximum value at d(p-GaAs) = 0.5-1.5 μm. Solar cells designed for concentrated light conversion must have a thicker p-GaAs layer, to reduce the sheet resistance. For L_n = 5-6 µm, d (p-GaAs) maybe increased to 2 µm with the photocurrent values still remaining fairly high. High efficiency radiation recombination effects may give rise to secondary photons that may be re-absorbed, leading to larger diffusion lengths. This photon recycling effect may increase the photoresponse in cells with a larger thickness of active region.

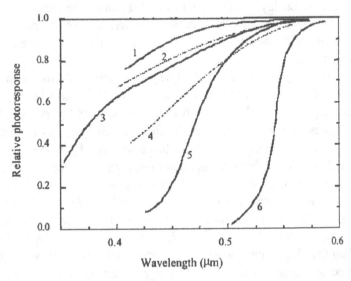

Figure 3. Experimental quantum yield in the short-wavelength range for solar cells based on a n-GaAs/p-GaAs/p-Al$_x$Ga$_{1-x}$As structure at different thickness (D) of the wide band gap Al$_x$Ga$_{1-x}$As layer (µm): 1 = 0.03; 2 = 0.2; 3 = 0.07; 4 = 0.5; 5 = 0.5; 6 = 10 with constant (curves 1, 3, 5, 6) and graded (2,4) compositions.

2.3.2. *Current-Voltage Characteristics of AlGaAs/GaAs Solar Cells*

Due to a high probability of radiation recombination under the concentrated light conditions in GaAs p-n junctions, the reverse saturation current amounts to $i_0 = 10^{-19}$ A/cm^2, while the minimum value is $i_0 = 10^{-20}$ A/cm^2. The close values of the actual and theoretical i_0 currents is a necessary condition for achieving open-circuit voltage and fill factor values close to their theoretical limits. We have pointed out that the current through the load connected in parallel with an illuminated solar cell can be represented as a difference between the photocurrent and the dark current (i_d) in the junction. For $i_d \gg i_o$, the latter represents the sum of the injection and recombination components: $i_d = i_{01}\exp(qU/kT) + i_{02}\exp(qU/2kT)$, where i_{01} and i_{02} are the injection and recombination components of the saturation current, respectively.

At a current density of $i = 3\times10^{-2}$ A/cm^2, corresponding to non-concentrated sunlight, solar cells based on GaAs p-n junctions exhibit a "combined" mechanism of current production (Figure 4, curve 1), with a total saturation current equal to 10^{-13} to 10^{-15} A/cm^2. Minimum values of i_o in this case are found in structures fabricated from high quality n-GaAs grown by liquid-phase epitaxy or MOCVD.

The recombination component of the saturation current is proportional to the intrinsic carrier concentration, which decreases with increasing band gap energy. In n-GaAs/n-Al$_{0.4}$Ga$_{0.6}$As/p-GaAs/p-Al$_{0.8}$Ga$_{0.2}$As structures with a thin Al$_{0.4}$Ga$_{0.6}$As layer in the space charge region (Figure 4b), the value of the recombination saturation current is appreciably lower, actually below $i_{02} = 10^{-15}$ A/cm^2 [31]. The transition from a recombination to an injection mechanism of current flow occurs here (Figure 4, curve 2) at a current density two orders of magnitude lower than in a GaAs p-n junction structure. In this case, due to the predominant injection from Al$_{0.4}$Ga$_{0.6}$As into p-GaAs, the injection component of the saturation current decreases by more than an order of magnitude, down $i_{01} = 10^{-20}$ A/cm^2. As a result, the respective solar cells exhibit an injection mechanism of current under solar illumination ($i_{ph} = 3\times10^{-2}$ A/cm^2) with a reverse saturation current of $i_{01} = 10^{-20}$ A/cm^2, permitting a high open-circuit voltage and fill factor to be obtained. The difference between curves 1 and 2 in Figure 4 increases with decreasing current. Therefore, the advantage of structures with a wide band gap "insert" is more pronounced under low illumination. For example, a 12% efficiency was measured in a cell based on this structure at $P_s = 10^{-3}$ W/cm and T = 100°C. This is 1.5 times higher than in a n-GaAs/p-GaAs/p-AlGaAs cell.

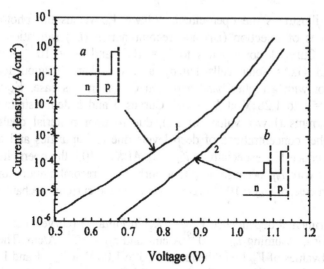

Figure 4. Experimental dark current-voltage characteristics of solar cells based on a heterostructure with a GaAs p-n junction (1) and on a structure with an Al$_{0.4}$Ga$_{0.6}$As layer in the space-charge region (2).

The saturation current component due to injection decreases, and V_{oc} increases for higher doping levels in the n- and p-regions. However, leakage and tunnel currents are more likely at higher doping levels. For these reasons, the optimum free carrier concentrations near the p-n junction are $n = (1-5)\times10^{17}$ cm^{-3} and $p = (1-2)\times10^{18}$ cm^{-3}. In diffused junctions, the field in the p-region, arising from the acceptor concentration gradient, causes an increase in L_n at a hole concentration of 10^{19} cm^{-3} near the heterointerface, which should result in a higher V_{oc}.

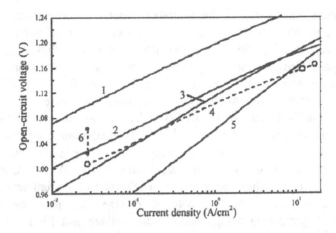

Figure 5. Theoretical (1-3,5) and experimental (4) dependences at 300 K of the open-circuit voltage on photocurrent density for different values of the injection (i_{01}) and recombination (i_{02}) components of the saturation current, i, (A/cm^2): 1 - $i_{01} = 10^{-20}$, $i_{02} = 10^{-12}$; 2 - $i_{01} = 10^{-19}$, $i_{02} = 10^{-11}$; 3 - $i_{01} = 10^{-19}$, $i_{02} = 10^{-10}$; 4 - experiment; 5 - $i_{01} = 10^{-19}$, $i_{02} = 10^{-6}$; 6 - measured V_{oc} values at $i_{ph} = 3 \times 10^{-2}$ A/cm^2 ($K_s = 1$).

Figure 5 illustrates the open-circuit voltage V_{oc} versus the photocurrent for different values of injection (i_{01}) and recombination (i_{02}) saturation currents at T = 300 K. Curve 1 corresponds to $i_0 = 10^{-20}$ and $i_{02} = 10^{-12}$ A/cm^2, values achievable in GaAs solar cells having a space charge region located in a semiconductor with a higher band gap than GaAs. In this case, V_{oc} varies from 1.1 V at $K_s = 1$ to 1.28 V at $K_s = 10^3$. Curves 3 and 5 demonstrate that higher saturation currents (lower values of L_D), due to poor material quality and the resulting higher concentrations of deep levels due to impurities and defects, can appreciably reduce V_{oc}, especially at $K_s < 10$. At $K_s > 100$, the open-circuit voltage varies mainly with i_{01}, because despite a fairly large recombination component of the saturation current ($i_{02} = 10^{-10}$ A/cm^2), the injection current mechanism becomes dominant.

The experimental dependence $V_{oc} = f(K_s)$ in Figure 5 (curve 4) can be fitted to the calculations, assuming $i_{01} = 10^{-19}$ A/cm^2 and $i_{02} = 10^{-10}$ A/cm. The maximum experimental values of V_{oc} lie in the range 1.03 to 1.06 V at $K_s = 1$ and 1.1 to 1.18 at $K_s = 10^3$. The larger discrepancies between the experimental (4) and theoretical curves (2 and 3) at $K_s > 10^2$ are likely to be associated with a higher equilibrium temperature of solar cells possessing higher concentration ratios.

The fill factor (FF) increases with a lower reverse saturation current and a higher concentration ratio. Figure 6 (curves 1-3) illustrates the effect of the injection (i_{01}) and recombination (i_{02}) components of the reverse saturation current on FF for concentrated sunlight (neglecting the ohmic losses). As is clear from curves 2 and 3, an increase in i_{02} from 10^{-11} to 10^{-10} A/cm^2 at $K_s = 1$ ($i_{ph} = 3 \times 10^{-2}$ A/cm^2) causes FF to decrease from 0.86 to 0.8. At higher concentration ratios, the difference between curves 2 and 3 becomes smaller, demonstrating a weaker influence of reverse saturation current on FF at $K_s > 10^2$.

The experimental curves 4-5 in Figure 6 are markedly affected by ohmic losses, but at low concentration ratios this influence is insignificant. The maximum FF values obtained in un-concentrated sunlight ($i_{ph} = 3 \times 10^{-2}$ A/cm^2) are 0.84 to 0.87. With increasing K_s, FF rises to 0.86-0.89, and then rapidly falls off due to greater ohmic losses.

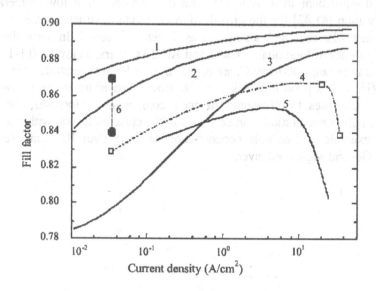

Figure 6. Calculated (1-3) and experimental (4-6) dependences of the fill factor on the photocurrent density for GaAs solar cells at different values of injection (i_{01}) and recombination (i_{02}) saturation currents, (A/cm^2): 1 - $i_{01} = 10^{-20}$, $i_{02} = 10^{-12}$; 2 - $i_{01} = 10^{-19}$, $i_{02} = 10^{-11}$; 3 - $i_{01} = 10^{-19}$, $i_{02} = 10^{-10}$.

2.3.3. *Efficiency of AlGaAs/GaAs Concentrator Solar Cells*

GaAs solar cells, in which the current is due to the diffusion mechanism, have a maximum calculated efficiency rising from 29% at $K_s = 1$ (AM1.5, 27°C) to 35% at $K_s = 1000$. In practice, the achievable efficiency rises from 25% (AM1.5) at $K_s = 1$ to 28% at $K_s = 100$-300, mainly due to the increasing open-circuit voltage (Figure 5). The reduced efficiency with further increase in K_s is caused by the fill factor decrease, due to greater ohmic losses.

Figure 2 *(c)* shows the structure of a GaAs solar cell with the highest one-sun and concentrator efficiencies for single-junction cells. The first cell structures of this kind were prepared by metal-organic chemical vapor deposition (MOCVD). The one-sun AM0 efficiency of 21.7% [40] and the concentrator AM0 efficiency of 24.5% at 170 suns have been obtained in cells based on these structures, with an n-Al$_x$Ga$_{1-x}$As layer as the back surface barrier. The better performance of these cells relative to earlier AlGaAs/GaAs cells is primarily due to improved MOCVD epitaxial growth and cell processing. An increase in the growth temperature from 700-740°C to 775-800°C significantly reduces the recombination in the cell bulk and at the interface. The long-wavelength response is substantially higher at higher growth temperatures, due to a lower recombination in the base. The open-circuit voltage, which is limited by the base recombination, has also improved. Today, the MOCVD method is widely used for the preparation of multilayer AlGaAs/GaAs and GaInP/GaAs structures for one-sun and concentrator applications.

Liquid phase epitaxy can provide good crystal quality epitaxial layers with less sophisticated equipment than MOCVD. The development of a low-temperature LPE modification [41,42] for the growth of AlGaAs/GaAs structures has yielded high efficiency solar cells with a structure as shown in Figure 7. In particular, the crystallization rates of high quality GaAs and AlGaAs layers, as low as 0.1-1 nm/s in the temperature range 400-550°C, are of the same order of magnitude as in MBE and MOCVD, with a good tractability of the growth procedure in the nanometer scaling of thicknesses by changing the melt composition. Moreover, the free electron and hole concentrations can be varied in the epitaxial layers, within a wide range. For example, a free hole concentration of more than 10^{10} cm^{-3} can be obtained in Ge- and Mg-doped layers.

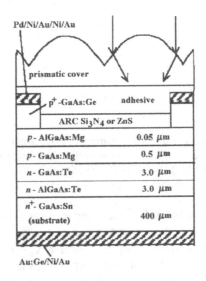

Figure 7. Schematic diagram of an LPE grown AlGaAs/GaAs solar cell with a back surface field layer made of AlGaAs [41].

Low temperature growth of multilayer AlGaAs/GaAs structures is performed using a "piston boat" [33]. The main advantages of this method over the conventional "sliding boat" technique are as follows. Before reaching the substrate, the melt is squeezed through a slot by a piston, to provide effective mechanical cleaning of any oxide film. The melt substitution is performed by replacing one melt by another, to reduce the number of micro-defects. Crystallization on the substrate can be carried out from a very thin melt layer, which provides a low growth rate and structural planarity. The epitaxial structure shown in Figure 7 consists of a n-Al$_x$Ga$_{1-x}$As Te-doped back-surface barrier, a n-GaAs Te-doped 3 μm thick base (n = 1.3×10^{17} cm^{-3}), a p-GaAs Mg-doped 0.4-0.5 mm thick front photoactive layer, a p-AlGaAs (x = 0.8-0.9) Mg-doped window, and a p-GaAs Ge-doped cap layer. The cap layer was locally removed between the photolitographically shaped 0.1-02 mm spaced contact grid fingers. An antireflection coating was deposited by magnetron sputtering or thermal evaporation. Silicone prismatic covers optically eliminated the grid line obscuration loss in the concentrator cells.

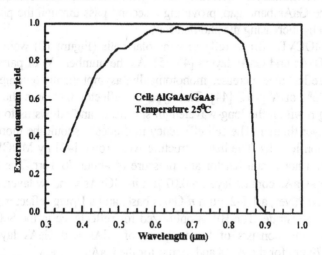

Figure 8. Spectral response of an AlGaAs/GaAs cell prepared by low-temperature LPE.

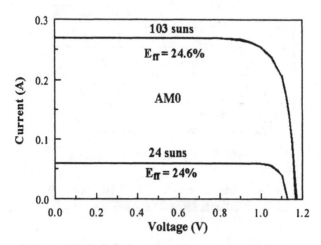

Figure 9. Current versus voltage for an AlGaAs/GaAs-based
concentrator cell prepared by low-temperature LPE.

Owing to the high crystal quality of the LPE material and optimized optical parameters, a high quantum yield is obtainable in a wide spectral range (Figure 8). The conversion efficiencies for solar cells with a prismatic cover (Figure 9) are 24.6% (AM0, 103 suns, 25°C) [41] and 27.6% (AM1.5D, 140 suns).

2.3.4. *Solar Cells with an Internal Bragg Reflector*

The Bragg reflector (BR) is widely used in lasers and other optical devices. The principles of multilayer dielectric reflectors are well known. By using a multiple layer composed of materials with different refractive indices, a nearly 100% reflectance can be achieved over a restricted wavelength range. The thickness of each of the two materials is chosen for quarter-wavelength reflection for the given wavelength. This multilayer dielectric stack will selectively reflect unabsorbed photons near the GaAs band gap, providing a second pass through the photoactive region and thereby increasing the photocurrent.

Epitaxial (MOCVD) Bragg reflectors in solar cells (Figure 10) were designed for pairs of AlGaAs and GaAs layers [43-45]. As the number (N) of pairs is made larger, the BR reflectance increases monotonically, asymptotically tending to unity, and reaching 96% at $N = 12$ [44]. This type of reflector increases the effective absorption length within the long-wavelength spectrum, and allows us to make the n-GaAs base layer thinner. The cell efficiency in this case should be more tolerant to lower diffusion lengths. The heterostructure was grown [44] by MOCVD using equipment with a horizontal reactor at a pressure of about 76 Torr. The solar cell consists of a top-GaAs contact layer, a 0.07 μm n-AlGaAs window layer, a 0.5 μm GaAs photoactive layer, a 1.5-2.5 μm n-GaAs base, and a Bragg reflector grown on an n-type GaAs substrate. The BR is optimized for reflectance in the 800-900 nm spectral region, and consists of twelve pairs of AlAs and GaAs layers, with thicknesses of 72 nm for the AlAs and 59 nm for the GaAs.

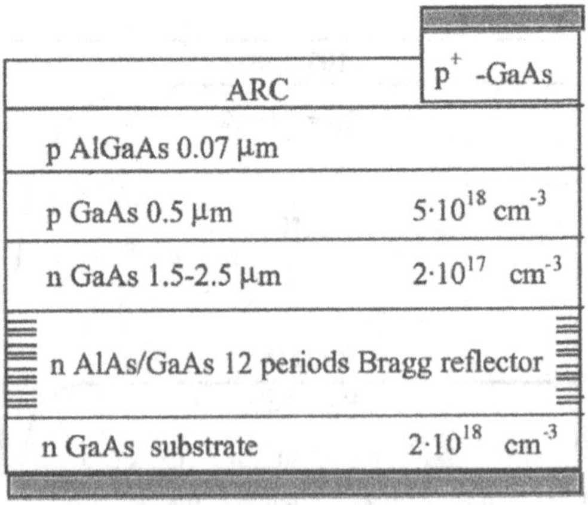

ARC	p^+ -GaAs	
p AlGaAs 0.07 μm		
p GaAs 0.5 μm	$5 \cdot 10^{18}$ cm^{-3}	
n GaAs 1.5-2.5 μm	$2 \cdot 10^{17}$ cm^{-3}	
n AlAs/GaAs 12 periods Bragg reflector		
n GaAs substrate	$2 \cdot 10^{18}$ cm^{-3}	

Figure 10. Schematic cross section of an AlGaAs/GaAs solar cell
with an internal Bragg reflector.

The photocurrent density registered in solar cells based on structures with a prismatic cover was as high as 32.7 mA/cm^{-2} (AM0, 25°C). This value is fairly good, considering the small thickness of the n-GaAs base layer. The red response of a cell with BR, and with a 1.5 μm n-GaAs layer is nearly the same as for a cell with a 3-4 μm *n*-GaAs layer without a BR. This improves the cell radiation resistance. An AM0 efficiency of 23.4% (18 suns, 25°C) was registered for cells with a prismatic cover [45].

Thus, high efficiency, irradiation and temperature stable AlGaAs/GaAs solar cells have been developed during the last decade [40-49]. The record confirmed efficiencies in concentrator cells are 24.5-24.6% at 100 suns under AM0 spectra [40, 41], 27.6% at 140 suns, AM1.5, 26.2% at 1000 suns [49] and 23% at 5800 suns [48].

3. Cascade Solar Cells

The high efficiency solar cells described in the preceding section have already passed the research and development stage, and find ever-wider applications in space and terrestrial photovoltaic installations. Today, such units are capable of competing with arrays of low-cost cells operating in un-concentrated sunlight. Despite the fairly high cost of concentrator cells, their contribution to the cost of the electrical power produced is not dominant, and decreases in proportion to the concentration ratio. For this reason, the greater sophistication and higher cost of the solar cells are justified, as long as their efficiency rises. If, for example, the cost of a cell is one tenth of that of a power unit, a 2-3% increase in the efficiency will be sufficient to make the generated power cheaper, even if the cost of the cells should rise by as much as 50%. This accounts for the great research effort being made to improve cascade cells so as to obtain a substantially higher efficiency [50-64]. For example, a tandem cell based on III-V heterostructures (AlGaAs/GaAs, GaInP/GaAs or GaAs/GaSb) with a 33% efficiency at $K_s = 1000$ and a cost of $10/cm^2 may be more cost-effective than a Si concentrator cell with a 26% efficiency at $K_s = 100$ and a cost of $1.0/cm^2.

3.1. TYPES OF CASCADE SOLAR CELL

Still more solar energy is utilized in cascade cells, representing a set of solar cells with different band gaps. Here, the solar spectrum is either split into several beams by selective mirrors or passed through a sequence of cells, whose band gap energy decreases along the solar beam path. Two types of tandem solar cell are represented schematically in Figure 11. The top cell in each tandem unit is based on a heterostructure with a band gap E_{g1}. In the bottom cell, the band gap of the photoactive layer is E_{g2}.

The first type (Figure 11a,b) is designed as a monolithic p-n-n^+-p^+-p-n heterostructure. A series connection of the upper and lower cells is provided by a reverse-biased tunnel p-n junction, to produce the tandem. In the second type, the cells are connected by a network of ohmic contacts (Figure 11c,d).

The short-wavelength sunlight component with photon energies $h\nu > E_{g1}$ is absorbed in the lower bandgap layer of the top cell. Photons with $h\nu < E_{g1}$ will reach the bottom cell and those with $h\nu > E_{g2}$ will be absorbed in the vicinity of the second *p-n* junction. Thus, the short-wavelength portion of the solar spectrum will be converted within the upper cell and the long-wavelength component within the bottom cell. In this design, the solar energy loss can be reduced through optimization of each cell for a respective spectral range, thus increasing the efficiency.

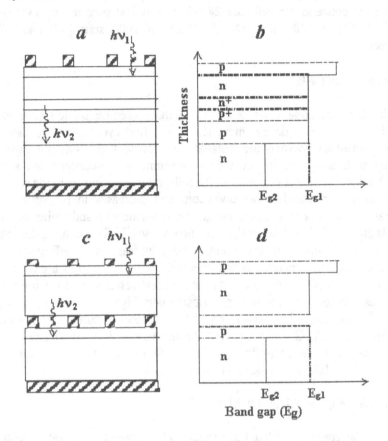

Figure 11. Two types of tandem solar cell with connection via a tunnel p^+-n^+ junction (*a, b*) or a contact grid (*c,d*).

The main difficulty in the implementation of the tandem design shown in Figure 11 is to connect the top and bottom cells without introducing additional ohmic and optical losses. The "tunnel" connection involves considerable complications in the fabrication technology and extra ohmic losses due to the current through the tunnel *p-n* junction. In the other design (Figure 11*c,d*) additional optical losses are associated with light passing from the upper to the lower cell. These can be reduced by application of anti-reflection coatings to the back surface of the top cell and to the top surface of the bottom one. This is done by introducing an optical medium with a large refractive index between the cells, using a prismatic cover, or by reducing the area of the metallization grid connecting the cells in cascade.

3.2. III-V COMPOUNDS FOR TANDEM SOLAR CELLS

Figure 12 is a 3-dimensional diagram representing the efficiency variation with the band gaps E_{g1} and E_{g2}. A series of curves is seen on the surface, representing the function $\eta = f(E_{g1}, E_{g2})$. The first set of lines are the $\eta = f(E_{g1})$ curves calculated for E_{g2} values varying between 0.8 and 1.6 eV, with a 0.1 eV step. The second set are the $\eta = f(E_{g2})$ curves for fixed E_{g1} values in the range 1.4-2.4 eV, with the same step. The non-monotonic character of the curves is due to the atmospheric absorption bands within the AM2.3 terrestrial solar spectrum. Efficiencies of more than 40% are obtained for $E_{g1} = 1.6$-1.8 eV and $E_{g2} = 0.9$-1.1 eV.

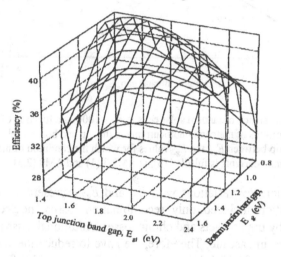

Figure 12. Spatial representation of the calculated efficiency for tandem solar cells as a function of E_{g1} and E_{g2} at AM2.3, $K_s = 1000$.

The dependence of the efficiency on E_{g1} and E_{g2} presented in Figure 12 refers to tandem cells, with each cell connected independently. These curves have been calculated neglecting the requirement that the photocurrents through the upper and lower cells should be matched. The condition described as "current matching" should be observed in tandems with a series connection of top and bottom cells. Figure 13 gives the calculated efficiency as a function of E_{g2} at $K_s = 1000$ (curves 1-2) under the conditions of matched photocurrents for different air mass values: AM2.3 (curve 1) and AM0 (curve 2). The same figure (curve 3) demonstrates the relationship between E_{g1} and E_{g2} for AM0 under the current matching condition.

The smooth character of the efficiency variation with E_{g1} and E_{g2} facilitates the choice of semiconductor for tandem cell fabrication. Even a considerable deviation of E_g from the optimum value causes but a small decrease in the efficiency. For example, the fairly high efficiency predicted for $E_{g2} = 1.4$ eV and for $E_{g1} = 1.4$ eV makes GaAs an attractive material for the fabrication of both the bottom and the top cell. Note that the efficiency values given in Figure 13 differ only slightly from the maximum value in Figure 12, calculated with a photocurrent mismatch.

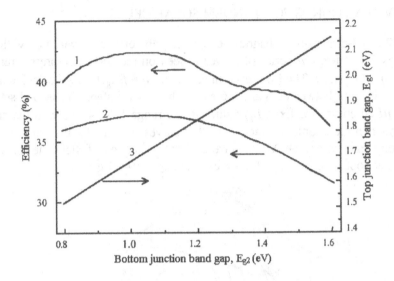

Figure 13. Efficiency of tandem solar cells at $K_s = 1000$ as a function of E_{g2} (1,2) under conditions of matched photocurrents in the top and bottom cells, and a relationship between E_{g1} and E_{g2} values for which the condition of matched photocurrents is fulfilled (3), at AM2.3 (1) and AM0 (2,3).

The above efficiency variations with E_{g1} and E_{g2} represent a theoretical limit, since the calculation included only recombination losses, neglecting reflection losses, screening by the contacts and ohmic losses. Additional losses arise when the cells are connected in cascade. Therefore, we have to reduce the values in Figures 12 and 13 by a factor of 1.10-1.15, to find practically achievable efficiencies.

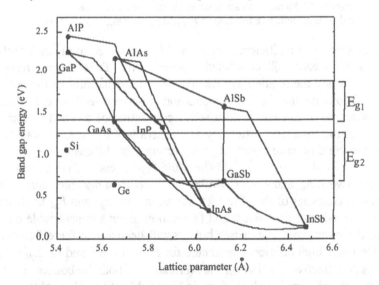

Figure 14. Dependences of the band gaps of Ge, Si and III-V compounds and their solid solutions on the lattice parameter of these materials.

Figure 14 illustrates the band gaps and lattice parameters for a number of III-V compounds and solid solutions. The marked E_{g1} and E_{g2} bands represent optimum E_g values for the bottom and top cells. One can see that Ge, GaSb, Si, InGaAs and GaAsSb can be used for the fabrication of bottom cells, while GaAs, AlGaAs, GaPAs, GaInP, AlGaAsSb and AlInPAs for top cells. GaAs substrates can be employed for the growth of GaInAs and GaAsSb epilayers of a desired composition (E_{g2} = 1.1 eV). The layer growth by liquid-phase epitaxy or MOCVD is performed in several stages, each at a higher In or Sb content. In this way, GaInAs solid solution layers of satisfactory quality have been grown [63], and used for subsequent epitaxial growth of lattice-matched top cells in the AlGaInP/GaInP system. Lattice-matched $Al_xGa_{1-x}As$/GaAs and $GaInP_2$/GaAs heterostructures provide a fairly high efficiency for tandems with GaAs bottom cells and AlGaAs (E_g = 1.9-1.95 eV) or GaInP (E_g = 1.85-1.9 eV) top cells.

As germanium has a lattice parameter close to that of gallium arsenide, it can profitably be used for the fabrication of the bottom element in tandem Ge/GaAs solar cells. In that case, the theoretical efficiency is almost the same as in tandem cells based on GaAs/AlGaAs or GaAs/GaInP structures. A still higher efficiency (η = 40%, AM0, K_s = 1000) has been calculated for triple-junction Ge/GaAs/AlGaAs or Ge/GaAs/GaInP cascade cells.

The InP/InGaAsP system is of special interest. Indium phosphide and lattice-matched InGaAs and InGaAsP solid solutions have been used to fabricate both top (on InP) and bottom cells [50,51]. The recombination losses of minority carriers generated near the InP surface by short-wavelength light are lower, due to a low surface recombination rate in InP. However, the efficiency of cells with an InP $p-n$ junction is less than in cells based on AlGaAs/GaAs or GaInP/GaAs heterostructures. An efficiency increase in InP cells, especially for concentrated sunlight, is hampered by the lack of lattice-matched materials having a sufficiently high band gap, such as AlGaAs solid solutions, which serve as a perfect window in GaAs cells. For this reason, a high efficiency can be expected only for moderate concentration ratios (K_s = 50-100) in InP cells.

As seen in Figure 14, silicon is a material with an optimum band gap for the fabrication of bottom cells. Wide band gap cells can, however, be made only from such materials as AlGaAs, GaInP, and GaPAs solid solutions, which are not lattice-matched with silicon. For this reason, mechanically stacked cells connected through contact networks have long been considered as the only design possible for silicon cells. The top cell in a stack can be made as a thin film structure glued to a glass plate, with a gallium arsenide substrate etched off or removed by other means. The ultimate thickness of the structure is 10-20 μm. Mechanically stacked cells based on $Si/Al_xGa_{1-x}As$ with x = 0.15-0.2 can provide over 30% efficiencies. Lately, significant advances have been made in the fabrication of GaAs, GaPAs, and $Al_xGa_{1-x}As$ epitaxial layers on single crystal Si substrates, using the MBE and MOCVD techniques. The results obtained hold a promise for high efficiency monolithic cascade cells, which should be less costly than those with heterostructures on GaAs or Ge substrates.

3.3. ACHIEVED EFFICIENCIES IN MECHANICALLY STACKED AND MONOLITHIC CASCADE SOLAR CELLS

Despite a large number of theoretical studies on cascade solar cells, their efficiency has remained low for a long time, since the ohmic and optical losses in available designs were unacceptably large. It is only recently that monolithic and mechanically stacked tandem cells with over 30% efficiency have been fabricated. For mechanically stacked cells, the maximum efficiency obtained [52] is about 35% at AM1.5 and $K_s = 100$ with GaSb bottom and AlGaAs/GaAs top cells. For AM0 concentrated sunlight (100x), the attainable efficiency is about 30% [52, 53]. AlGaAs/GaAs cells for tandems were made transparent to the infrared spectrum in stacked cells, so the n-GaAs substrate doping density was reduced to 10^{17} cm^{-3}. The anti-reflection coatings on the front and back sides of GaAs cells were optimized to reduce the reflection losses in a broader spectral region - throughout the visible energy range as far as the GaSb band edge at 1700 nm.

GaSb cells for tandems have been fabricated using Zn-diffusion [52-54]. A silicon nitride dielectric deposited on the GaSb wafer top surface served as a diffusion mask, so that a junction was formed only in the region intended for illumination. Single crystal n-GaSb wafers ($n = 3\text{-}5\text{x}10^{17}$ cm^{-3}) were used as substrates, and the source for diffusion was a mixture of Zn, Ga and Sb. The diffusion temperature range was 450-600°C, and the typical diffused p-GaAs layer thickness for solar cells ranged between 0.2 and 0.4 µm. The surface Zn concentration was about 10^{20} cm^{-3}. The maximum photocurrent densities in GaSb cells for the AM0 spectrum are about 30 mA/cm^2 behind a GaAs filter, and efficiencies of 6 to 7% have been achieved behind a transparent GaAs cell.

A GaAs and InGaAs/InP cell pair can also be used for the preparation of mechanically stacked tandem cells [50,53]. Lattice-matched GaInAs or GaInPAs layers were grown on an InP substrate by LPE or by MOCVD epitaxy. In the first case, the cells had no contact grid on the top surface, because of a low sheet resistance provided by a thick (350 µm) InP infrared-transparent substrate window. A 6% efficiency (AM0, 100x) was obtained in cells illuminated through a GaAs filter, and about 30% efficiency (AM0, 100 suns) was achieved in a tandem GaAs/InGaAs stack [53].

High efficiency monolithic cascade cells were fabricated on GaAs/AlGaAs [55,56] and GaAs/GaInP [57-64] heterostructures. GaInP/GaAs tandem solar cells are quite suitable for high-efficiency photovoltaic technology. The structures consist of a GaInP top cell connected in series by a low-resistively, grown-in tunnel junction to a GaAs bottom cell. With the ultimate objective of lower cost, GaInP/GaAs/Ge triple-junction solar cells on germanium substrates have also been developed [58,59,64]. The record efficiency of 32.5% (AM1.5D, 300 suns) was achieved in GaInP/GaAs/Ge triple junction cells [64].

To conclude, single $p\text{-}n$ junction and cascade solar cells operating on concentrated sunlight are being used in designs of efficient solar systems for supplying electrical power. Their cell area, and hence the total cost, are reduced in proportion to the sunlight concentration ratio. This opens up new prospects for a considerable reduction in the cost of the power produced. These considerations allow us to predict the integration of heterostructure cells into a large-scale solar power industry, in the near future.

4. References

1. Vasil'ev, A.M. and Landsman, A.P. (1971) *Poluprovodnikovyefotopreobrasavateli (Semiconductor photoconverters)*, Sov. Radio, Moscow (in Russian).
2. Hovel, H.J. (1975) *Semiconductors and semimetals*, Vol. 11, *Solar Cells*, Acad. Press, New York, London.
3. Koltun, M.M. (1985) *Optica i metrologia solnechnykh elementov (Optics and metrology of solar cells)*, Nauka, Moscow (in Russian).
4. Chopra, K.L. (1983) *Thin film solar cells*, Plenum Press, New York - London.
5. Grilikhes, V.A., Orlov, P.P. and Popov, L.B. (1984) *Solnechnaya energya i kosmicheski polioty (Solar energy and space flights)*, Nauka, Moscow (in Russian).
6. Fahrenbruch, A.L. and Bube, R.H. (1983) *Fundamentals of solar cells*, Acad. Press, New York - London.
7. Rauschenbach, H.S. (1980) *Solar cell array design handbook. The principles and technology of photovoltaic energy conversion*, Litton Educational Publishing, Inc., New York.
8. Alferov, Zh.I. (1978) Photovoltaic solar energy conversion, in: E.B. Etingof (ed.), *Future of Science*, Znanie, Moscow, pp.92-101 (in Russian).
9. *Current topics in photovoltaics* (1985) T.J. Coutts and J.D. Meakin (eds.), Acad. Press, London.
10. *Solar cells and their applications* (1995) L.D. Partain (ed.), John Wiley & Sons.
11. Flood, D. and Brandhorst, H. (1987) Space solar cells, in T.J. Coutts and J.D. Meakin (eds.), *Current topics in photovoltaics*. Vol. 2, Acad. Press, New York - London, pp. 143-202.
12. Andreev, V.M., Grilikhes V.A. and Rumyantsev, V.D. (1997) *Photovoltaic conversion of concentrated sunlight*, John Wiley & Sons, Chichester.
13. Luque, A. (1989) *Solar cells and optics for photovoltaic concentration*, Adam Hilger, Bristol-Philadelphia.
14. Sinton, R.A. (1995) Terrestrial silicon concentrator solar cells, in L.D. Partain (ed.), *Solar cells and their applications*, John Wiley & Sons, Inc., pp. 79-98.
15. Klausmeier-Brown, M.E. (1995) Status, prospects and economics of terrestrial, single junction GaAs concentrator cells, in L.D. Partain (ed.), *Solar cells and their applications*, John Wiley & Sons, Inc., pp. 125-142.
16. Fraas, L.M. (1995) High-efficiency III-V multijunction solar cells, in L.D. Partain (ed.), *Solar cells and their applications*, John Wiley & Sons, Inc., pp. 143-162.
17. Alferov, Zh.L, Andreev, V.M., Zimogorova, N.S. and Tret'iakov, D.N. (1969) Photovoltaic properties of heterojunction AlGaAs-GaAs, *Fir. i Tekhn. Polupr.* 3, 1633-1637. Translated into English in *Soviet Physics Semicond.* 3, No. 11.
18. Alferov, Zh.L, Andreev, V.M., Kagan, M.B., Protasov, I.I. and Trofim, V.G. (1970) Solar cells based on heterojunction p-AlGaAs-n-GaAs, *Fiz. i Tekhn. Polupr.* 4, 2378-2379. Translated into English in *Soviet Physics Semicond.* 4, No. 12.
19. Andreev, V.M., Glolovner, TM., Kagan, MB., Koroleva, N.S., Lubochevskaya, T.A., Nuller, TA. and Tret'yakov, D.N. (1973) Investigation of high efficiency AlGaAs-GaAs solar cells, *Fiz. i Tekhn. Pohipr.* 7, 2289-2296. Translated into English in *Sov. Phys. Semicond.* 7, No.12.

20. Andreev, V.M., Kagan, M.B., Luboshevskaia, T.L., Nuller, TA. and Tret'yakov, D.N. (1974) Comparison of different heterophotoconverters for achievement of highest efficiency, *Fiz. i Tekhn. Pohipr.* **8**, 1328-1334. Translated into English in *Sov. Phys. Semicond.* **8**, No. 7.

21. Hovel, H.J. and Woodall, J.M. (1972) High-efficiency AlGaAs-GaAs solar cells, *Appl. Phys. Lett.* **21**, 379-381.

22. Vinogradova, E.B., Kagan, MB., Koroleva, N.S., Lubochevskaya, TA. and Nuller, T.A. (1974) Temperature and intensity dependences of solar cells illumination prepared on the basis of AlGaAs-GaAs heterojunctions, *Zh. Tekhn. Fiz.* **44**, 2229-2238. Translated into English in *J. Techn. Phys.* **44**, No.10.

23. James, L.W. and Moon, R.L. (1975) GaAs concentrator solar cell, *Appl. Phys. Lett.* **26**, 467-470.

24. Woodall, J.M. and Hovel, H.J. (1975) Outlook for GaAs terrestrial photovoltaics, *J. Vac. Sci. Technol.* **12**, 1000-1009.

25. Alferov, Zh.I, Andreev, V.M., Daletskii, G.S., Kagan, M.B., Lidorenko, N.S. and Tuchkevich, V.M. (1977) Investigation of high efficiency AlAs-GaAs heteroconverters, *Proc. World Electrotechn. Congress,* Moscow, Section 5A, report 04.

26. Alferov, Zh.I., Andreev, V.M., Garbuzov, D.Z. and Rumyantsev, V.D. (1977) Hetero-photocell with intermediate radiation conversion, *Fiz. i Tekhn. Polupr.,* **9**, 1765-1770. Translated into English in *Sov. Phys. Semicond.* **9**, No.6.

27. Woodall, J.M. and Hovel, H.J. (1977) An isothermal etchback-regrowth method for high efficiency AlGaAs-GaAs solar cells, *Appl. Phys. Lett.* **30**, 492-493.

28. Hamaker, H.C., Ford, C.W., Werthen, J.C., Virshup, C.F., Kaminar, N.F., King, D.L. and Gee, J.M. (1985) 26% efficient magnesium doped AlGaAs/GaAs solar concentrator cells, *Appl. Phys. Lett.* **47**, 762-764.

29. Dingle, R., Wiegmann, W. and Henry. C.H. (1974) Quantum states of confined carriers in very thin $Al_xGa_{1-x}As$-GaAs-$Al_xGa_{1-x}As$ heterostructures, *Phys. Rev. Lett.* **33**, 827-830.

30. Batey, J. and Wright, S.L. (1986) Energy band alignment in GaAs: (Al,Ga)As heterostructures: The dependence on alloy composition, *J. Appl. Phys.* **59**, 200-209.

31. Andreev, V.M., Egorov, B.V., Lantratov, V.M. and Troshkov, S.I. (1985) Heterojunction solar cells with low value of saturation current, *Fiz. i Tekhn. Polupr.* **19**, 276-281. Translated into English in *Sov. Phys. Semicond.* **19**, No.2.

32. Alferov, Zh.I., Andreev, V.M., Konnikov, S.G., Larionov, V.R. and Shelovanova, G.N. (1975) Liquid phase epitaxy of $Al_xGa_{1-x}As$ -GaAs heterostructures, *Kristall und Technik* **10**, 103-110.

33. Alferov, Zh.I., Andreev, V.M., Konnikov, S.G., Larionov, V.R. and Pushny, B.V. (1976) Investigation of new LPE method of obtaining Al-Ga-As heterostructures, *Kristall und Technik* **11**, 1013-1020.

34. Andreev, V.M. (1983) Heterostructure solar energy converters, in M. Herman (ed.), *Optoelectronic Materials and Devices,* Polish Sci. Publ., Warszawa, pp.479-495.

35. Andreev, V.M., Dolginov, L.M. and Tret'iakov, D.N. (1975) *Zhidkostnaya epitaxia v tekhnologii poluprovodnikovykh priborov (Liquid phase epitaxy in semiconductor device technology),* Sov. Radio, Moscow (in Russian).

36. Panish, M.B. and Sumski, S. (1969) Ga-Al-As: phase, thermodynamic and optical properties, *J Phys. Chem. Solid.* **30**, 129-137.

37. Andreev, V.M., Egorov, B.V., Syrbu, A.V., Trofim, V. and Yakovlev, V.P. (1980) Liquid phase epitaxy of AlGaAs heterostructures on profiled substrates, *Kristall und Technik.* **15**, 379-385.

38. Dupuis, R.D., Dapkus, P.D., Vingling, R.D. and Moundy, L.A. (1977) High-efficiency GaAlAs/GaAs heterostructure solar cells grown by metal organic chemical vapor deposition, *Appl. Phys. Lett.* **31**, 201-203.

39. Nelson, N.J., Jonson, K.K., Moon, R.L., Vander Plas, HA. and James, L.W. (1978) Organometallic-sourced VPE AlGaAs/GaAs concentrator solar cells having conversion efficiencies of 19%, *Appl. Phys. Lett.* **33**, 26-27.

40. Tobin, S.P., Vernon, S.M., Woitczuk, S.J., Baigar, C., Sanfacon, M.M. and Dixon, T.M. (1990) Advances in high-efficiency GaAs solar cells, *Conf. Record 21st IEEE Photovoltaic Specialists Conference*, Las Vegas, pp. 158-162.

41. Andreev, V.M., Kazantsev, A.B., Khvostikov, V.P., Paleeva, E.V., Rumyantsev, V.D. and Shvarts, M.Z. (1994) High-efficiency (24.6%, AM0) LPE grown AlGaAs/GaAs concentrator solar cells and modules, *Conf. Record First World Conference on Photovoltaic Energy Conversion*, Hawaii, pp. 2096-2099.

42. Khvostikov, V.P., Larionov, V.R., Paleeva, E.V.. Sorokina. S.V., Chosta, O.I., Shvarts, M.Z. and Ziznogorova, N.S. (1995) Space concentrator solar cells based on multilayer LPE grown AlGaAs/GaAs heterostructure", *Proc. 4th European Space Power Conference*, Poitiers, France, pp. 359-362.

43. Tobin, S.P., Vernon, S.M., Sanfacon, M.M. and Mastrovito, A. (1991) Enhanced light absorption in GaAs solar cells with internal Bragg reflector, *Conf. Record 22nd IEEE Photovoltaic Specialists Conference*, Las Vegas, pp. 147-152.

44. Andreev, V.M., Komin, V.V., Kochnev, I.V., Lantratov, V.M. and Shvarts, M.Z. (1994) High efficiency AlGaAs-GaAs solar cells with internal Bragg reflector, *Conf. Record First World Conference on Photovoltaic Energy Conversion*, Hawaii, pp. 1894-1897.

45. Andreev, V.M., Kalinovsky, V.S., Komin., V.V., Kochnev, I.V., Lantratov, V.M. and Shvarts, M.Z. (1995) High efficiency radiation stable AlGaAs/GaAs solar cells with internal Bragg reflector, *Proc. 4th European Space Power Conference*, Poitiers, France, pp. 367-370.

46. Boettcher, R.J., Borden, P.G. and Gregory, P.E. (1981) The temperature dependence of the efficiency of an AlGaAs/GaAs solar cell operating at high concentration, *IEEE Electron. Dev. Lett.* **EDL-2**, 88-89.

47. Andreev, V.M., Khvostikov, V.P., Larionov, V.R., Rumyantsev, V.D., Paleeva, E.V. and Shvarts, M.Z. (1998) Very high concentrator AlGaAs/GaAs solar cells, *Proc. 2nd World Conference on Photovoltaic Solar Energy Conversion*, Vienna, pp. 3719-3722.

48. Andreev, V.M., Khvostikov, V.P., Larionov, V.R., Rumyantsev, V.D., Paleeva, E.V. and Shvarts M.Z. (1999) 5800 suns AlGaAs/GaAs concentrator solar cells, *Tech. Digest of 11th Photovoltaic Sci. and Eng. Conf.*, Sapporo, Japan, pp. 147-148.

49. Ortiz, E., Algora, C., Rey-Stalle, I., Khvostikov, V.P. and Andreev V.M. (2000) Experimental improvement of concentrator LPE GaAs solar cells for operation at 1000 suns with an efficiency 26.2%, *Conf. Record 28th IEEE PVSC*, Anchorage, USA, pp. 1122-1125.

50. Wanlass, M.W., Ward, J.S., Emery, K.A., Gessert, T.A., Osterwald, C.R. and Coutts, T.J. (1991) High performance concentrator tandem solar cells based on IR-sensitive bottom cells, *Solar Cells* **30**, 363-371.

51. Andreev, V.M., Karma, L.B., Rumyantsev, V.D., Shvarts, M.Z. and Tabarov, T.S. (1994) Narrow gap InGaAs/InP solar cells illuminated through transparent InP substrate, in *Proc. 12th Photovoltaic Solar Energy Conf.*, Amsterdam, pp. 1398-1400.

52. Fraas, L.M., Avery, J.E., Martin, J., Sundaram, V.S., Giard, G., Dinh, V.T., Davenport, T.M., Yerkes, J.W. and O'Neil, M.J. (1990) Over 35-percent efficient GaAs/GaSb tandem solar cells, *IEEE Trans. Electron. Dev.* **37**, 443-449.

53. Andreev, V.M., Karlina, L.B., Kazantsev, A.B., Khvostikov, V.P., Rumyantsev, V.D., Sorokina, S.V. and Shvarts, M.Z. (1994) Concentrator tandem solar cells based on AlGaAs/GaAsInP/InGaAs (or GaSb) structures, *Conf. Record First World Conference on Photovoltaic Energy Conversion*, Hawaii, pp. 1721-1724.

54. Belt, A.W., Keser, S., Stollwerck, G., Sulima, O.V. and Wettling, W. (1996) GaSb-based (thermo) photovoltaic cells with Zn diffused emitters, *Conf. Record 25th IEEE Photovoltaic Specialists Conference*, Washington DC, pp. 133-136.

55. Chung, B-C., Virshup, G.F., Hikido, S. and Kaminar, N.R. (1989) 27.6% efficiency (1 sun, air mass 1.5) monolithic $Al_xGa_{1-x}As/GaAs$ two junction cascade solar cell with prismatic cover glass, *Appl. Phys. Lett.* **55**, 1741-1743.

56. Andreev, V.M., Khvostikov, V.P., Paleeva, E.V., Rumyantsev, V.D. and Shvarts, M.Z. (1997) Monolithic two-junction AlGaAs/GaAs solar cells, *Proc. 26th IEEE Photovoltaic Specialists Conference*, Anaheim, USA, pp. 927-930.

57. Bertness, K.A., Kurtz, S.R., Friedman, D.J., Kibbler, A.E., Kramer, C. and Olson, J.M. (1994) 29.5%-efficient GaInP/GaAs tandem solar cells, *Appl. Phys. Lett.* **65**, 989-991.

58. Bertness, K.A., Kurtz, S.R., Friedman, D.J., Kibbler, A.E., Kramer, C. and Olson, J.M. (1994) High-efficiency GaInP/GaAs tandem solar cells for space and terrestrial applications, *Conf. Record 1st World Conference on Photovoltaic Energy Conversion*, Hawaii, pp. 1671-1678.

59. Friedman, D.J., Kurtz, S.R., Bertness, K.A., Kibbler, A.E., Kramer, C. and Olson, J.M. (1994) GaInP/GaAs monolithic tandem concentrator cells, *Conf. Record 1st World Conference on Photovoltaic Energy Conversion*, Hawaii, pp. 1829-1832.

60. Friedman, D.J., Kurtz, S.R., Sinha, K., McMahon, W.E., Kramer, C.M. and Olson, J.M. (1996) On-sun concentrator performance of (GaInP/GaAs tandem cells, *Conf. Record 25th IEEE Photovoltaic Specialists Conference*, Washington DC, pp. 73-75.

61. Chiang, P.K., Krut, D.D., Cavicchi, B.T., Bertness, K.A., Kurtz, S.R. and Olson, J.M. (1994) Large area GaInP/GaAs/Ge multijunction solar cells for space application, *Conf. Record 1st World Conference on Photovoltaic Energy Conversion*, Hawaii, pp. 2120-2123.

62. Chiang, P.K., Ermer, I.H., Niskikawa, W.I., Krut, D.D., Joslin, D.E., Eldredge, J.W. and Cavicchi, B.T. (1996) Experimental results of $GaInP_2/GaAs/Ge$ Triple junction cell development for space power systems, *Conf. Record 25th IEEE Photovoltaic Specialists Conference*, Washington DC, pp. 183-186.

63. Bett, A.W., Dimroth, F., Large, G., Meusel, M., Beckert, R., Hein, M., Riesen, S.V. and Schubert, U. (2000) 30% monolithic tandem concentrator solar cells for concentrations exceeding 1000 suns, *Conf. Record 28th IEEE Photovoltaic Specialists Conference*, Anchorage, USA, pp. 961-964.

64. Cotal, H.L., Lillington, D.R., Ermer, J.H., King, K.R., Kahan, N.H. Kurtz, S.R., Friedman, D.J., Olson, J.M., Ward, J.S., Duda, A., Emery, K.A. and Moriarty, T. (2000) Triple-junction solar cell efficiencies above 32%: the promise and challenges of their application in high-concentration-ratio PV systems, *Conf. Record 28th IEEE Photovoltaic Specialists Conference*, Anchorage, USA, pp. 955-960.

MICRO-/POLY-CRYSTALLINE SILICON MATERIALS FOR THIN FILM PHOTOVOLTAIC DEVICES: DEPOSITION PROCESSES AND GROWTH MECHANISMS

J.K. RATH

Utrecht University, Debye Institute, Surface, Interface and Devices, P.O. Box 80.000, 3508 TA Utrecht, The Netherlands

1. Introduction

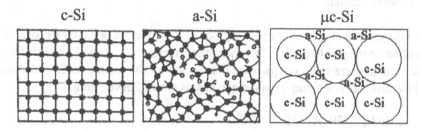

Figure 1. Structure of crystalline, amorphous and microcrystalline silicon

In recent years, poly-crystalline silicon (Poly-Si) has drawn a lot of attention because it can utilise the advantages of crystalline silicon (c-Si) and amorphous silicon (a-Si), which have already consumed a lot of investment and effort for many years. Figure 1 shows the structure of these types of material. C-Si is a material with periodically arranged Si atoms. Defects are mostly structural, while the concentration of electronic defects is very small (less than 10^{14} cm^{-3}, even for doped materials). This owes a lot to the development of the material for the IC industry. Amorphous silicon, on the other hand, has a totally disordered structure (only short range order present) and contains a lot of three fold coordinated Si atoms and dangling bonds. Dangling bonds create states in the middle of the band gap of a-Si, and act as efficient recombination centres for photogenerated carriers. However, none of these types of material have been able to satisfy the needs of solar cell industry. On the one hand, c-Si shows high efficiencies, but the cost of production is high. On the other hand, amorphous silicon is cheap but suffers from a low efficiency and light induced degradation.

Poly-Si achieves the best of both these processes, i.e., high efficiency, stability and large area deposition. Thin film silicon (f-Si) should be differentiated from high temperature (> 1400°C) bulk poly-Si. This f-Si category of material can be formed

157

J.M. Marshall and D. Dimova-Malinovska (eds.),

Photovoltaic and Photoactive Materials - Properties, Technology and Applications, 157–170.

© 2002 *Kluwer Academic Publishers.*

by two approaches: high (> 600°C) and low (< 600°C) temperature. The criterion for differentiation is guided by the types of substrate that can be used (for example a glass substrate implies a low growth temperature). However there is one more criterion: the hydrogen effusion maximum at about 600°C. Due to this, the low temperature processes can yield material (also called microcrystalline silicon (μc-Si)) with passivated grain boundaries during growth, whereas high temperature films invariably need a second step for hydrogen passivation.

Examples of high temperature poly-Si growth techniques are thermal chemical vapour deposition, liquid phase epitaxial (LPE) growth and solid phase crystallisation (SPC). Examples of low temperature approaches are plasma enhanced chemical vapour deposition (PECVD) and hot wire chemical vapour deposition (HWCVD). In this article, we discuss the low temperature approaches.

2. Growth Models

The nucleation and growth process of poly-Si is a complex one, and various models have been proposed to explain the phenomenon. However, as will be clear from the next section, there is no unique model to satisfy all the experimental observations, and the deposition techniques and mechanisms are still developing.

2.1. SURFACE GROWTH MODELS

Several actions take place on the surface of a growing film. The various types of silicon-containing species reaching the surface have to stick to it, and are subject to abstraction, diffusion and insertion reactions. Two kinds of surface growth model have been proposed.

2.1.1. *The Chemical Transport Model*
This model [1] is based on a mechanism whereby nucleation occurs in a regime where there is an equilibrium between deposition and etching at the growing surface. There are two types of species needed for growth: (1) a growth precursor and (2) an etchant species. As for the former, Veprek [1] differs from many other researchers. In this model, SiH_2 is seen as the main precursor. Two steps are important for the growth process:

(i) Adsorption of silicon-containing precursors on the growing surface, dehydrogenation and formation of Si solid.

(ii) An etching process whereby there is reaction of the Si solid by atomic hydrogen to form SiH_4.

Deposition of microcrystalline material occurs under the condition of a partial chemical equilibrium (PCE) state, in which there is a balance between plasma induced formation and decomposition of silane, given by the equation

$$Si\ (s) + (4/m)\ H_m => SiH_4\ (g).$$

If the growth rate (the difference between the etching and deposition rates) is small compared to the etch rate, microcrystalline films are formed without noticeable amorphous tissue. This condition defines the PCE. In fact, the growth rate, which is a function of the departure from partial chemical equilibrium, defines the microcrystalline to amorphous transition.

It should be mentioned here that microcrystalline films can also be made under chemical equilibrium. However, this generally leads to poor quality material. Such a case occurs for the condition of high depletion, with a sufficiently large dwell time (residence time) and $\tau_{1/2} = \tau_{dwell}$. This model explains the following experimental observations that favour microcrystallinity: decreasing the SiH_4 to hydrogen ratio, increasing the deposition temperature and decreasing the level of ion bombardment. Under long dwelling time conditions, there is no detectable amorphous component. The rate-limiting step is ion-induced dehydrogenation from the growing surface.

Validity of the model. In this process, the role of ions at the growing surface is essential. The growth process containing only radical precursors and the dehydrogenation process are not understood. The observed high deposition rates (for example in HWCVD) cannot be explained by such a model.

A variation of the above model is the selective etching model [2], which is based on the difference in the etching rates of the amorphous tissue and the dense regions, leading to selective etching of the former and the growth of the dense crystalline one.

2.1.2. *The Surface Diffusion Model*
This model [3] is based on the diffusion of Si species on the growing surface. Here again, hydrogen on the growing surface plays a crucial part. However, its role is mainly to improve the diffusion of the Si species. The etching and abstraction reactions for crystallinity are of less importance. When a precursor reaches a growing surface, it can either stick or abstract. If it sticks, the admolecule will go through the following processes:

(i) Diffusion: the molecule will diffuse on the growing surface with a thermally activated diffusion coefficient $D_s = a^2 \exp(-E_s/kT)$, where a is the distance between the adsorption sites and E_s is the activation energy of diffusion. The species finds a suitable site and chemisorbs.

(ii) Recombination: hydrogen atoms from adjacent species react and H_2 is eliminated, leaving a Si solid.

The rate-limiting step here is the diffusion coefficient on the growing surface. From the diffusion equation, it is clear that at low temperatures the small diffusion coefficient causes less crystallinity. However at high temperatures, another effect comes into the picture. The hydrogen elimination causes the surface to be more reactive, thereby increasing the activation energy of diffusion. This decrease of D_s

due to hydrogen elimination causes a crystalline to amorphous transition at 500°C. By this model, Matsuda [3] explained his observation of a maximum in the crystalline volume fraction at around 350°C, above which it steadily decreases. The hydrogen ions impinging on the growing surface cause lattice distortion, which reduces the grain size. There is a critical hydrogen ion concentration, above which the crystalline to amorphous transition takes place. Matsuda proposed a triode configuration for the deposition system (to reduce ion bombardment), and achieved an increased grain size at high positive bias.

Validity of the model: The model cannot explain the deposition of polycrystalline material at temperatures higher than 500°C. In fact, many reports show that the grain size and crystalline volume fraction are large at temperatures as high as 500-600°C. The model also cannot explain microcrystalline deposition at temperatures as low as 25-50°C. In fact, the roll of hydrogen in abstraction and etching, which may be crucial in the nucleation process, especially at low temperatures, is not taken into account.

2. 2. THE GROWTH ZONE MODEL

2.2.1. *The Hydrogen Chemical Potential Model* [4]

The equilibrium structure of a material is determined by a minimum in the free energy. Amorphous or even microcrystalline silicon is not in an equilibrium solid phase. The model assumes that structures in small volumes may represent a minimum free energy, subject to long-range disorder of the network. Equilibrium is attained by the mobile hydrogen. At the normal deposition temperature (~ 200°C) which is well above the glass transition temperature, hydrogen is mobile enough to allow changes in the bonding configuration, e.g. by breaking Si-Si bonds and terminating dangling bonds. Below the glass transition temperature, hydrogen is immobile and the structure is frozen.

In the equilibrium picture, the hydrogen concentration in the plasma is described by a hydrogen chemical potential;

$$\mu_{Hg} = E_H + kT\ln(N_H/N_{H0}) \qquad (1)$$

where E_H is the energy of the atom in the vacuum, N_H is the hydrogen concentration in the plasma, and N_{H0} is the effective density of states. A similar expression can be defined for the hydrogen chemical potential (μ_{Hs}) for the hydrogen concentration in the grown film. The first step is that the chemical potential in the gas phase tends to be in equilibrium with the chemical potential in the solid. This means that the amount of hydrogen entering the material is in equilibrium with the amount diffusing out. The disorder in the material is reduced at a high hydrogen chemical potential. However, there is a minimum disorder necessary to sustain an amorphous network. Street [4] has presented a plot depicting the minimum sustainable disorder for a-Si:H at various hydrogen concentrations. For a sufficiently high hydrogen chemical potential, a situation arises where the degree of structural order determined by μ_H far exceeds the available order in an amorphous network, and the instability changes the phase to a microcrystalline deposition.

Validity of the model. The role of hydrogen is a centrally essential part of this model, which finally determines the eventual structure of the material. In that sense, the hydrogen chemical potential of the plasma is a critical parameter. This aspect is perfectly compatible with a microcrystalline deposition growth process in which the role of hydrogen is well established. The model predicts that the material structure is not influenced by the growth process, but rather by the thermodynamic parameters. The fact that wide varieties of microcrystalline materials can be made for similar thermodynamic parameters, by choosing the deposition process, is incompatible with this model. In fact, there is no unique microcrystalline structure. Moreover, the role of ionic species in the deposition process (which is a very crucial parameter in determining the resulting structure) is not taken into account. The model does not distinguish between atomic hydrogen and hydrogen ions.

2.2.2. *Layer-by-layer growth* [5,6]

In this deposition method, the roles of deposition and treatment are kept separate. Thus, the effect of treatment, the role of hydrogen, etching and abstraction can be separated from the deposition process. The standard deposition in LBL proceeds as follows:

(1) a-Si deposition for a time T_d.

(2) Pumping out SiH_4 and inserting H_2 gas into the chamber.

(3) Hydrogen plasma treatment for a time T_t.

This cycle is repeated several times, to obtain the required film. The treatment gas can be a reactive element such as hydrogen or inert noble gases such as Ar and He.

The structural relaxation caused by atomic hydrogen is the mechanism of phase transformation from an amorphous to a crystalline structure within the subsurface. The permeated hydrogen breaks some Si-Si bonds and causes structural relaxation, including long range ordering. The structural transformation by the LBL process proceeds in the following way. A porous layer is formed in the subsurface region with an increase in treatment time. This highly porous and hydrogen rich phase seems to be a necessary condition for the nucleation of the crystalline phase. This is also the reason that microcrystalline silicon easily grows on a porous/disordered surface (a-Si:H made at low temperature, or a-SiC:H), whereas it is difficult to make μc-Si on device quality amorphous silicon [7].

A few observations can be made concerning the LBL process. The crystalline volume fraction increases (almost exponentially) with the number of LBL cycles. The effect of hydrogen radicals is not only the etching the surface but also to cause reactions well below the surface of the film, up to a few tens of a nanometre in depth. This modifies the subsurface growth zone while the deposition proceeds. A nice experiment gives evidence that such a physical process is taking place. One LBL cycle was defined in such a way that the a-Si:H deposited during T_d was etched away during the T_t hydrogen treatment period. This means that there is no effective deposition at the end of one cycle. After the deposition of a thick amorphous silicon layer, a series of such LBL cycles was made. At the end of these cycles, the thickness of the original Si film remained unchanged, as the effective deposition rate of an LBL cycle was zero. However, it was observed that the Si layer became microcrystalline, after an intermediate phase of a porous layer.

Validity of the model: The role of ions, especially on the growing surface, is not taken into account in this process. In general, the species for the growth has no specific role in the final outcome of the microcrystallinity.

3. Effect of Temperature on Crystallinity

As mentioned above, there are several options for making poly-Si films, based on the deposition temperature. The low temperature process, upon which we concentrate in this article, has many possibilities. It is therefore necessary to define the temperature regime required to make the poly-Si film, because a lot of aspects of solar cell fabrication depend on this: the structure of the i-layer, the choice of substrate, the selection of doped layers, the type of configuration (substrate or superstrate), the type of metal oxide TCO etc. As mentioned in describing the surface diffusion model, Matsuda [3] showed from experiment and explained through his model that there is an optimum growth temperature ($\sim 350^\circ$C), i.e. the temperature at which the maximum crystalline volume fraction is expected. At higher temperatures, the crystallinity gradually decreases, and above 500°C (around the temperature of the hydrogen effusion maximum) no crystallinity is expected. However, this observation is seriously contested by Veprek [8], who showed that not only can poly-Si films be made at temperatures above 500°C, but also that the crystallinity is higher at high temperatures. Moreover he could reproduce the results of Matsuda, by carefully inserting an oxygen impurity into the material via the gas phase. It was confirmed in recent studies by Matsuda's group that indeed, in an impurity free atmosphere, highly crystalline poly-Si films are made at high temperatures. The above results suggest two things:

(i) In an ideal deposition environment such as that in an ultra high vacuum deposition system, it is at high temperatures that materials with a high crystalline fraction and grain size can be made.

(ii) To get the highest crystallinity, there is an optimum temperature that depends on the base pressure, type and volume of the chamber.

If the above criterion is taken into consideration, one would believe that given a high quality ultra high vacuum multichamber with a load lock, everyone would be inclined to work at high substrate temperatures. However this is not the case. One needs to consider the transition temperature above which hydrogen elimination from the Si matrix takes place. This is the normal temperature used for making device quality a-Si:H films. Below this temperature, better passivation of the grain boundaries is expected and the poly-Si will have a large hydrogen concentration. On the other hand, above this temperature large grain sized poly-Si is expected. This has led to two regimes of temperature in which device quality Si films are made for solar cells:

(i) High Temperature, 400-550°C: Films have large grains, compact poly-Si, thin grain boundary, no or negligible amorphous component, and a small impurity (oxygen) concentration. Examples are the poly-Si films and cells made by Kaneka Co, Japan and Utrecht University. The efficiency (η) of Kaneka solar cells is 10.7%. Multi-junction cells with 14.1% initial and 12% stabilised efficiencies have been grown. These films are called poly-silicon.

(ii) Low Temperature, 100-350°C : The grain sizes are small, typically around 20 nm. The crystalline volume fraction is high, but the non-crystalline fraction/ amorphous fraction is not negligible. The material has a high hydrogen content. Examples are IMT Neuchatel (η for single junction: 8.5%, micromorph cell: 13.1% initial, 12% stabilised), KFA Julich (η: 8.2%), Canon Co. (η: 9.3%). These films are called microcrystalline silicon.

The above results clearly suggest that both regimes of growth can lead to high efficiencies.

4. Deposition Techniques

The deposition models, notwithstanding their diversity, suggest that hydrogen plays a crucial role in the nucleation process. For this reason, the hydrogen dilution of silane gas is commonly used in low temperature chemical vapour deposition processes for making poly-Si films. A low silane flow in a silane/hydrogen gas mixture restricts the deposition rate.

There are many deposition systems in which to make poly-Si films. It has been found that a certain technology is suitable for a certain application. For example, laser induced crystallisation has been most successful in thin film transistor (TFT) applications. Solid phase crystallisation is also very suitable for this purpose. However, poly-Si films made in a one step process are still not so successful for TFTs, because of the problem of the incubation phase. The reverse is true for the case of solar cell applications. The above deposition techniques are not suitable here, mainly because of the high fabrication time. Solar cells need thick intrinsic layers which can be made by techniques with high deposition rates. The two techniques which are most successful in this respect are plasma enhanced chemical vapour deposition (PECVD) and hot-wire chemical vapour deposition (HWCVD). The following section will expand further on these two important techniques, upon which poly-Si research at Utrecht University is based.

4.1. PECVD

The most common configuration for PECVD is a parallel plate capacitor structure. An rf voltage is applied between the electrodes, and the standard frequency is 13.56 MHz. In recent developments, especially relevant to the case of microcrystalline silicon deposition, the use of very high frequencies (VHF) in the range 20-110 MHz, microwaves (2.56 GHz), and a remote plasma using a microwave power source (ECR) have become more popular. They create a high hydrogen density in the plasma, which is beneficial for microcrystalline silicon growth. In plasma deposition, electron impacts with SiH_4 molecules in the chamber lead to dissociation processes. The SiH_3 radical is considered to be the most important species for microcrystalline deposition. However, ions play a large role in determining the final nature of the film.

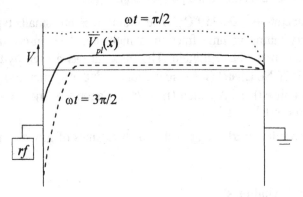

Figure 2. The plasma potential between the electrodes [9]

Figure 2 shows the potential profile in a plasma volume and at the electrodes [9]. The plasma potential is slightly positive with respect to the electrodes, and due to this, positive ions bombard the electrodes. However, due to a difference in the sizes of the electrodes, a negative DC bias is generated at the powered one. This is the reason why this electrode is called the cathode and the grounded one is called the anode, where the substrate is attached for deposition with a lower level of bombardment by positive ions. However, high energy ion bombardment has an adverse effect on the crystallinity [1], and a triode configuration can be used to completely eliminate ions reaching the substrate [3]. Numerous experiments have recently shown that control of the ion energy and density is the key to achieving poly-Si films of high quality, especially at a high growth rate. In this respect, not only a modified plasma configuration (e.g. the triode configuration), but also changes in process parameters as well as deposition at a high frequency are promising. A detailed review on these aspects is provided elsewhere [10].

4.2. HWCVD

This technique relies on the catalytic decomposition of silane gas by a metal. Thus, a filament is a basic component in this system, in the same way that an rf electrode is the basic component in PECVD. In HWCVD, the gas in the presence of a heated filament is decomposed to radicals, which diffuse to and are deposited on the substrate. Although electron emission from the filament can lead to ionised species, the effect is negligible. This is a very advantageous situation compared to PECVD, as there are no ionic species to disturb the nucleation process. There are two configurations to carry out such a deposition:

(i) Filament perpendicular to substrate: The configuration is also called the "hot wire cell" (TIT, Tokyo) [11]. In this configuration, the filament length is parallel to the gas flow direction and perpendicular to the substrate plane. Two things happen because of this. Since the gas flow is through the filament along its full length, the decomposition is very efficient. It can be achieved with a small amount of the total heated component, thereby reducing the heat radiation. Konagai *et al.* [11] have achieved deposition rates as high as 3nm/s by such a method, whereas poly-Si can be made at as low a temperature as 200°C, without hydrogen dilution.

(ii) Filament parallel to substrate: This configuration is widely used, and the best amorphous silicon and poly-Si films and solar cells have been achieved by it. However, this does not mean that this configuration is the most suitable for achieving good poly-Si films. Rather, it points out that it is the easiest to implement, so that many groups use it. However, as can be realised, many of the advantages of the perpendicular configuration are lost in this process.

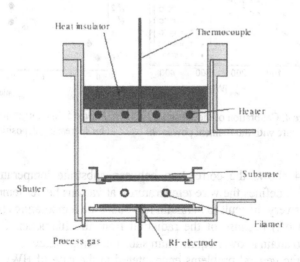

Figure 3. Cross section of the HW assembly

166

Figure 3 shows the configuration used at Utrecht University. Two coiled tungsten (W) filaments, 0.5 mm wide and 6 cm long, are placed parallel to each other, with a 4 cm gap in between. The filaments are placed parallel to the substrate at a distance of 3 cm. As can be seen, an rf field is also sometimes applied, to deposit films at an increased rate in a combined rf-HWCVD process. A shutter is placed between the filament and the substrate. Before the deposition, the filament is first heated to the desired temperature with the shutter in place. This is to prevent the substrate from receiving unwanted species from the filament during the heating up process. The filament is kept at some annealing temperature to de-gas it, and is then brought to the deposition filament temperature (T_w) and stabilized. After stabilization, gas is let in and the shutter is opened to deposit the film. The substrate is heated up by the radiation from the filament. The exact determination of the substrate temperature is a crucial step, as this critically defines the final quality of the poly-Si. There are two components to the substrate temperature: heating by the heater attached to the substrate holder (back side) and heating of the front side from the filament. Calibration has been made to estimate the actual substrate temperature for a certain heater and filament temperature [12]. The relations are

$$T^4_{sub}=0.375.T^4_{heater}+0.612.T^4_{walls}+6.3 \times 10^8.P_{fil} \qquad \text{(shutter open)}$$

$$T^4_{sub}=0.5.T^4_{heater}+0.408.T^4_{walls}+4.2 \times 10^8.P_{fil} \qquad \text{(shutter closed)}$$

where P_{fil} is the power supplied to the filament.

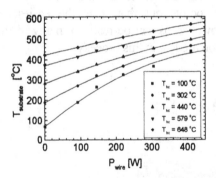

Figure 4. Calibration of the substrate temperature with the filament power on.

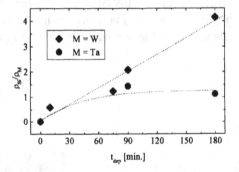

Figure 5. Si content in a tungsten wire for increasing deposition periods [13]

Figure 4 shows the correlation between substrate temperature and filament power (which defines the wire temperature), at various heater temperatures. As one can see, it is very difficult to maintain low substrate temperatures (even if the heater is switched off) because of the radiation from the filaments. Cooling has to be employed to attain a lower equilibrium substrate temperature.

One of the general problems encountered in the case of HWCVD deposition is filament aging, which leads to breakage and the need for replacements at regular intervals. This is a particularly severe problem at low filament temperatures. The choice of filament material also is important in achieving its longevity. For example, it was found that a tantalum filament has a longer lifetime than a tungsten one. Studies at Utrecht University (UU) [13] have shown that filament fragility is connected to Si incorporation into the material. Figure 5 shows the Si/W and Si/Ta ratios (measured by XPS) as a function of deposition time. It is clear that Si incorporation is much less for Ta, and that it saturates quickly. Si incorporation has a serious consequence, especially in the case of poly-Si deposition where the deposition time is invariable long because of the thick layer needed for a solar cell. It is necessary to ascertain that the properties of the filament and the catalytic reaction process do not change during the course of a single deposition.

It is unfortunate that the best device quality poly-Si, showing the best solar cell results, has been achieved in the case of a W filament (instead of a Ta one, with a high longevity). It is for this reason that other types of filament material, such as Rh, have been tried, but without much success.

4.2.1. *Material Studies*

Poly-Si (such as that made at UU) is deposited from a silane and hydrogen gas mixture. HWCVD poly-Si deposition at UU shows a general trend [14]. Variations in the gas flow rate lead to structural changes in crystal orientation, crystalline volume fraction, crystal size and shape. High pressure leads to an increased deposition rate, though with a simultaneous increase in the defect density (N_d). The defect density is most sensitive to the wire temperature (T_w), and a low defect density and thin grain boundary can be achieved at an optimum wire temperature of about 1800°C. The gas phase reaction that leads to poly-Si deposition is as follows. Above 1500°C, silane molecules crack catalytically in the presence of heated filaments, giving Si and H atoms i.e., $SiH_4 \Rightarrow Si + 4H$ ($T_{fil} > 1500$ °C). The other reaction is $H_2 \Rightarrow H + H$. However, for poly-Si deposition, these species do not reach the surface, but rather go through gas phase reactions. Hydrogen reacts with silane to create radicals: $H + SiH_4 \Rightarrow SiH_3 + H_2$, and the main species SiH_3, along with abundant atomic hydrogen, facilitates poly-Si growth. On the other hand, Si is lost in a gas phase reaction, $Si + SiH_4 \Rightarrow Si_2H_4^*$.

5. Homogeneity

5.1. HWCVD

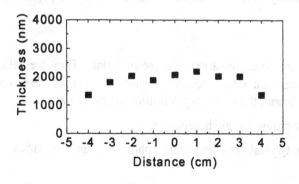

Figure 6. Thickness profile of a poly-Si film perpendicular to the length of the wires [15]

If a line source is used (filament in the parallel configuration) the flux density of species at a distance L (perpendicular to the filament line) is proportional to $1/L^2$. This gives a Lorentzian shape of the film in the direction perpendicular to the filament. This would lead to very inhomogeneous growth, depending upon how the filament is mounted. To overcome this, multiple heaters are used to improve the homogeneity. However, this leads to an increase in the heated filament area, which increases the radiation to the substrate. It also increases the gas depletion. The latter is beneficial for poly-Si growth, but the former restricts the lowest substrate temperature that can be maintained. Hence, most groups make a compromise between substrate temperature and homogeneity. Utrecht University uses two wires.

Figure 6 shows the thickness of a poly-Si film grown on Corning 7059 glass (with W filaments), in the direction perpendicular to the filament [15]. As can be observed, the film is uniform across 6 cm of a 10x10 cm substrate. The homogeneity of the poly-Si is expected to be enhanced compared to amorphous silicon, because poly-Si is deposited at a higher chamber pressure and a significant gas phase reaction is expected. Scaling up to large area has been demonstrated by the University of Kaisersluiten, where a homogeneity of $\pm$ 2.5% over a 20 cm diameter is achieved for a deposition on a 30x30 cm substrate, by using multiple wires with a distance of 4 cm between them [16].

5.2. PECVD

The homogeneity of deposition suffers in PECVD, when the wavelength of the plasma excitation approaches the dimensions of the chamber. Inhomogeneity problems occur in the electrode voltage distribution when the electrode dimension becomes comparable to a quarter of the free space associated with the excitation frequency [17]. For example, λ is 5.53 m at 13.56 MHz, but 0.75 m at 100 MHz. This poses problems at VHF and higher frequencies, which are becoming popular among poly-Si material growers. For this reason, Kaneka Corporation and IVP Julich are concentrating on 13.56 MHz depositions. However, a proper combination of pressure and flow rate can overcome this problem. Using such a process, a homogeneity of $\pm$ 5% has been achieved in UU for a 30x40 cm area a-Si deposition by 13.56 MHz PECVD. Use of a modulated plasma has also yielded a homogeneity of $\pm$ 5% over a 10x10 cm area by VHF CVD deposition of a-Si:H at UU [9].

6. Nomenclature

There is a lot of confusion in the presentation of these materials. They are called poly-Si, microcrystalline silicon (μc-Si) and sometimes also nano-crystalline silicon (nc-Si). To make the terms clearer, the following definition is put forward:

- Poly-silicon: Crystalline grains + grain boundaries,
- Microcrystalline silicon: Crystalline grains + amorphous + grain boundaries.

For μc-Si or poly-Si to be called device quality, they have to satisfy the following characteristics:

(1) intrinsic conduction,

(2) reduced oxygen incorporation,

(3) small number of grain boundary defects,

(4) high deposition rate,

(5) initial growth (interface property) on the substrate.

The first three properties will be discussed in the second of this three paper series [18], where it will be shown how these conditions are satisfied for the case of poly-Si made by HWCVD at UU. The fourth and fifth conditions are relevant to cell properties and fabrication processes, and will be discussed in the third paper of the series [19]. The properties of such a device quality poly-Si film are given in Table 1.

TABLE 1. Physical properties of device quality poly-Si material made by HWCVD at Utrecht University [14]. σ_{ph}: white light photoconductivity, σ_d: dark conductivity.

Material Properties	
Crystalline volume fraction (%; Raman)	95
Average grain size (XRD) (nm)	70
Deposition rate (Å/s)	5.5
H content (at.-%)	0.47
Diffusion length (SSPG) (nm)	568
σ_{ph} ($\Omega^{-1}cm^{-1}$) (100mW/cm^2 AM1.5)	2×10^{-5}
σ_d ($\Omega^{-1}cm^{-1}$)	1.5×10^{-7}
Band gap (eV)	1.1
Activation energy (eV)	0.54
Defect density (ESR) (cm^{-3})	7.8×10^{16}
Activation energy (eV) (Hall mobility)	0.012
Roughness (AFM) (nm)	150
SIMS (oxygen content)	$\sim 3 \times 10^{18}/cm^3$
IR (oxygen in the film)	very low

7. References

1. Veprek, S., Sarrott, F.-A. and Iqbal, Z. (1987) *Phys. Rev.* **36**, 3344.
2. Solomon, I., Drevillon, B., Shirai, H. and Layadi, N. (1991) *J. Non-Cryst. Solids* **164-166**, 989.
3. Matsuda, A. (1983) *J. Non-Cryst. Solids* **59/60**, 767.
4. Street, R.A. (1991) *Phys. Rev. B* **43**, 2454.
5. Roca i Cabarrocas, P., Layadi, N., Heitz, T., Drevillon, B. and Solomon, I. (1995) *Appl. Phys. Lett.* **86**, 3609.
6. Akasaka, T. and Shimizu, I. (1996) *J. Non-Cryst. Solids* **198-200**, 883.

7. Rath, J.K. and Schropp, R.E.I. (1998) *Solar Energy Materials and Solar Cells* **53**, 189.
8. Veprek, S., Sarrott, F.-A. and Ruckschlos, M. (1991) *J. Non-Cryst. Solids* **137-138**, 733.
9. Biebreicher, A. (2000) *Ph.D. Thesis*, Utrecht University.
10. Rath, J.K. *Solar Energy Materials and Solar Cells* (to be published).
11. Konagai, M., Tsushima, T., Kim, M., Asakusa, K., Yamada, A., Kudriavtsev, Y., Villegas, A. and Asomoza, R. (2001) *Thin Solid Films* **395**, 152.
12. Feenstra, K.F.F. (1998) *Ph.D. Thesis*, Utrecht University.
13. van Veenendaal, P.A.T.T., Gijzeman, O.L.J., Rath, J.K. and Schropp, R.E.I. (2000) *Thin Solid Films* **395**, 194.
14. Rath, J.K., Meiling, H. and Schropp, R.E.I. (1997) *Jpn. J. Appl. Phys.* **36**, 5436.
15. Rath, J.K., van Cleef, M.W.M. and Schropp, R.E.I. (1997) *Proc. 14th EPVSEC, Barcelona*, p. 597.
16. Ledermann, A., Weber, U., Mukherjee, C. and Schroeder, B. (2000) *Thin Solid Films* **395**, 61.
17. Schimdt, J.P.M. (1992) *Mat Res. Soc. Symp. Proc.* **219**, 631.
18. Rath, J.K. (2002) Micro-/poly-crystalline silicon materials for thin film photovoltaic devices: physical properties, *This volume*.
19. Rath, J.K. (2002) Micro-/poly-crystalline silicon materials for thin film photovoltaic devices: application in solar cells, *This volume*.

MICRO-/POLY-CRYSTALLINE SILICON MATERIALS FOR THIN FILM PHOTOVOLTAIC DEVICES: PHYSICAL PROPERTIES

J.K. RATH
Utrecht University, Debye Institute, Surface, Interface and Devices,
P.O. Box 80.000, 3508 TA Utrecht, The Netherlands

1. Introduction

This paper discusses some of the physical properties of poly-Si films that determine their device performance. For a specific case, the compact poly-Si films grown by HWCVD at Utrecht University (UU) will be described. The causes for the naturally enhanced optical absorption of poly-Si, which makes it possible to use it as a thin film photovoltaic material, will be probed, as will the oxygen incorporation which creates donor states that have adverse effects on solar cell performance. It will be shown that extremely intrinsic poly-Si material can be made by HWCVD.

2. Physical Properties

2.1. OPTICAL PROPERTIES

Optical absorption, light scattering (internal and surface) and their effects on band gap and subgap absorption are important in understanding carrier generation in a poly-Si film. Here, it should be emphasized that the indirect band gap of microcrystalline silicon is 1.1 eV. The band gap is estimated from a plot of $\sqrt{\alpha}$ against hv (where α is the optical absorption coefficient and hv is the photon energy). To estimate the band gap, the absorption spectra at different energies (hv) are obtained by photothermal deflection spectroscopy (PDS) [1] or the constant photocurrent method (CPM) in the low energy region, in contrast to the case of amorphous silicon where (due to the high absorption and high optical gap) the absorption coefficient (α) is measured by simple reflection/transmission. For amorphous silicon, the optical gap can be estimated either from a Tauc plot of $(\alpha hv)^{1/2}$ against hv or, for a cubic gap, from $(\alpha hv)^{1/3}$ vs. hv.

There are a few differences between the absorption spectrum and the parameters that are derived from it for amorphous silicon and microcrystalline silicon. An enhanced optical absorption has been observed in microcrystalline silicon films. Meier et al [2] showed that although microcrystalline silicon film has the same optical gap as c-Si, the absorption coefficient is almost an order of magnitude higher, due to which a layer only about 2μm thick is necessary for the microcrystalline cell, compared to the more than 100 μm thick wafer used in a c-Si

J.M. Marshall and D. Dimova-Malinovska (eds.),
Photovoltaic and Photoactive Materials - Properties, Technology and Applications, 171–182.
© 2002 *Kluwer Academic Publishers.*

cell. The enhanced light absorption in these microcrystalline films and solar cells has been attributed mainly to a longer optical path, occurring as a result of efficiently diffused light scattering at the textured film surface.

The absorption coefficient of microcrystalline silicon films has been modelled using the effective medium approximation. It is modelled either as a two-phase system of a-Si and c-Si [3] or a three-phase system of a-Si, c-Si and grain boundaries [4]. Diehl *et al.* [4] showed that elastic light scattering is not primarily responsible for the enhanced absorption, especially in the range 1.4 to 2 eV. The absorption in this region is a superposition of those of the crystalline grains, the grain boundary regions and the amorphous phase, weighted by the corresponding volume fractions. The grain boundary absorption is enhanced due to relaxation of the k-selection rule, and is also influenced by the hydrogen at the grain boundaries.

Three distinct regions of the absorption spectra can be identified: (1) the region between 1.2 and 1.4 eV, (2) absorption above 1.5 eV and (3) the subgap region below 1.1 eV. It has been reported in one study [5] that absorption below 1.4 eV is predominantly due to surface scattering (due to surface texture) with, in the case of porous materials, a contribution from bulk scattering at the voids also being present. However, by polishing the material or using scattering modeling, the true optical absorption can be obtained. This is the same for microcrystalline silicon as for c-Si, between 1.2 and 1.4 eV. The absorption above 1.5 eV has been attributed mainly to the amorphous component, although other contributions have also been speculated upon. The sub-bandgap absorption is correlated with defect absorption. The Urbach energy, E_0, defined via $\alpha = \alpha_0 \exp (E/E_0)$, can be obtained from the exponential region of the absorption, where it is manifested as a linear region in the log α vs. hv plot. For microcrystalline silicon, this inverse slope is not always clearly visible. Vanecek *et al.* [5] observed such a linear region, attributed to the Urbach tail, and correlated it with a valence band tail width of between 50 and 60 meV.

Studies of HWCVD poly-Si samples give a different picture [6]. Figure 1 shows atomic force microscopic (AFM) images over an area of 10x10 cm for such a poly-Si specimen made at Utrecht University (UU). The as-deposited film has a

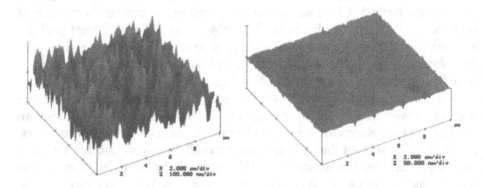

Figure 1. AFM images of a poly-Si film made by HWCVD at Utrecht University, in the unpolished (left) and polished (right) states [6].

naturally textured surface, with a root mean square (rms) roughness of about 50 nm. This is considered to play a big role in the enhanced optical absorption. Figure 1 shows that polishing reduces the rms surface roughness from 50 to about 5 nm.

Angular resolved light scattering measurements performed on as-deposited and polished poly-Si show that the as-deposited film has a significant diffused reflection, which correlates to the surface texture of the poly-Si. Polishing the film reduces the diffused reflection as a result of a decreasing surface roughness. Polishing does not affect the angular dependence of the transmittance. This is a result of the high absorption in the bulk of the film, due to which the diffused part is mostly absorbed in the film.

The thickness dependence of the optical absorption was recorded during successive polishing steps. The material employed here was a profiled layer made using a 20 nm seed layer, on top of which device quality poly-Si was deposited. In this way, the amorphous incubation phase was avoided, while maintaining the bulk film properties. A cross sectional transmission electron microscopy (XTEM) image of this material is provided as Figure 4 in an accompanying paper in this volume [7]. The XTEM image clearly reveals that there is no noticeable amorphous phase either at substrate/film interface or in the bulk of the film.

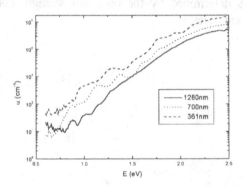

Figure 2. Optical absorption recorded by PDS for a poly-Si film, at different thicknesses obtained by polishing [6].

Figure 2 shows that polishing actually increases the optical absorption throughout the energy range studied. This thickness dependence is attributed to the structure and disorder in the initial stages of deposition. In this part of the film, enhanced absorption is expected due to a greater relaxation of the k-selection rule.

2.2. HYDROGEN

2.2.1. *Vibrational Spectra*

The hydrogen present in microcrystalline silicon is predominantly bonded at grain boundaries, and the presence of hydrogen inside the grains is considered unlikely due to the low solubility of H in the crystalline matrix. This is reflected in the IR spectra. The IR spectrum of hydrogenated microcrystalline silicon films normally shows absorption at 2100 cm^{-1}. This is actually a doublet [8], and the spacing between the sharp peaks is about 20 cm^{-1}. The doublet has been attributed to stretching modes of silicon hydrogen bonds. When the silane concentration in a SiH$_4$/H$_2$ mixture was reduced below 7.5%, Kroll et al. [9] saw a gradual disappearance of the stretching modes at 2000 cm^{-1} and 2090 cm^{-1}, and the growth of stretching peaks around 2100 cm^{-1}. This effect was correlated to a transition from an amorphous to a microcrystalline matrix. They attributed the different stretching modes around 2100 cm^{-1} to mono and/or dihydride bonds on the (100) and (111) surfaces of the silicon crystallites. Comparing the peak positions with the vibrational frequencies of Si-H bonds on the c-Si planes, Stryahilev et al. [10] concluded that the low frequency peak is indeed assigned to a grain boundary with index (111). However, they argued that the high frequency contribution can neither be attributed to SiH$_2$ (as it was not confirmed by an annealing experiment) nor to SiH bonds on any of the low index planes. Thereby, they rejected the proposition of assigning the doublet to different bonding configurations [9] or to a contribution from Si-H bonds on different crystalline planes [11]. Instead, they proposed that the splitting of the absorption bands in the IR spectra of μc-Si films should be attributed to optically anisotropic Si-H monolayers residing at the μc-Si grain boundaries, which are preferentially oriented along the (111) surface. This frequency difference between the peaks is determined by the oscillator strength component directed normal to the grain boundaries.

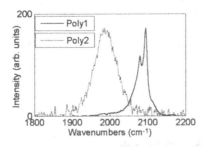

Figure3. FTIR spectra of Poly1 and Poly2 films [12].

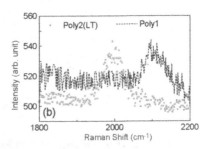

Figure 4. Raman spectra from the top side of Poly1 and Poly2 films [12].

Rath et al. [1] observed that the Si-H stretching vibrations depend sensitively upon the structure, which is determined by the hydrogen dilution. Whereas poly-Si films made at high hydrogen dilutions (Poly1) showed a typical doublet (2080 and 2100 cm^{-1}), films made at low hydrogen dilutions (Poly2) showed only one peak, at 1993 cm^{-1} (see figure 3). The doublet for the high hydrogen dilution poly-Si was attributed to Si-H monohydrides on the (111) and (110) surfaces of the grain boundaries. No bending modes corresponding to dihydrides have been observed, which rules out their involvement. However the case of absorption around 2000 cm^{-1} is very special to poly-Si films made by this particular deposition technique, as such peaks are generally not observed in microcrystalline films.

A comparison of XTEM and Raman studies confirmed that the 2000 cm^{-1} mode is indeed from a completely crystalline region, not any amorphous tissue. The 2000 cm^{-1} vibration is due to Si-H bonds at completely coalescent grains (within thin grain boundaries). Oxygen incorporation is greatly reduced in poly-Si films having such structures. From the above result, one might still suggest that some amorphous region in the film might be involved in the 2000 cm^{-1} mode. Raman spectroscopy solves this problem [12]. The poly-Si at the top is a densely packed crystalline region, characterised by a Raman spectrum (transverse optic (TO) band of Si-Si vibration) at 520 cm^{-1}. We used Raman measurements to detect the stretching modes. As the beam penetration depth is only about 100 nm, we could be certain that we were detecting only the top highly crystalline region. Figure 4 shows the Raman spectrum in the energy range of silicon-hydrogen vibrations of the samples. It is observed that Poly2 has only one band at 2000 cm^{-1}, whereas Poly1 has a band at 2100 cm^{-1}, consistent with the IR result. These data give definite proof that the 2000 cm^{-1} mode in Poly2 is indeed due to Si-H bonds in the crystalline region, and not to Si-H bonds in any amorphous network.

2.2.2. Effusion

To understand the bonding configuration, hydrogen effusion experiments are conducted. A sample is heated from room temperature to around 1000°C at a constant rate, and the hydrogen effused out of the sample is detected by a quadrupole analyser. The effusion processes are of two types; (i) a surface desorption process of hydrogen into an interconnected network of voids, generally observed as the low temperature (LT) maximum in amorphous silicon samples made at a low substrate temperature or in microcrystalline silicon films, and (ii) diffusion of atomic hydrogen in compact material. The diffusion coefficient, D, is defined as $D = D_0 \exp(-E_D/kT)$, where D_0 is the diffusion prefactor and E_D is a diffusion energy. The temperature, T_M, of the effusion maximum is related to D by the expression [13]

$$\ln(D/E_D) = \ln(d^2\beta/\pi^2 kT_M^2) = \ln(D_0/E_D) - E_D/kT_M, \qquad (1)$$

where β is the heating rate (20 K/min) and d is the thickness of the film. From a measurement of T_M as a function of d or β, the diffusion coefficient can be evaluated. In our case, we fixed β and varied the thickness. For effusion experiments, films on c-Si substrates are generally used.

Effusion curves for microcrystalline silicon films normally show a weak double maximum at about 400 and 500°C. The peak positions are thickness independent, but shift to lower temperatures with boron doping. This double peak is attributed to the partial reconstruction of the material during the evolution process [14]. It is also typical for molecular hydrogen desorption from a c-Si surface. The similarity in the behavior leads to the conclusion that the hydrogen evolution process in μc-Si:H is mainly due to desorption from the grain boundaries, followed by rapid out-diffusion of H_2 along the grains. Hence, H_2 surface desorption from voids or grain boundaries is the rate-limiting step for hydrogen effusion, and the out-diffusion of H_2 is fast due to the presence of an interconnected void network. However the effusion peak positions depend on the exact microstructure. For more compact material, the peak position is at about 600°C, which is typical for hydrogen diffusion. This is the case for Poly2 films made by HWCVD at UU. For such materials, the effusion rate is limited by the diffusion of atomic hydrogen, as is well known for compact a-Si:H films of about 1μm thickness. For the case of ECR CVD, in addition to the high temperature peak at 600°C, a shoulder at 400°C has been observed [15]. A peak at around 300°C is also observed for μc-Si:H films made at low temperature (T_s = 150°C), and has been attributed to desorption of water.

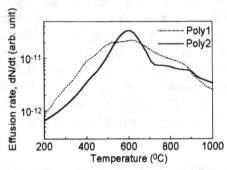

Figure 5. Hydrogen effusion as a function of effusion temperature for Poly1 and Poly2 films [13].

The hydrogen effusion spectrum of the poly-Si films made by HWCVD at UU is presented in Figure 5. Poly1 shows a broadly distributed effusion maximum. The low temperature (LT) contribution is attributed to molecular hydrogen desorption through interconnected voids present in the material. This analysis is supported by the structure of a Poly1 film (seen from cross-sectional transmission electron microscopy (XTEM) (see Figure 13 below)) which showed a large number of interconnected voids. The Poly2 film shows only one maximum, between 550 and 650°C. This maximum is similar to that for the effusion of H in a-Si:H.

To ascertain the effusion mechanism giving such a maximum, and to determine the Si-H bonding configuration, experiments were done on samples of various thickness, i.e., 180, 480, 1060, 1470 and 2350 nm. The shift of T_M with increasing thickness (following equation 1) confirmed the diffusive nature of the hydrogen. This dependence is similar to that for hydrogen at compact Si-H sites in a-Si:H, which is also diffusion limited. The slope of the function $\ln(d^2/T_M^2) = A + B/T_M$, fitted to the thickness dependence data, gives an activation energy, E_D, of 2.36 eV. This is significantly different than the 1.4-1.6 eV values observed for standard amorphous silicon. However, this diffusion energy agrees with that for the high temperature maximum (2.3 eV, within error) in the data for low substrate temperature deposited (low-T_S) a-Si:H. This would suggest that a polymeric type of a-Si:H may be present in the poly-Si, through which hydrogen diffuses. However, the absence of any low-temperature (LT) maximum in the effusion profile of the Poly2 sample indicates that the hydrogen environments are not similar to those in low-T_S a-Si:H samples. To clarify these differences, diffusion measurements of deuterium-implanted poly-Si were made.

2.2.3. Diffusion

In a diffusion experiment, deuterium is implanted into Si:H at a depth (typically 0.4 μm) determined by the energy of the implantation. The implantation dose is generally kept well below the hydrogen concentration in the sample. In a typical case, for experiments made at KFA Julich on samples from UU, deuterium was implanted at an energy of 30 keV. The implantation dose (1×10^{16} cm^{-2}) was chosen to keep the deuterium concentration smaller than the hydrogen content. For secondary ion mass spectroscopy (SIMS) profiling, an oxygen (O_2^+) sputtering beam at normal incidence was used, and positive secondary ions were detected.

The reason for using deuterium in diffusion experiments is that this species can be differentiated from the native hydrogen background, and thus easily tracked. Moreover, deuterium diffusion reflects the properties of hydrogen, because the two isotopes have similar diffusion coefficients. The material is annealed at different temperatures and the diffused deuterium profile is recorded by secondary ion mass spectrometry (SIMS). The diffusion profile of D⁻ implanted material is fitted to a complimentary error function (erfc), and from the relation $D = D_0 \exp(-E_D/kT)$, the diffusion parameters are calculated. Although the coefficient ($D \sim 10^{-14}$ cm^{2}s^{-1}) for hydrogen diffusion in these HWCVD poly-Si (Poly2) films [16] is similar to that for hydrogen implanted in c-Si, the diffusion prefactor, ($D_0 \sim 10$ cm^{2}s^{-1}), and activation energy ($E_D \sim 2.1$eV) are higher than those for c-Si implanted with a comparable hydrogen content. However, they are similar to those for amorphous silicon with a comparable hydrogen content (either implanted or as grown), as shown in Figure 6.

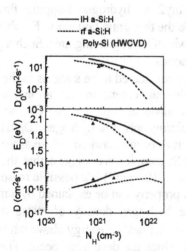

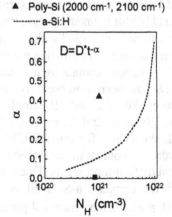

Figure 6. Diffusion parameters of H in poly-Si compared to (a) H in c-Si and (b) H in a-Si:H [16]

Figure 7. Variation of the diffusion time dependence with hydrogen content [16].

Using the D_0 and E_D values for poly-Si films obtained from SIMS, we can calculate the T_M values using equation (1). The value obtained was $T_M = 630°C$, which agrees well with that determined from the effusion experiment at 640°C.

Another interesting observation concerns the time dependence of the diffusion coefficient, $D = D^*t^{-\alpha}$. This is attributed to the loss of hydrogen which moves into voids during diffusion. Figure 7 shows that the time dependence coefficient, α, of a device quality Poly2 film is very low. It is comparable to or even lower than that for the case of compact a-Si:H material with a similar hydrogen concentration. The absence of the time dependence of diffusion in Poly2 material essentially proves that there are very few voids present. However, another poly-Si film, with a similar hydrogen content but showing the 2100 cm^{-1} mode in the IR vibration spectrum, had a significant time dependence of the hydrogen diffusion. This implies that the void structure (not the hydrogen content) controls the time dependence of the diffusion coefficient.

In a polycrystalline film, Si-H bonds can reside in three configurations: (i) dispersed in the interior of crystals, (ii) at the grain boundaries and on the surfaces of crystal columns, and (iii) in amorphous regions. The third case can be excluded as the Poly2 material on a c-Si substrate (the sample used for the diffusion experiment) has no noticeable amorphous region either at the substrate/film interface or in the bulk. The hydrogen migration is anisotropic; it is predominant at the grain boundaries and negligible inside the c-Si grains (as the diffusion parameters of hydrogen are different in Poly2 than in c-Si). In amorphous silicon, hydrogen diffusion through compact sites is initiated by excited mobile hydrogen from a Si-H bond, and the thermally induced mobile hydrogen hops through sites analogous to the Si-Si bond centered sites in c-Si. In this process, the Si-Si bonds are continuously broken and reconstructed. In compact poly-Si, the diffusion of hydrogen through the thin grain boundary should follow this model. We attribute the behaviour of hydrogen diffusion in Poly2 to hydrogen hopping through deformed Si-Si bonds (at compact sites) inside the thin grain boundary. For this, we propose a (110) tilt boundary region through which hydrogen migrates. Such a grain boundary can be constructed without broken bonds.

The high temperature effusion maximum is considered to be same as in the case of an a-Si:H sample containing a smaller hydrogen content. In fact, an increase of E_D and D_0 has been observed for samples made at high (> 500°C) substrate temperatures. For such a-Si:H samples containing only 1% hydrogen, D_0 and E_D values of 120 cm^{2}s^{-1} and 2.1 eV respectively have been measured. These agree very well with the values for our HWCVD poly-Si sample. We suggest that the low initial hydrogen content in our poly-Si films (8.3x10^{20} cm^{-3}) is a possible reason for the present diffusion constants. This diffusion property can be explained by a model proposed for compact a-Si:H films. In the hydrogen density of states model (Figure 8), the hydrogen chemical potential, μ_H, defines the energy above which the shallow unfilled traps are located, and below which the deep traps occur. The trap depth, $E_{tr}-\mu_H$, where E_{tr} is the energy of hydrogen transport, can be obtained from the relation, $D = D_{H0} \exp(-(E_{tr}-\mu_H)/kT)$, where D_{H0} is the microscopic diffusion coefficient. Assuming $D_{H0} = 10^{-3}$ cm^{2}s^{-1}, we obtained a trap depth of 1.5 eV. This is slightly higher than the value for standard amorphous silicon ($\sim$10% hydrogen), but agrees very well with that for a-Si:H with a similar low hydrogen content ($\sim$1%).

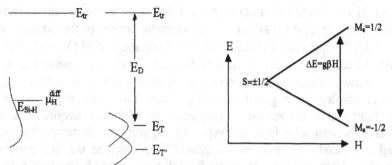

Figure 8. Hydrogen density of states.   *Figure 9*. ESR transition of a dangling bond state

The high activation energy of our Poly2 film can be explained by a statistical shift of the hydrogen chemical potential, given by the relation $E_{tr}-\mu_H(T) = E_{tr} - \mu_H(0) - \gamma T$, where $E_D = E_{tr} - \mu_H(0)$. The simultaneous increase in D_0 and E_D is due to a Meyer-Neldel type of behaviour. The situation is similar to the statistical shift of the Fermi level in the electronic density of states in amorphous silicon, which gives rise to Meyer-Neldel behaviour for electronic transport. The Meyer-Neldel rule for electronic transport is given by the relation; $\ln(\sigma_0)=B + E_a/E_{NMR}$, where σ_0 is the conductivity prefactor and E_a is the activation energy of electrical conductivity. A similar expression between the diffusion prefactor and the activation energy of hydrogen diffusion is obeyed for HWCVD poly-Si films.

The diffusion energy for hydrogen in Poly2 is a confirmation of the compact nature of the film. The Si-H bonds with a high binding energy, being located at compact sites, have a vibrational mode at 2000 cm^{-1}. Although it is, at present, not possible to ascertain whether such sites are in the inter-columnar region or the crystal interior, the similarities of the diffusion coefficient to those of amorphous films indicates that strained Si-Si bonds are involved in the migration. The inter-columnar regions are sites at which the probability of being occupied by H is greater than for other possible sites. The proximity of crystal columns facilitates strained bonds between the columns, creating a compact network. The hydrogen in the isolated Si-H bonds in this region migrates through these strained bonds.

2.3. DEFECT PAIRING

The dangling bond (db) is considered to be a characteristic defect in microcrystalline silicon, as is the case for amorphous silicon. This defect is believed to occur predominantly at the grain boundaries, although the presence of dbs inside the crystalline matrix (small in number) is not ruled out. In addition, dbs inside the amorphous tissue are present, as in normal a-Si:H. However, probing such defects and ascertaining their locations (especially in which matrix they reside) is a difficult task by any experiment that records only global defects, such as CPM or PDS. Moreover, in the absence of a DOS picture of μc-Si and the assignments of the optical transitions involved, the CPM or PDS sub-gap absorption is still not very helpful even for estimating the total defect density. Electron spin resonance (ESR) is a very useful technique for characterising local defects. The silicon dangling bond is a three-fold coordinated Si atom forming a defect with one unpaired electron. Due to this single spin, it behaves as a paramagnetic centre. In the presence of a magnetic field, the spin can be parallel ($M_s = +1/2$) or antiparallel ($M_s = -1/2$) to the magnetic field, leading to spin polarization (Figure 9). Under microwave excitation, a transition between these two levels occurs as result of spin flip when the condition $h\nu = g\beta H$ is satisfied. Here, g is the Lande g-factor, β the Bohr magnetron, H the magnetic field and ν the excitation frequency. In ESR, the frequency is kept constant and the magnetic field is swept. All experiments described here were conducted at X-band. This transition is generally recorded as the first derivative spectrum of the intensity versus magnetic field. Recent reports have given a wealth of data on various defects and their locations in microcrystalline silicon. Here, we concentrate on the specific case of HWCVD poly-Si made at UU, and show that in addition to some common defects, new types of clustered defects are detected.

Generally, for the defects in standard amorphous silicon, the line shape of the ESR dangling bond signal can be simulated as a Gaussian one caused by a number of non interacting isolated defects. In ESR terminology, this is a dilute spin system. However for the case of HWCVD poly-Si, the ESR spectrum cannot always be simulated as such a simple line shape. Thus, a deconvolution procedure was applied. For device quality poly-Si made at low hydrogen dilution (Poly2), a single line at g = 2.0055 was observed. The line shape could indeed be simulated as a Gaussian, with the g tensor reflecting a dilute spin system. This is not surprising, due to the small spin density ($\sim 7.8 \times 10^{16}$ cm^{-3}) in such material. However, for materials made under the same deposition conditions except for an increased chamber pressure, the line shape changes appreciably and a narrowing is observed.

The deconvolution showed two types of defect: a Gaussian line identical to that observed in low defect material, and a narrow line with an axially symmetric g tensor. From careful observation via temperature dependent ESR measurements, it was concluded that Heisenberg-type temperature independent exchange narrowing is taking place [17]. The magnetic susceptibility (χ) (proportional to the intensity of ESR line) gives a picture of the type of defects involved. For low defect density material, the spins follow a simple Curie behaviour, i.e., $\chi = C/T$ where C is a constant and T the temperature. In such a case, $1/\chi$ varies as a straight line going through T = 0. For highly defective material, this is not the case. Deconvolution showed two types of defect distribution: (1) isolated defects (db), following the

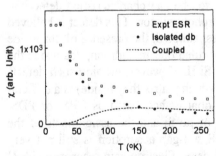

Figure 10. Temperature dependence of the magnetic susceptibility [18].

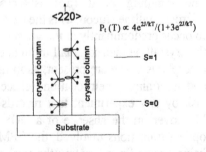

Figure 11. Grain boundary in a (220) oriented structure, showing isolated and paired dangling bond defects [18].

simple Curie behaviour, and (2) paired spins experiencing antiferromagnetic coupling through the Heisenberg exchange interaction [18]. This leads to Curie-Weiss behaviour (at high temperatures): $\chi = C_1/(T-\theta)$, with $\theta = -136.4°K$. Figure 10 shows these two contributions to the magnetic susceptibility.

For coupled defects, the paramagnetic state is different than in the case of isolated defects. A paired defect can be parallel or antiparallel, which splits the level into singlet (S=0) and triplet (S=1) states (Figure 11). The triple is paramagnetic, which splits it in a magnetic field into $M_s = 1, 0$ and -1. Transitions between these states give the ESR signal. Only at high temperatures, where the triplets are populated with probability $P_t(T)$, will they contribute to the ESR signal. At low temperatures, only the ground state (singlet) is populated. Being non-paramagnetic, it does not contribute to the ESR signal. This is manifested in the temperature dependence of the susceptibility of the paired spin in Figure 10. Structural considerations which could yield such defect pairing can be obtained from the XTEM image and the results of hydrogen diffusion experiments. For (220) oriented grains (as in the case of Poly2), the dangling bonds are perpendicular to the column face in the (111) plane, and face each other. If the grain boundary is narrow, the orbitals of these oppositely facing dangling bonds will overlap (Heisenberg interaction) leading to antiferromagnetic coupling and a split of the energy levels into singlets and triplets.

2.4. OXYGEN INCORPORATION

Figure 12 shows the hydrogen and oxygen depth profiles of poly-silicon films, as measured by secondary ion mass spectroscopy (SIMS) [19]. For such experiments, the sample is on a c-Si substrate. For the case of a Poly1 layer, it is observed that the oxygen homogeneously penetrates the film at a concentration of 2×10^{21} cm^{-3}, within a few days of exposure to air. In the Poly2 layer, the oxygen profile (even after one year of exposure to air) shows a sharp decrease at a depth of 50 nm from the surface. The oxygen concentration inside the film is nearly constant at about 3×10^{18} cm^{-3}. It is also seen that the hydrogen content in Poly1 (3.5×10^{21}/cm^3) is much higher than in Poly2 (6×10^{20}/cm^3). In fact, the ratio of hydrogen to oxygen in Poly1 is only 2:1, suggesting incorporation of water vapour after deposition. In sharp contrast, in the *profiled* layer (a double layer of Poly2 on top of Poly1), the hydrogen content in the top Poly2 layer is the same as that in the bottom Poly1 layer, except for a slight drop at the interface. The oxygen content in the Poly2 region of this *profiled* layer is again very low in the bulk, as in case of a single Poly2 layer. The oxygen content rises sharply in the bottom Poly1 layer, although the value (2×10^{20} cm^{-3}) is an order of magnitude lower than in the single Poly1 layer. We attribute this to oxygen incorporation during growth, facilitated by the structure. The IR spectrum of Poly1 shows a strong Si-O$_x$ band (~1050 cm^{-1}), which is absent in the Poly2 film. For a *profiled* film (Poly2 on Poly1), there was also no detectable Si-O$_x$ IR absorption. It was also observed that for Poly2 material, even when deposited in a high oxygen atmosphere (adding oxygen to the chamber) there was no noticeable oxygen in the film (from the IR spectrum). The bulk value of the oxygen concentration in Poly 2 is comparable to that in c-Si grown by the Czochralski technique, and to that in a microcrystalline film deposited by VHF CVD using an extra gas purifier [2]. The difference in the oxygen content between

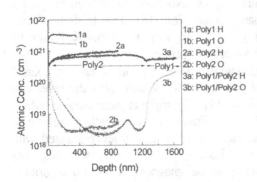

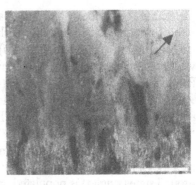

Figure 12. SIMS depth profile of oxygen and hydrogen in three films: Poly1, Poly2 and Poly2/Poly1. The depths of the layers are as shown.

Figure 13. Voids in poly-Si films, measured by TEM in under-focussed conditions. Part of the Poly1 layer and the Poly2 layer above it are shown.

Poly1 and Poly2 is attributed to the different porosities, leading to the intrusion of water vapour in Poly1. Figure 13 shows the voids in these two layer systems, measured by XTEM in defocused (here under focus) conditions [20]. From the above observations, we conclude that a compact structure not only prevents oxygen incorporation after deposition, but also inhibits its incorporation during growth.

3. References

1. Rath, J.K., Meiling, H. and Schropp, R.E.I. (1997) *Jpn. J.Appl. Phys.* **36**, 5436.
2. Meier, J., *et al.*(1996) *Mat. Res. Soc. Symp. Proc.* **420**, 3.
3. Yoo Yeong Cho, Yoo Yeong and Lim, K.M. (1997) *Jap. J. Appl. Phys.* **36**, 1094.
4. Diehl, F., Schroder, B. and Oechsner, H. J. (1998) *Appl. Phys.* **84**, 3416; Diehl, F., Scheib, M., Schroder, B. and Oechsner, H. (1998) *J. Non-Cryst. Sol.* **227-230**, 973.
5. Vanecek, M., Poruba, A., Remes, Z., Beck, N. and Nesladek, M. (1998) *J. Non-Cryst. Solids* **227-230**, 967.
6. van Veenendaal, P.T.T., Rath, J.K. and Schropp, R.E.I. (2000) *Proc. 16th European Photovoltaic Solar Energy Conference* (Glasgow, May 2000), p. 458.
7. Rath, J.K. (2002) Micro-/poly-crystalline silicon materials for thin film photovoltaic devices: application in solar cells, *This volume*, Figure 4.
8. Ito, T., Yasumatsu, T., Watabe, H., Iwami, M. and Hiraki, A. (1990) *Mat. Res. Soc. Symp. Proc.* **164**, 205.
9. Kroll, U. *et al.* (1996) *J.Appl. Phys.* **80**, 4971.
10. Saito, T. and Hiraki, H. (1985) *Jpn. J. Appl. Phys.* **24**, L491.
11. Stryahilev, D., Diehl, F. and Schroder, B. (1999) *J. Non-Cryst. Solids* **266-269**, 166.
12. Rath, J.K.and Schropp, R.E.I. (1999) *Mat. Res. Soc. Sym. Proc.* **557**, 573.
13. Rath, J.K., Schropp, R.E.I. and Beyer, W. (1999) *J. Non-Cryst. Solids* **266-269**, 548.
14. Finger, F., Prasad, K., Dubail, S., Shah, A., Tang, X.-M., Weber, J. and Beyer, W. (1991) *Mat. Res. Soc. Symp. Proc.* **219**, 383.
15. Beyer, W., Hapke, P. and Zastrow, U., (1997) *Mat. Res. Soc. Symp. Proc.* **467**, 343.
16. Rath, J.K., Schropp, R.E.I. and Beyer, W. (2001) *Solid State Phenomena* **80-81**, 109.
17. Rath, J.K., Barbon, A. and Schropp, R.E.I. (1998) *J. Non-Cryst. Solids* **227-230**, 1277.
18. Rath, J.K., Barbon, A. and Schropp, R.E.I. (1999) *J. Non-Cryst. Solids* **266-269**, 548.
19. Schropp, R.E.I., Alkemade, P.F.A. and Rath, J.K. (2001) *Solar Energy Materials and Solar Cells* **65**, 541-547.
20. Rath, J.K., Tichelaar, F.D. and Schropp, R.E.I. (1999) Solid State Phenomena **67-68**, 465.

MICRO-/POLY-CRYSTALLINE SILICON MATERIALS FOR THIN FILM PHOTOVOLTAIC DEVICES: APPLICATION IN SOLAR CELLS

J.K. RATH

Utrecht University, Debye Institute, Surface, Interface and Devices,
P.O. Box 80.000, 3508 TA Utrecht, The Netherlands

1. Solar cells

1.1 INTRODUCTION

A solar cell is a diode, which converts light energy to electrical energy. Normally, for example in the case of c-Si, it is a p-n junction. The electron and hole formed due to absorption of a photon (light) are separated from each other and collected across a junction. For the c-Si case, where the diffusion length is large enough, the carrier transport is predominantly by diffusion. However, for amorphous and microcrystalline silicon, a p-i-n cell is used, in which an intrinsic layer is sandwiched between n and p doped layers. The intrinsic layer is the carrier-generating region, whereas the doped layers define the field within the solar cell.

The reason for using a p-i-n cell is that for amorphous silicon and poly-Si films, the diffusion lengths of both the minority and the majority carrier are too small to be collected by diffusion. Hence, carriers are collected through drift induced by the field across the junction. The doped layers are highly defective, as has been explained by the thermal equilibrium model. Due to this, doped layers do not generate photo-induced carriers and are called dead layers. Moreover, carrier trapping at the defect sites in the doped layers reduces the field inside these regions. In a p-i-n cell, the depletion region is maintained in the less defective i-layer bulk, due to less trapping in the smaller defect density bulk. Moreover, the reduced concentration of defects allows efficient carrier generation in the i-layer. The carriers are separated by the high built-in field inside the i-layer, and collected by drift to the doped contacts. However, for the case of poly-crystalline silicon, if the grain sizes are large, then depending on the grain orientation with respect to the transport path, the diffusion length can be high. In such a case, collection of carriers in a poly-Si cell can have both drift and diffusion components. This will be discussed in more detail at a later stage, for the case of a HWCVD poly-Si cell.

Figure 1 shows schematic diagrams of two types of p-i-n solar cell. The doped regions, being dead layers (no carrier generation), are very thin, so that they absorb as little as possible, to avoid optical losses.

J.M. Marshall and D. Dimova-Malinovska (eds.),
Photovoltaic and Photoactive Materials - Properties, Technology and Applications, 183–196.
© 2002 *Kluwer Academic Publishers.*

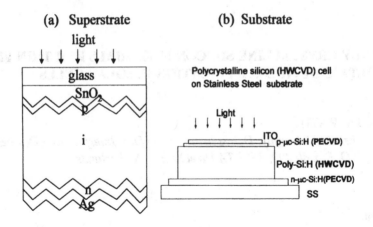

Figure 1. Types of solar cell structure

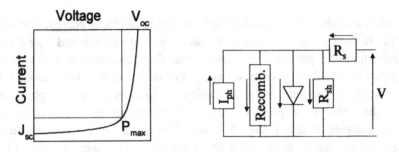

Figure 2. Schematic I-V characteristics and equivalent circuit of a solar cell

Figure 2 shows the current voltage characteristics of a solar cell. The open circuit voltage (V_{oc}), short circuit current density (J_{sc}) and maximum power (P_{max}) are defined. A parameter that characterises the quality of the i-layer (recombination) and the cell performance (series and shunt resistance) is the fill factor, defined as $FF = P_{max}/(J_{sc}*V_{oc})$. The efficiency is calculated from the relation $\eta = P_{max}/P_L = (FF* J_{sc}*V_{oc})/(100 \text{ mW/cm}^2)$.

1.2. TECHNOLOGY OF THIN FILM PHOTOVOLTAIC DEVICES, AND THE CHOICE OF MATERIAL

A material is suitable for solar cell applications if it satisfies two basic properties: efficient light absorption, and a good photosensitivity. Two aspects define the first criterion: (1) the band gap and (2) whether it is direct or indirect. The band gap should be tuned to the solar spectrum. For too low a value, the photoexcited carriers lose energy in excess of the band gap of the material. Too high a gap will lead to less absorption. 1.5 eV is considered an ideal band gap for efficient absorption of the solar spectrum, because this energy matches the peak of the solar spectrum. The second aspect of the optical properties makes materials with direct gaps more favourable. From these points of view GaAs (1.4 eV), CdTe (1.5 eV) and CuInSe (1.2 eV) can be considered as the most suitable, as these materials are direct gap semiconductors, in addition to having fairly optimum band gaps. From the same logic, c-Si should not be among the suitable candidates. This material has band gap (1.12 eV) that is too low. Being an indirect band gap material, absorption in the visible part of the solar spectrum is also low. Due to this, thick wafers must be used to completely absorb the solar spectrum. However, recent studies have shown that a c-Si wafer as thin as 50 μm can be used for a solar cell. This needs additional optical enhancements. However, c-Si has been used extensively for solar cells and modules, and is still the major material for industrial photovoltaic production for terrestrial applications. The choice for this material has come mainly because of the experience in the IC industry and because in the past the PV industry used reject materials from semiconductor industry. However, from the cost point of view and to increase throughput, thin film materials have been explored. From a thermo-dynamical limit, there are a number of materials which can, in an ideal case, deliver efficiencies between 25 to 30%. However, most of the thin films have achieved less than 50% of their thermodynamic limits.

TABLE 1. The estimated maximum efficiency, η_L, and characteristic years, c, for various types of solar cell.

Technology	η_L	c	a_0
c-Si	29	30	1948
Thin film silicon	30	19	1989
CIS/CIGS	29	30	1969
a-Si	18	20	1968
Organic cells	18	25	1995
New material	42	25	2000

TABLE 2: Reported record efficiencies of various types of solar cell.

Material	Efficiency	
c-Si	24.7%	[24]
a-Si/a-SiGe/a-SiGe	13%	[16]
Poly-Si	10.87%	[23]
a-Si/poly-Si/poly-Si	12%	[12]
CIS	18.8%	[25]
CdTe	16%	[26]
Dye Sensitised TiO_2	11%	[27]

A calculation has been performed concerning the potential thermodynamic limits and achieved efficiencies of various materials used in solar cells [1]. From this, it is clear that c-Si has almost reached its limit, whereas thin film silicon is at less than 30% of the limit. CuInSe and CdTe also have scope for development. Only in the future will it become clear which of these materials will succeed. However, there are already some predictions, based on developments so far and the present status of many of these materials. One such is by the Fraunhofer Institute for Solar Energy Systems, Germany [2]. The available data of solar cell efficiency have been fitted to a function

$$\eta(t) = \eta_L(1-\exp((a_0-a)/c)), \qquad (1)$$

where $\eta(t)$ is the time-dependent efficiency, η_L is the limiting asymptotic maximum efficiency, a_0 is the year at which $\eta(t)$ is zero, a is the calendar year and c is a characteristic development time. Table 1 shows the maximum achievable efficiencies predicted. From this study, it is clear that thin film silicon is the most promising candidate, not only because of the maximum efficiency limit that may be achieved (~30%) but because of a short characteristic time for development. The new material mentioned in the table is for the cells of the future, called third generation solar cells. At present, thin film cells have made a lot of progress and are already challenging the monopoly of c-Si based cells and modules. Table 2 shows the best efficiencies obtained by various types of low temperature (< 600°C) thin film solar cells. The efficiency for c-Si is provided for comparison.

Low temperature poly-crystalline (microcrystalline, μc) silicon achieved recognition with the development of a single junction solar cell delivering 4.6% efficiency [3]. The utility gained further recognition as part of the so-called micromorph solar cell, which combines an amorphous silicon top cell with a μc-Si bottom cell. A combination of the 1.7 and 1.1 eV band gaps provided an efficient way of trapping a wide range of the solar spectrum. However, the contribution of Kaneka Co. Japan in the development of single junction poly-Si cells and multijunction cells combining amorphous and poly-Si devices has been very impressive [4]. For the first time, industrial application looks viable, and production on the commercial scale has already proved the importance of low temperature poly-Si. Because the process temperature is less than 550°C, Kaneka has been successful in fabricating its cells on glass substrates. Both n-i-p (substrate) and p-i-n cell (superstrate) structures have been developed. The best efficiency has been achieved with a substrate structure, STAR (naturally Surface Texture and enhanced Absorption with back Reflector). It consists of a glass substrate coated with a back reflector (generally a combination of a reflecting metal such as Ag or Al and TCO (transmitting conducting oxide)). The back reflector has been texturised to achieve optical enhancement by the scattering of light. The poly-Si n-i-p structure lies on top of this. There is texture on the top of the Si layer due to the natural roughness of poly-Si, and this causes scattering at the front surface. The cell is completed by an ITO (indium tin oxide) top contact, plus grid lines to reduce the series resistance.

1.3. TYPES OF SOLAR CELL

Generally there are two kinds of solar cell:

(1) The superstrate type. In this construction p, i and n layers are grown successively on the substrate, as shown in Figure 1a. The substrate is transparent (glass), coated with TCO, and the light enters through it.

(2) The substrate type. In such a structure (Figure 1b) n, i and p layers are successively grown on top of the substrate. Light enters from the top side, and the substrate can be opaque.

The next section will present the process used at Utrecht University (UU) to fabricate an n-i-p solar cells on stainless steel (SS) substrates using HWCVD poly-Si, and will also show their characteristics.

2. Solar cell Fabrication

2.1. THE SINGLE JUNCTION CELL

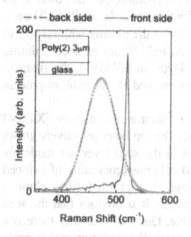

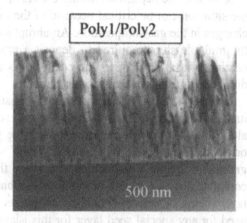

Figure 3. Raman spectra of a poly-Si film, from the front and back sides [5].

Figure 4. A profiled layer made by depositing device quality poly-Si on a seed layer of high nucleus density [5].

Poly-Si films grown by hot-wire CVD at a high hydrogen dilution of SiH_4 (Poly1) show a polycrystalline initial growth, but with a high defect density and randomly oriented grains. The crystal columns consist of many small grains. As described in the previous paper in this series, oxygen penetrates deep into the film through large voids between the columns perpendicular to the substrate. On the other hand, films grown with a low hydrogen dilution (Poly2) showed device quality (purely intrinsic nature, only (220) oriented growth), but with an incubation phase during which amorphous initial growth takes place.

Figure 3 shows the Raman spectra of these poly-Si films. The Ar ion laser beam used had a wavelength of 514.5nm. At this wavelength, the penetration depth of the probe beam is only ~50 nm for amorphous silicon material and ~100 nm for the crystalline regions. From the spectra of a film probed from the front and back sides respectively, it is clear that for Poly2 the region at the back side is amorphous, whereas the top region is highly crystalline. From the characterisation studies presented in the previous paper in this series, we know that the presence of only a 2000 cm⁻¹ IR peak, and high temperature (~ 600°C) hydrogen evolution proves the existence of a compact structure which inhibits oxygen incorporation into the film.

In order to use device quality intrinsic poly-Si, while avoiding an incubation phase, we proposed [5] a new approach to integrate these two growth regimes and make *profiled* poly-Si layers. In the first of these schemes (for which we will show solar cell results), a Poly1 layer of fixed thickness acts as a seed layer, on top of which the Poly2 is deposited. The second approach is to start the deposition at high dilution and continuously ramp the hydrogen flow to end the deposition at a low dilution. Experience so far has shown that both the schemes yield similar results, although the latter scheme has a lot of parameters available for varying the structure, and may lead to better efficiencies.

The above approaches are possible in HWCVD deposition, because there is no need to stop the deposition during the change of dilution steps. In the PECVD case, the situation can be critical because of the instability of the plasma with respect to changes in the gas composition. An abrupt switch on and off may be needed during the multiple steps required to deposit a profiled layer. In HWCVD, the filament heating (which determines the catalysis) is always on, and the deposition process during any gas phase transition is smooth.

Figure 4 shows a cross sectional transmission electron microscopy (XTEM) image of a profiled layer made by the first scheme. The top layer selectively grows along the (220) direction, even though the grains in the seed layer are randomly oriented. The top layer is compact, with V shaped columns (consisting of twinned grains) of height approaching the layer thickness. The profiled films show a complete absence of any amorphous incubation phase. It is obvious that the seed layer (being defective) should be as thin as possible. One may ask why there is a need for any special seed layer for this i-layer in a cell. This question makes sense when one considers that in a solar cell the i-layer is deposited on a microcrystalline n-layer (not glass), which could itself act as a seed layer. Experiments showed that the mechanism of nucleation is not that simple [6]. A cell grown on n-type c-Si, made in the structure n^{+}c-Si/poly-Si/p-μc-Si, had a good efficiency (3.15%), whereas a cell made by depositing a poly-Si i-layer layer on top of a PECVD n-type μc-Si showed very poor results. Although the exact mechanism is not yet clear, the change of deposition process and the type of seed layer structure have significant effects on the transition phase to the subsequent poly-Si layer. It must be noted that the Poly1 seed layer is randomly oriented, as are the grains in the n-μc-Si layer. However, the subtle difference in structure plays a big role in the transition regime.

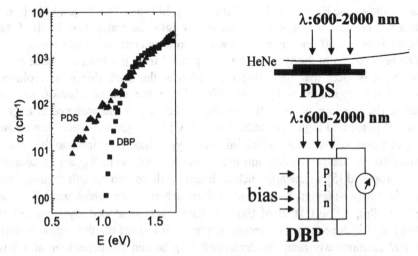

Figure 5. PDS and DBP spectra of a poly-Si film [9].

Cells incorporating the *profiled* poly-Si:H films were made on stainless steel (SS) substrates, in the configuration SS/n-μc-Si:H(PECVD)/i-poly-Si:H(HWCVD)/ p-μc-Si:H(PECVD)/ITO. A *profiled* layer (a device quality poly-Si layer on top of a thin seed layer) of thickness 1.22 μm was used instead of homogeneous poly-Si as the i-layer in these cells. An efficiency of 4.41% was achieved [7]. The other parameters were $V_{oc} = 0.34$ V, $J_{sc} = 19.45$ mA/cm^2 and FF = 0.59. Such a high current density in an i-layer only 1.22 μm thick is due to the large optical absorption in these films as described in the second paper in this series. A detailed study was made to understand the working of these cells. Density of states measurements by ESR and photothermal deflection spectroscopy (PDS) on single films, space charge limited current measurements (SCLC) on n-i-n structures, and dual beam photo-conductivity (DBP) measurements in cell configurations have given evidence about the location of the defects [8]. They have shown a distinction between a global defect density and those defects involved in the transport path.

Figure 5 compares of the optical absorption coefficients obtained by PDS and DBP [9]. Absorption at the low energies is a measure of the defects in the material. In PDS, a laser beam grazing the film surface is deflected due to the temperature gradient of the refractive index of a fluid on top of the film, which is heated due to absorption of monochromatic light. Thus, the absorption coefficient of the material is proportional to the defection at any particular wavelength. It is clear that such an experiment records the global defect absorption in the material. DBP was done by measuring the spectral dependence of the primary photoconductivity in a cell. In such an experiment, the carrier transport (which is normal to the substrate) is through the crystal columns in the i-layer of the cell and thus records only the defects connected to this transport path. Whereas PDS (which measures global defects) shows a high density of defects in the material, manifested as subband gap absorption in the low energy region, the DBP spectrum shows a sharp edge, as in c-Si. It is inferred that the transport of photogenerated carriers in the cell bypasses the defects, leading to the low sub-bandgap absorption obtained by DBP. This makes it possible for cells to operate even though the total defect density is 6.8×10^{16} cm^{-3}.

Computer simulation using the AMPS program has been employed to simulate the poly-Si cell, fit the cell parameters, test the sensitivity to various parameters and obtain predictions for cell optimisation [8]. The moderate efficiency of the cell can be explained as follows. Cracks in the n-layer (observed by XTEM) are speculated to affect the cell parameters. Moreover, the lack of a diffusion barrier between the stainless steel and the n-layer leads to the diffusion of impurities such as Cr from the substrate to the top of the n-layer. This was confirmed by X-ray photo-luminescence (XPS). This problem can be overcome by using a ZnO layer as a diffusion barrier. The third and most important reason is the absence of a back reflector in the cell. Stainless steel is a poor reflector and, in such a cell, a special back reflector such as Ag/ZnO helps to increase the current by more than 30%. A comparison of an a-Si:H cell on Asahi TCO/p-i-n/Ag with SS/n-i-p/ITO/grid and without back reflector has shown that the n-i-p cell has almost 30% less efficiency. Hence, with a proper back reflector the efficiency is expected to reach around 6%.

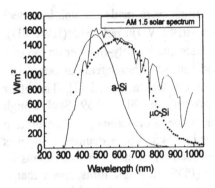

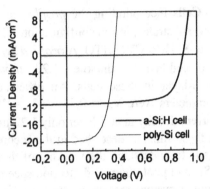

Figure 6. Spectral responses of a-Si and poly-Si solar cells.

Figure 7. I-V characteristics of poly-Si and a-Si solar cells.

2.2. TANDEM CELL

In order to use different parts of the solar spectrum effectively, multijunction solar cells with different band gap i-layers are used. Figure 6 shows the spectral response of two kinds of cell, namely amorphous silicon and polycrystalline silicon. The former, with a band gap of 1.75 eV, absorbs light in the short wavelength range of the visible spectrum, whereas the latter absorbs in the long wavelengths. The corresponding I-V characteristics of single junction cells made with these materials are shown in Figure 7. The a-Si:H cell has a high voltage but less current, whereas the poly-Si cell has a high current but a small voltage. The best way to make a tandem cell is to use a-Si as the top cell and poly-Si as the bottom one. Figure 8 shows such a structure [10]. All the i-layers were made by HWCVD, whereas the doped layers were made by PECVD. An efficiency of 8.1% was achieved by such a cell (Figure 9).

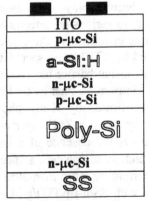

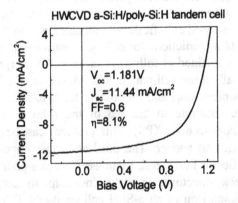

Figure 8. Cross-section of an a-Si/poly-Si tandem cell made at UU.

Figure 9. I-V characteristics of an a-Si/poly-Si HWCVD tandem cell made at UU [10].

3. Temperature Dependence of the Cell Parameters

One of striking difference between the performances of c-Si and a-Si cells is the temperature dependence of the cell parameters. For microcrystalline cells, it is found that the temperature dependence of the FF for small V_{oc} cells is similar to that for c-Si (high temperature coefficient), whereas the temperature coefficient for high V_{oc} cells is similar to that for a-Si (low temperature coefficient) [11]. It has been speculated that this behaviour is linked to type of carrier transport, i.e, drift limited transport (as in a-Si) is less affected by temperature than diffusion limited transport (as in c-Si). For the case of micromorph cells, the temperature dependence is less than for c-Si or μc-Si, but follows that of a-Si. On the other hand, Kaneka Co. poly-Si cells showed a temperature dependence similar to that of c-Si [12]. This suggests that the Kaneka cells feature electronic transport with a similar behaviour to that in c-Si. This is due to the fact that the material has large grains, forming a compact structure. The above observation again shows the fundamental difference between large grained (poly-Si) and small grained (μc-Si) cells. However, even for Kaneka cells, hybrid (Poly-Si/Poly-Si/a-Si) and tandem (poly-Si/a-Si) cells show a temperature coefficient similar to that for amorphous silicon. The small temperature coefficient of the hybrid cell gives it a special advantage for use at high temperatures, such as in desert and tropical areas.

4. What Limits Poly-Si Cell Performance

Among other parameters, it is the open circuit voltage that limits the performance of poly-Si cells. V_{oc} is limited by recombination at the grain boundaries, and a correlation between V_{oc} and effective diffusion length, L_{poly}, of poly-Si can be made. L_{poly} is given by the expression $L_{poly} = L_{mono}/\sqrt{(1+(2SL_{mono}/v_d g))}$, where L_{mono} is the diffusion length and v_d the diffusion velocity in single crystalline grains, g is the grain size and S is the recombination velocity. This is given by $S = S_0 \exp(q\psi/kT)$, where $q\psi$ is the barrier at the grain boundary and $S_0 = v_{th}\sigma N_{it}$ where v_{th} is the thermal velocity of free carriers, σ is the capture cross-section and N_{it} is the concentration of the interface states per unit area. The barrier height depends on the charges and the grain size.

A detailed review [13] of the performance of various high temperature poly-Si cells showed that the open circuit voltage decreases monotonically with decreasing grain size. However low temperature poly-Si cells show much higher open circuit voltages, although the grain sizes are typically less 100 nm. This is attributed to

(i) the intrinsic layer,

(ii) a low oxygen content,

(iii) a high hydrogen content

(iv) selective growth of grain boundaries with small numbers of broken bonds.

The intrinsic layer, together with the small grain size, reduces the band bending at the grain boundaries, whereas low oxygen content, high hydrogen content and

192

special nature of the grain boundaries reduce the grain boundary defect density, together with the recombination velocity. Selective growth of grain boundaries occurs in (220) oriented materials, where the grain-boundary is of (110) tilt type, which leaves bonds unbroken. A review of the results of various low temperature cells [13] (among which the UU cell is an example) shows that most of the high efficiency cells have a preferential (220) oriented structure. Commenting on the grain size dependence of the open circuit voltage, Kocka *et al.* [14] suggested a guiding rule that an optimum grain size for a high open circuit voltage can be achieved at $\sigma_0 > 100 \ \Omega^{-1} \ cm^{-1}$ and $E_a > 0.5$ eV, where σ_0 and E_a are the conductivity pre-factor and activation energy respectively. They showed that at high crystalline volume fractions, σ_0 and E_a are reduced due to a change of the transport path, leading to a lower open circuit voltage. In principle, this could explain the solar cell results of many groups using microcrystalline layers of small grain size and moderate crystalline volume fraction (at the transition from the crystalline to the amorphous phase). However, the results for UU poly-Si cells give a different picture [15]. UU uses materials that are so called poly-Si, which means that they have a negligible amorphous phase and a crystalline volume fraction of > 90%, with a large grain size. According to the Kocka *et al.* [14], this should lead to a low σ_0 and E_a, of about $10 \ \Omega^{-1} \ cm^{-1}$ and 0.45 eV respectively. This would have resulted in a low V_{oc}. In practice, it is observed that a σ_0 of $1 \times 10^4 \ \Omega^{-1} \ cm^{-1}$ and an E_a of 0.582 eV are obtained for the UU poly-Si material, which is also consistent with a high V_{oc}. It has been proposed [15] that grain environment plays a major role in the open circuit voltage, and that low temperature deposition leads to a better grain environment.

5. Use of a-SiGe:H or Poly-Si as a Bottom Cell

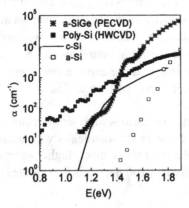

Figure 10. Comparison of the optical absorption coefficients of various thin films

As a low band gap material, poly-Si has to compete with a-SiGe:H for applications in multi-band gap cells. It has to be remembered that the highest initial ($\eta = 14.6\%$) as well as stabilised ($\eta = 13.0\%$) efficiency in a thin film amorphous silicon based solar cell was achieved with a-Si/a-SiGe/a-SiGe triple junction cell [16]. The material a-SiGe:H has many advantages. Its band gap can be suitably tailored by

TABLE 3. Comparison between poly-Si and a-SiGe cells.
SJ = single junction, LBG = low band gap

Parameters	Poly-Si	a-SiGe	Issues
Thickness	(~2μm)	100-200 nm	R_d
Cell structure	Texturisation	Grading p/i, n/i	Back reflector
Sub. Temp.	200 °C-500 °C	150-200 °C	Substrate, stable n and p-layer
Stability to Light (SJ)	stable	degrades	Microstructure
a-Si:H/LBG Tandem cell	Thick top cell, degradation in top cell	Thin top cell, degradation in bottom cell	Degradation
Efficiency(UU) Single Tandem	4.41% (n-i-p) 8.1% (n-i-p)	7.6% (p-i-n) 9.1% (p-i-n)	ITO,Grid, Shunt

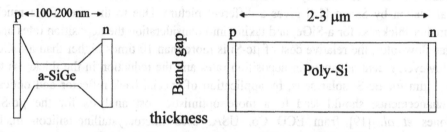

Figure 11. Band engineering for a-SiGe and poly-Si cells.

varying the Ge content in the film. It is a direct band gap material like amorphous silicon, due to which the high absorption allows a very thin layer 100-150 nm to generate more than 20 mA/cm^2 current in a solar cell. In contrast to this 1-3μm thick film is needed for poly-Si, because of a much lower absorption. Figure 10 shows the optical absorption of a-SiGe:H, poly-Si films, compared to a-Si:H and c-Si. It is clear that the poly-Si has a much higher absorption than c-Si, which allows it to be much less thick (almost two orders lower compared to c-Si) for adequate light trapping in a solar cell. However, in the visible region, a-SiGe:H still has a much higher absorption than poly-Si.

Table 3 lists some of the advantages and disadvantages of a-SiGe:H and poly-Si. The two materials also differ a lot in the device design. The a-SiGe cell in particular is made with a lot of band gap engineering. Figure 11 shows the band gap profile of both cells. In the case of the poly-Si one, both the p and n layer are also poly/μc-Si doped layers, having similar band gaps to the intrinsic middle layer. This makes for a simple device design. However, for a-SiGe:H, the doped layers are p-type amorphous silicon carbide and n-type amorphous silicon, with band gaps of 2.0 and 1.75 eV respectively. This leads to band gap mismatches at the p/i and n/i interfaces. In order to facilitate good carrier collection, U-type, V-type and E-type band profiles are used [17].

Generally, for the a-SiGe materials of band gap ~ 1.45 eV used in devices, deposition rates are lower than 0.1 nm/s, due to the high hydrogen dilution used to make device quality material. On the other hand, the deposition rate of μc-Si material for the best efficiency is ~ 0.5 nm/s. However, μc-Si loses this advantage of a higher deposition rate, due to the thicker (~ 2-3 μm) μc-Si used in a cell, compared to only ~ 150 nm for an a-SiGe layer needed to generate a similar current and efficiency. This is due to a much higher absorption in a-SiGe than in μc-Si (see Figure 10). Thus, the deposition rate for μc-Si should be at least 10 times higher than that for a-SiGe. This condition is still not satisfied at present, as far as devices are concerned. However recent advances in the concept of the high pressure depletion method combined with very high frequencies (VHF), such as microwave plasma and HWCVD, have shown that deposition rates of more than 2nm/s can be achieved; which has to translate into device performance.

Another consideration is the cost of the gas. a-SiGe is made from a $SiH_4/GeH_4/H_2$ gas mixture, whereas μc-Si is made from only a SiH_4/H_2 mixture. GeH_4 gas is almost 100 times more expensive than SiH_4, which puts a question mark on the cost viability of solar cells based on a-SiGe. However an actual calculation by Sanyo [18] gives a different picture. Due to the need for a much smaller thickness for a-SiGe, and taking into consideration the deposition rates and gas flow rates, the relative cost of μc-Si is more than 10 times higher than a-SiGe. However, recent advances in deposition rates and the reduction in the thickness to ~ 1 μm for μc-Si solar sells, by application of special back reflector and optical enhancements, should lead to a more optimistic cost analysis for the μc-Si. Jones *et al.* [19] from ECD Co. USA, used microcrystalline silicon made by microwave PECVD for the bottom cells of tandem and triple junction cells with initial efficiencies of 9.8% and 11.4% respectively. The μc-Si was deposited at 1.5 nm/s.

These cells have been compared with high quality a-SiGe:H ones, as typically used as the bottom unit in triple-junction a-Si/a-SiGe/a-SiGe solar cells. These silicon-germanium based cells were prepared at United Solar, using a standard PECVD technique and an i-layer deposited at 0.1 nm/s. The lower J_{sc} values for μc-Si cells [19] imply that these do not absorb as much red light as an a-SiGe cell. However, they absorb more light above 900 nm, due to their low band gap. One positive aspect of μc-Si cells is that P_{max} degrades by less than 5%, whereas the value is 17% for an a-SiGe cell. They believe that μc-Si cells can best be used as replacements for the lowest band gap a-SiGe layer in triple junction cells. In this case, the differences in the band gaps of μc-Si and a-SiGe are much less, and the demands for the higher wavelength response (800-100nm) are greater.

In a tandem cell, the currents in the top and bottom cells have to be matched. As the current in a μc-Si cell is higher than in an a-SiGe one, the current flowing in the top amorphous cell of an a-Si/μc-Si tandem cell is higher than that in an a-Si/a-SiGe one. This leads to a thicker top i-layer (a-Si) in a micromorph tandem, and higher degradation in the top cell, which limits its performance. On the other hand, in an a-Si/a-SiGe tandem, the degradation in a thinner or less current generating top cell is moderate, and the cell performance is limited by degradation in the bottom cell. The thickness of the top cell in an a-Si/a-SiGe device is half of that in a-Si/μc-Si (typically 160 and 320 nm respectively). Moreover, the comparatively low current in an a-Si/a-SiGe cell allows the use of a wide band gap proto-crystalline (high stability) top cell. In contrast, in a micromorph cell, a standard amorphous top cell is used to keep the thickness low [20]. As the power generated in the top cell is 2/3 of the total for the tandem structure, the degradation of the top device is a crucial aspect in the choice of cell structure and material. Moreover, recent studies by the UU group has shown that very stable a-SiGe cells, which show negligible light induced degradation, can be made [21,22]. However, considering all these aspects, recent decisions by certain groups such as Canon Co. Japan (a former sister company of ECD, USA) are pointer to the fact that poly-Si has a lot of advantages, such as fabrication simplicity. After years of experience in a-SiGe, Canon has now preferred to concentrate on poly-Si [23].

6. References

1. Kazmerski, L.L. (2001) *Tech. Digest Int. PVSEC-12*, Jeju, Korea, p.11.
2. Goetzberger, A., Luther, J. and Willeke, G. (2001) *Tech. Digest Int. PVSEC-12*, Jeju, Korea, p. 5.
3. Meier, J., Fluckiger, R., Keppner, H. and Shah, A. (1994) *Appl. Phys. Lett.* **65**, 860.
4. Yamomoto, K., Yoshimi, M., Tawada, Y., Okamoto, Y. and Nakajima, A. (2000) *J. Non-Cryst. Solids* **266-269**, 1082.
5. Rath, J.K., Tichelaar, F.D., Meiling, H. and Schropp, R.E.I. (1998) *Mat. Res. Soc. Symp. Proc.* **507**, 879.
6. Rath, J.K., Feenstra, K.F., van der Werf, C.H.M., Hartman, Z. and Schropp, R.E.I. (1998) *Proc. 2nd World Conference and Exhibition on Photovoltaic Solar Energy Conversion, Vienna*, p. 1665.
7. Rath, J.K., Tichelaar, F.D. and Schropp, S.E.I. (1999) *Solid State Phenomena* **67-68**, 465.
8. Rath, J.K., Rubinelli, F.A., van Veghel, M., van der Werf, C.H.M., Hartman, Z. and Schropp, R.E.I. (2000) *Proc. 16th European Photovoltaic Conference and Exhibition, Glasgow*, p. 462.
9. Rath, J.K., Barbon, A. and Schropp, R.E.I. (1998) J. Non-Cryst. Solids **227-230**, 1277.
10. Rath, J.K., Stannowski, B., van Veenendaal, P.A.T.T., van Veen, M.K. and Schropp, R.E.I. (2001) *Thin Solid Films* **395**, 320-329.
11. Meier, J., Keppner, H., Dubail, S., Kroll, U., Torres, P., Pernet, P., Ziegler, Y., Selvan, J.A.A., Cuperus, J., Fischer, D. and Shah, A. (1998) *Mat. Res. Soc. Symp.* **507**, 139.
12. Yamamoto, K. (1999) *IEEE Trans. Elecron. Dev.* **46**, 2041.
13. Werner, J. (2000) *Tech. Digest 13th Sunshine Workshop on Thin Film Solar Cells*, M.Konagai (ed.) (NEDO, Tokyo) page 41-48.
14. Kocka, J., Struchlikova, H., Stuchlik, J., Rezek, B., Mates, T. and Fejfar, A. (2001) *Tech. Digest Int'l PVSEC 12*, Korea, p 469.
15. van Veenendaal, P.A.T.T., van der Werf, C.H.M., Rath, J.K. and Schropp, R.E.I. (in press) *J.Non-Cryst. Solids* (ICAMS19).
16. Yang, J., Banerjee, A. and Guha, S. (1997) *Appl. Phys. Lett.* **70**, 2975.

17. Zambrano, R.J., Rubinelli, F.A., Rath, J.K. and Schropp, R.E.I. (2002) *J. Non-Cryst. Solids* **229-302,** 1131.
18. Shima, M., Isomura, M., Maruyama, E., Okamoto, S., Haku, H., Wakisaka, K., Kiyama, S. and Tsuda, S. (1998) *Mat. Res. Soc. Symp. Proc.* **507,** 145.
19. Jones, S.R., Crucet, R. and Izu, M. (2000) *Conference Record of the 28th IEEE PhotoVoltaic Specialists Conference-2000,* Anchorage, p.134.
20. Platz, R., Meier, J., Fischer, D., Dubail, S. and Shah, A. (1997) Mat. Res. Soc. Symp. Proc. **467,** 699.
21. Rath, J.K., Gordijn, A., Tichelaar, F.D. and Schropp, R.E.I. (2001) *Tech. Digest Int'l PVSEC-12,* Korea, p. 567.
22. Gordijn, A., Zambrano, R.J., Rath, J.K. and Schropp, R.E.I. (2002) *IEEE Trans. Elec. Dev.* **49,** (in press).
23. Saito, K., Sano, M., Sakai, A., Hayasi, R. and Ogawa, K. (2001) *Tech. Digest Int. PVSEC-12,* Jeju, Korea, p. 429.
24. Zhao, J., Wang, A. and Green, M.A. (1999) *Progress in Photovoltaics,* **7,** 471.
25. Contreras, M.A., Egaas, B., Ramanathan, K., Hiltner, J.U. and Swartzlander, A. (1999) *Progress in Photovoltaics,* **7,** 311.
26. Ohyama, H., Aramoto, T., Kumazawa, S., Higuchi, H., Arita, T., Shibutani, S., Nishito, T., Nakajima, J., Tsuji, M., Hanfusa, A., Hibno, T., Omura, K. and Murozono, M. (1997) *Proc. 26th IEEE Photovoltaic Specialist Conf.,* p 343.
27. Gratzel, M.(1997) *AIP Conf. Proc.,* **404,** 119.

ORGANIC MATERIALS AND DEVICES FOR PHOTOVOLTAIC APPLICATIONS

JEAN-MICHEL NUNZI
ERT Cellules Solaires PhotoVoltaïques Plastiques,
Laboratoire POMA, UMR-CNRS 6136, Université d'Angers,
2 Boulevard Lavoisier, 49045 Angers, France

1. Introduction

The recent progress achieved using organic mono-crystal, multilayered thin film and interpenetrated network technologies permits anticipation of a very fast increase in the conversion yield of organic solar cells, in order to make them a competitive alternative to the various forms of silicon cell. Indeed, the past two years have seen a significant jump in the conversion yield of organic photovoltaic (PV) solar cells, rising from a 1% yield achieved 15 years ago [1] to a 5% yield one year ago [2]. This opens the prospect, on a typical five-year timescale, of organic PV cells with solar yields in excess of 10%.

The long term objective of such very active research is to reduce the cost of PV modules. This is a difficult but realistic objective. This is the objective of the research programme in Angers. The large number of newly established specific conferences demonstrates the increasing research activity in the field of organic PV cells [3]. In this review, we discuss some of the key technical aspects of the problem.

There is a need for organic solar cells, in order to permit users to buy cost-effective modules. The short term idea is not to replace silicon, a well developed and productive technology, nor to replace the future thin film technologies (a-silicon, CIS or photosensitised cells), but to develop a long term technology based on environmentally safe materials. Such low-cost technology could be used to cover a majority of roofs, using large-surface modules connected to the electricity distribution network. Plastic materials bear this potentiality.

Such an objective now becomes feasible, in a similar manner to the previous development of efficient organic displays in the electronics industry (see [4] for a review). Those displays were developed after 10 years of laboratory research, because they offer a low cost "easy" technology and a technically attractive alternative to liquid crystal displays. Organic light emitting diode (OLED) research [5] sets the guidelines for research into organic PV cells, and the OLED literature will be taken as an illustration throughout this paper.

J.M. Marshall and D. Dimova-Malinovska (eds.),
Photovoltaic and Photoactive Materials - Properties, Technology and Applications, 197–224.
© 2002 *Kluwer Academic Publishers.*

Progress in organic PV cells requires the development of a clear understanding of the peculiar physics of amorphous organic semiconductors and devices [6]. In this paper, we first review the organic material for PV applications. Emphasis is given to the structure/property relationships, which are the guidelines for a molecular engineering strategy aimed at optimising PV properties. We start with the basic understanding of organic semiconductors, and their essential properties such as charge transport. We give indications aimed at designing organic PV materials, and we review some classes of materials used in the different layers of a PV cell.

We then proceed to an electrical description of organic solar cells. A critical analysis of the physical processes leading to the photovoltaic effect in organic materials permits one to size the maximal and minimal yields achievable using different device structures. This also permits an assessment of the main physical parameters required for the achievement of a 10% solar energy conversion in trial devices. We finally describe several materials and device structures which are good candidates for such an objective.

2. Organic Materials for Photovoltaic Applications

There are several good reasons for using organic materials for photovoltaic solar cell applications. The most important rely on the peculiar advantages of organic materials:

- They can be processed easily, using spin coating or doctor blade techniques (wet-possessing) or evaporation through a mask (dry-processing).
- The amounts and necessary purity of organic materials are relatively small, and large scale production (chemistry) is easier than for inorganic materials.
- They can be tuned chemically, in order to adjust separately the band gap, valance and conduction band energies, charge transport, as well as the solubility or other structural properties.
- The vast variety of possible chemical structures of organic materials (polymers, oligomers, dendrimers, organo-minerals, dyes, pigments, liquid crystals, etc.) favours active research for alternative competitive materials with the desired PV properties. Such original materials and structures can then be covered by patents.

2.1 STRUCTURAL ASPECTS OF ORGANIC SEMICONDUCTORS

Organic semiconductors can feature very different structural aspects (Figure 1), depending on the mechanical properties required for processing:

- The polymers are made of a large (10-10^3) number of identical repeat units, all linked together by covalent bonds in a linear way. They are deposited in thin films from a solution (wet process). They can be soluble in various solvents, or insoluble if prepared from a precursor route. Polymers can also be attached together: crosslinked. Sol-gel technology provides an example

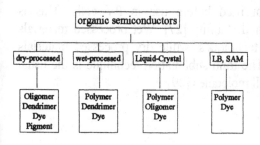

Figure 1. Organic semiconductors may be divided into different categories according to their mechanical, i.e. processing, properties: wet, dry, liquid crystalline, LB (Langmuir-Blodgett films) and SAM (self assembled monolayers). They may be divided further into monomers (dyes, pigments), oligomers, dendrimers and polymers [26].

of such cross linked polymers [7]. Thin films made from polymers are usually in a glassy state.

- Small fragments of polymers attached linearly and with a well defined length (2 to 12) are called oligomers [8]. The longest oligomers bear all the essential electrical and optical properties of polymers. They can be either soluble or insoluble. Thin films made from oligomers are usually in a polycrystalline state.

- Dendrimers are also made of identical repeat units linked together by covalent bonds, but in a three dimensional way like the branches of a star, forming an ellipsoid-like structure with a fractal surface [9]. They can be either soluble or insoluble. Thin films made from dendrimers are usually in a glassy state, with a high glass transition temperature (T_g).

- Pigments come from the paint industry. They are small molecules bearing specific properties relevant to light absorption and charge generation. They are processed by evaporation (sublimation) from a solid powder, in vacuum. Pigments usually form polycrystalline thin films.

- Dyes mainly have the same electronic structure and properties as pigments, but they bear "side groups" which make them soluble. They would not form thin films by themselves from a solvent, but they can be sublimated as pigments or be incorporated by mixing into a polymer host [10,11,12]. They can also be attached chemically to a host polymer [13,14].

- Liquid crystal materials (LCs) can be either polymers, oligomers, dyes or pigments. They can either be wet or dry processed. Their particularity is that the order which results from the liquid-crystal state (nematic, columnar, smectic, ...) permits separate adjustment of the light absorption, charge generation and transport [15] in the same layer. It also permits use of the light polarization properties in the case of small molecules [16,17] or polymers [18].

- Self assembled monolayers (SAM) [19], layer by layer deposited polymer films [20] and Langmuir-Blodgett films (LB) [21,22,23,24] permit the control of the molecular order, as for the liquid-crystal state. They can also be used to adjust the electrode properties: tuning the work function of electrodes for efficient electron or hole extraction [25].

All materials must be carefully purified before use in PV cells. This is mandatory to get optimum efficiency and stability [27]. Wet processed materials can be purified by solvent extraction techniques, while dry processed materials must be purified by sublimation [28]. The train sublimation technique is the usual way to improve the performance of small molecules [29].

2.2 ENERGY BANDS

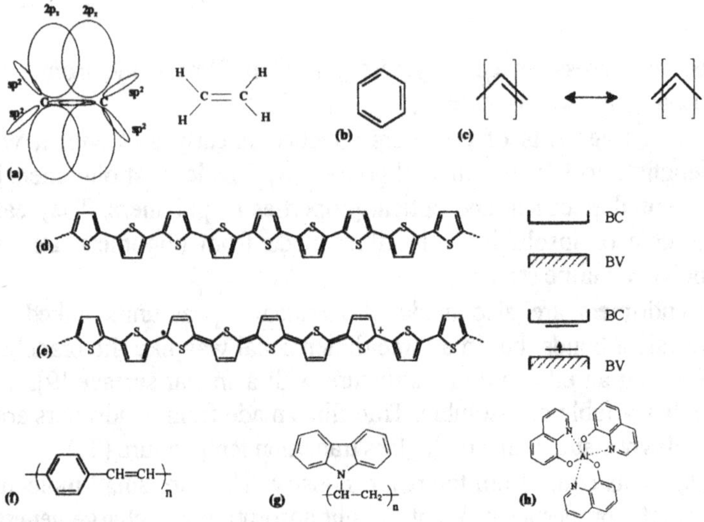

Figure 2. Model conjugated molecules and polymers: ethylene with its σ and π orbitals (a), benzene (b), a conjugated bond with no alternation (c), polythiophene (d), a charged polaron in polythiophene and corresponding energy levels (e), PPV (f), PVK (g), Alq₃ (h).

Polymers, oligomers, dendrimers, dyes, pigments, liquid crystals, organo-mineral hybrid materials, and all organic semiconductors share in common part of their electronic structure based on conjugated π electrons. By definition, a conjugated system is one featuring an alternation between single and double bonds. Ethene (figure 2a), butadiene and benzene (figure 2b) are basic representative elements of conjugated systems [30]. The essential property which comes from conjugation is that the π electrons are much more mobile than the σ electrons: they can jump from site to site between carbon atoms with a low potential energy barrier as compared to the ionisation potential (figure 2c). The π electron system shows all the essential electronic features of organic materials; that is light absorption and emission, charge generation and transport.

Each carbon atom in a conjugated system has 3 nearest neighbours with which it forms 3 equivalent σ bonds made from the trigonal sp^2 hybridisation of 3 valence atomic orbitals of the carbon atom: 2s, $2p_x$ and $2p_y$ for instance (figure 2a) [31]. For such a hybridisation state, the fourth orbital $2p_z$ lies perpendicular to the σ bond plane. It is the lateral overlap of these out-of-plane $2p_z$ atomic orbitals which gives the π bonds. In most molecules, unlike the case of figure 2c, double bonds are localised and the two extreme positions are usually not equivalent. A more general definition of a conjugated system is an ensemble of atoms whose p-orbitals overlap.

When the carbon chain length is increased, the molecule becomes a polymer. It has been proved experimentally, in the case of thiophene (figure 2d) [32] and phenyl-vinyl oligomers [33] for instance, and theoretically, in the case of phenyl-vinylene oligomers for instance [34], that oligomers with more than 5 to 8 repeat units bear all the essential electronic properties of infinite polymer chains in terms of the absorption and emission of light.

The electronic properties of polymers can be described in terms of semiconductor physics [35]. The particular framework of one dimensional periodic media is well suited to the basic understanding of an isolated polymer chain [36]. Polymers are bonded by strong covalent bonds. As the π-orbital overlap is weaker than the s-orbital overlap, the energy spacing (band gap) between the bonding and antibonding molecular orbitals is larger for the $\pi-\pi^*$ case than for the $\sigma-\sigma^*$ one. One can thus, to a first approximation, limit the band study to the $\pi-\pi^*$ molecular orbitals. These are respectively the HOMO (for Highest Occupied Molecular Orbital) and LUMO (for Lowest Unoccupied Molecular Orbital) states, in terms of molecular physics. They are also the usual valence (VB) and conduction bands (CB) of semiconductor physics, respectively (figure 2d). The σ-bonds then only contribute to the stability of the molecular structure [37].

As an example, the case of the infinite polyparaphenylene (PPP) chain is treated in a pedagogical manner by Moliton [38], within Hückel theory. PPP (with molecular formula $-(C_6H_4)_n-$) is a chain of benzene rings (figure 2b) attached in the para position. The problem is simply parameterized using the benzene resonance integral t_0 (or the exchange integral between two adjacent carbon atoms inside a ring) and the inter-ring resonance integral t_1. In the case of benzene, the HOMO – LUMO band-gap is $E_g = 2t_0 = 5.5eV$, as determined experimentally. In the case of the PPP polymer chain, the π and π^* orbitals split into two bands: the valence and conduction bands, with a band gap $E_g = 2t_0-4t_1 = 3eV$, as determined experimentally. Both band widths are given by Hückel theory as $4t_1 = 2.5eV$. This width is close to the experimentally determined ones, using electron energy loss spectroscopy. A more rigorous description of PPP energy levels confirms the validity of this simplified approach [39].

When the chain length is reduced, the maximum absorption of PPP shifts continuously to the blue, according to an experimental law $E_{max}= (3.36 + 3.16/n)$ eV [40], where n is the number of repeat units (benzene rings) of the oligomer. The maximum absorption, E_{max}, given here is larger than the band gap E_g. The same type of law applies to all the so-called alternated conjugated polymers (polymers in which double bonds are localized). This implies that in a "real" conjugated polymer with random length, corresponding to what most chemical synthetic routes deliver, the bands are broadened and the bad gap is apparently reduced.

In a real material, 3-dimensional interactions also play a major role in determining the transport properties, even dominating the transport which becomes an interchain hopping process.

Small molecules are bounded by weak interactions in the condensed state: Van der Waals forces. This yields a weak coupling between them, and the resonance integral t_l is thus small (a tenth of an eV at most) [41], resulting in narrow flat bands. The mobility is thus *a priori* smaller in small molecules, owing to a large effective mass $m* = (\hbar^2/2)(\partial^2 E/\partial k^2)^{-1}$. There can of course be exceptions to such a rule: the interdistance spacing can be small and molecular materials can in fact possess a rather large mobility. The first electrically pumped injection organic laser was indeed made from small molecules (a tetracene single crystal) [42].

2.3 TRANSPORT AND MOBILITY

Transport and mobility in organic materials require a knowledge of the charged species. A short physical review of properties is given by Schott [43]. The energy levels of the charges are usually determined by cyclic voltametry, for materials in solution. They can be characterized by XPS or UPS (X-ray and UV photo-electron spectroscopies) for solid materials. In small molecules, the charged species are localized spatially, they are simply the cation (positive) and anion (negative) radicals. In polymers, the electron-phonon coupling leads to so-called polarons, which are charges dressed by a reorganization of the lattice [44]. Polarons may be regarded as defects in conjugated polymer chains (figure 2e). Such a defect stabilises the charge, which is thus self-trapped as a consequence of lattice deformation. Hence, in the vast majority of organic semiconductors, transport bears all the characteristics of a hopping process in which the charge (cation or anion) propagates via side to side oxidation-reduction reactions (figure 3a). One must distinguish between intra-molecular charge transport along a conjugated polymer chain and intermolecular charge transport between adjacent molecules or polymer chains (figure 3b). The former, which is specific to conjugated polymers, is the most efficient.

The charge carrier mobility in organics is field dependent, especially in low mobility materials in which it usually follows phenomenologically a Poole-Frenkel law: $\mu \propto \exp(E^{1/2})$ [45]. The mobility can be experimentally determined by photo-current transients (time of flight) [46], field effect transistor saturation currents [47], space charge limited currents [48], or impedance spectroscopy [49]. The mobilities in organic semiconductors are usually rather small: from $10^{-2} \mathrm{cm}^2/\mathrm{Vs}$ in well ordered conjugated polymers (liquid crystalline polyfluorene) [50], down to $10^{-8} \mathrm{cm}^2/\mathrm{Vs}$ in guest-host polymer systems (dye doped poly-vinylcarbazole – PVK, figure 2g, for instance) [12]. The electron and hole mobilities differ by orders of magnitude in a single material; in small molecules such as the widely studied tris (8-hydroxyquinolinolato) aluminium - Alq$_3$ (figure 2h) [46], as well as in conjugated polymers such as the famous poly-paraphenylvinylene – PPV (figure 2f) [49]. The lowest mobilities are usually dispersive, which is the result of a distribution of mobilities [51]. The mobility can increase by up to two orders of magnitude with applied voltage, being eventually very large above 1MV/cm in conjugated polymers [52]. It is also increased by orders of magnitude when the molecular packing is improved. This is achieved by molecular ordering. The case of single crystals is obvious [42]: the electron mobility in fullerene C_{60} single crystals is 2.1 cm^2/Vs [53], but it is reduced by at least 3 orders of magnitude by imperfect purification and uncontrolled crystallization [54], as well as by oxygen traps [53]. With this in mind, the mobility in photovoltaic materials can be improved using liquid crystals made from molecules (columnar LCs [15]) or polymers (nematic LCs [50]). The mobility is also increased by orders of magnitude, up to 0.1 cm^2/Vs, between a random polymer (poly 3 alkylthiophene – P3AT) and its regioregular analogue [55,56]. The latter leads to superconductivity at low temperatures in a field-effect device [57].

Charge transport is also improved by purification or deposition conditions; for instance, mobility becomes non-dispersive in Alq_3 upon purification (oxygen induces traps) [58] and it becomes non-dispersive in soluble PPV derivatives upon selection of the solvent used for deposition [59]. The mobility is usually low and dispersive in randomly distributed polar molecules, but it is increased significantly when the dipoles are organized [60]. A record non-dispersive electron mobility of up to 2.10^{-4} cm^2/Vs was recently achieved in an air stable amorphous glassy molecular material [61]. Another record non-dispersive hole mobility of 10^{-2} cm^2/Vs was achieved in an amorphous glassy molecular material [62], together with the guidelines for developing such promising materials.

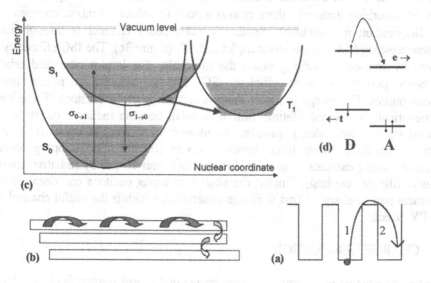

Figure 3. Photophysics of organic semiconducting materials: **(a)** a hopping process between molecules 1 and 2; **(b)** intra- (full arrows) and inter-molecular (broken arrows) charge transport; **(c)** absorption ($S_0 \rightarrow S_1$), luminescence ($S_1 \rightarrow S_0$) and inter-system crossing ($S_1 \rightarrow T_1$); **(d)** charge separation of an exciton into a free electron (e) – hole (h) pair at a donor (D) – acceptor (A) junction.

2.4 PHOTOPHYSICS

Light absorption by organic materials can also be rationalized using photophysics concepts [63]. It is from the optically excited state that the neutral excitation (exciton) can give rise to a free charge pair (figure 3d). The absorption and photoluminescence processes usually involve the same energy levels: the fundamental S_0 and first excited S_1 singlet states (figure 3c). Upon light absorption, molecules are excited from the fundamental to the excited state with a cross section $\sigma_{0\rightarrow1}$. Singlet-singlet transitions are very efficient ones, equivalent to the direct transitions in semiconductor crystals, leading to singlet excitons with a rather short lifetime (nanoseconds). The absorption spectrum extends inside the visible and near-infrared ranges. The exciton energy can then decay to the ground state radiatively, with a cross-section $\sigma_{1\rightarrow0}$ which is almost the same as $\sigma_{0\rightarrow1}$ (dipole coupling with the electromagnetic field): typically $10^{-16} cm^2$ in organic dyes and pigments. This is the usual luminescence which is a loss mechanism in photovoltaic cells. The exciton can also decay down to the ground state through vibrations (phonon emission in extended states). The speed of this decay is almost proportional to N!, where N is the number of vibration quanta ν_{vib} (stepladders in figure 3c) needed to decay down to the lower electronic state: $N \approx (\lambda_{max}.\nu_{vib})^{-1}$. This explains why UV photo-luminescence yields are often larger than IR ones. In order to improve the excited state lifetime in red absorbers, there is a need to inhibit large amplitude motions (ring rotation for instance, yielding the smaller possible N values). This can be achieved chemically (steric hindrance or linkages). For the near-IR absorbing materials, there is also a need to reduce vibration energies ν_{vib} (by fluorination for instance). Another exciton decay channel is through inter system crossing (ISC) to the lower triplet state T_1 (figure 3c). The ISC efficiency is driven by spin-orbit coupling inside the molecule. The largest permitted orbital moments permit the most efficient ISC. ISC is reduced in planar linear chromophores. The energy in triplet states is carried by triplet excitons. These have a dramatically enhanced lifetime (microseconds), because radiative decay to the ground state is forbidden (equivalent to indirect transitions in crystals). Triplet excitons can diffuse over large distances, up to 100 nm if no trapping occurs, although singlet excitons would not diffuse more than 10 nm by radiative energy transfer (dipole coupling). Finally, the singlet or triplet excitons can decay into a geminate pair of charges. This is charge generation, which is the useful channel for the PV effect.

2.5 CHARGE GENERATION

Exciton dissociation into a pair of charges occurs under large electric fields which can compete with the Coulomb interaction. The process is usually described by the so-called Onsager theory, which gives the efficiency of photo-dissociation ϕ of the exciton as a power series of power of the electric field E: $\phi = \phi_0 \exp(-r_c/r_0)[1 + \dfrac{r_c eE}{2!k_B T} + ...]$, with $r_C = e^2/4\pi\varepsilon k_B T$ being the critical distance below which the Coulomb energy is larger than the kinetic energy, and r_0 being the thermalisation distance (exciton size). Exciton dissociation can be equivalently promoted by charge transfer between donor and acceptor molecules (figure 3b), according to the Markus theory of electron transfer. A partner of the donor acceptor pair can simply be an impurity (O_2 for instance is an acceptor [6]). A junction at the interface with a metal or at the interface between a donor and an acceptor layer can also be a region of exciton dissociation. Such a junction effect can be phenomenologically interpreted in terms of an interfacial electric field, especially if the materials are the doped layers of a p-n junction. It is clear that the longest exciton lifetimes will permit the most efficient charge generation events.

2.6 MOLECULAR ENGINEERING

With the above considerations in mind, we can define some basic molecular engineering rules aimed at developing or selecting efficient PV materials. This includes tuning the band gap, choosing the acceptor or donor character (usually called p or n type by chemical engineers), optimising the charge mobility, plus selecting some structural aspects such as solubility, mechanical strength, etc. Each property is controlled by the assembly of different building blocks in a specific way (figure 4). The first element to choose is the deposition process: wet or dry. Practically, the dry process permits a larger purity of materials, for improved lifetimes. It is usually the domain of small molecules. The wet process permits the processing of larger surfaces, possibly at lower cost. It is usually the domain of polymers, although molecular glasses can be wet processed. In polymers, control of the size (degree of polymerisation and poly-dispersity) permits an improvement in the mechanical stability of the devices, and provides reproducibility of the fabrication procedure. From the electronic point of view, the length L of the conjugated skeleton permits some control of the band-gap energy (a+b/L law) in order to optimise sun-light capture, with a theoretical optimum around 1.5 eV. The use of donor or acceptor groups also permits control of the band gap (figure 4a) [64], its width if both donor and acceptor groups are used, and also the electron affinity if the molecule is made of donor groups (reduced work function) or acceptor groups (increased electron affinity).

Introduction of acceptors such as the oxadiazole (figure 4b) and triazole (figure 4c) groups permit to the electron affinity of the molecule to be increased (lowering the LUMO potential). Introduction of donors such as the carbazole (figure 4d) and amine (figure 4e) groups permit the ionization potential of the molecule to be decreased (increasing the HOMO potential). However, there is a lot of room for substitution around molecules, and this one must bear groups which permit solubilization in different solvents, bearing in mind that this can also be useful for chemistry, purification, and processing (solubilizing groups may be different for organic solvents, alcohols, water, fluorinated solvents, etc.). Such groups can also be used for linkage in order to make a non-conjugated polymer or a 3-dimensional

206

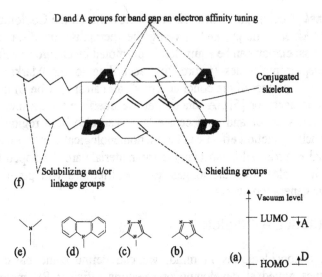

Figure 4. Molecular engineering of photovoltaic molecules: Effect of donor (D) and acceptor (A) groups on the HOMO and LUMO energies (a), examples of acceptor groups, Oxadiazole (b) and Triazole (c), donor groups, Carbazole (d) and Amine (e). Molecular building blocks (fragments) and their functionalities in EL materials (f).

crosslinked material with high mechanical and temperature stability. Such or similar groups can be used to protect reactive sites of the molecule, in order to increase stability in Red-Ox conditions (charge transport) or against oxygen attack. They can also be used to modify the planarity of the molecule, in order to adjust the conjugation (more planar), molecular packing or glassy behaviour. Importantly, organic molecules have donor or acceptor character (an intrinsic property). They can be independently p or n doped by impurities (an extrinsic character), and be efficient electron or hole transporters (a property determined by the orbital overlap in the solid state). An ideal donor should also permit efficient hole transport and eventually be n-doped. An ideal acceptor should also permit efficient electron transport and eventually be p-doped. Deviations from this rule would make poor PV cells.

2.7 MATERIALS

Excellent reviews on semiconductor materials are available today: general reviews [65], polymer specific reviews [66], molecule specific reviews [29], and reviews describing the state of the art of materials used in commercial or pre-commercial electronic devices [67]. Most of the materials currently used in electroluminescent diodes or in photocopying machines (xerographic materials) are now commercially available from different suppliers for different quantities and qualities (Covion [68], Syntec [69], American Dyes [70], H.W. Sands [71], Aldrich, etc.). Some of them can be readily used as PV materials. C60 (figure 5f) and its derivatives [72] are excellent acceptor molecules, with a sizable electron mobility. Perylen pigments (figure 5d) are also acceptors, with a larger near-IR absorption. The phtalocyanins,

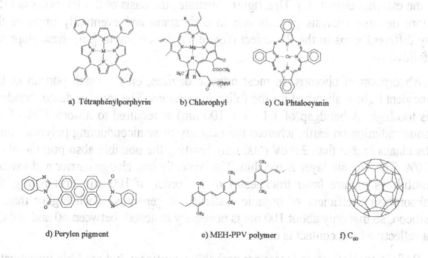

a) Tétraphénylporphyrin   b) Chlorophyl   c) Cu Phtalocyanin

d) Perylen pigment   e) MEH-PPV polymer   f) C_{60}

Figure 5. Model materials for photovoltaic cells: a), b), c) and e) are
Acceptors (A), d) and f) are Donors.

porphyrins and related molecules (figure 5a, b, c) are good donors with a sizable
electron mobility. Additionally, all the above molecules can be doped, in order to
improve charge transport and junction extension. MEH-PPV is a polymer whose
luminescence is efficiency quenched by charge transfer to C_{60}, making the so-called
"interpenetrated networks" [73]. Many photovoltaic molecules are described in
reference [74].

3. Physics of Organic Photovoltaic Solar Cells

3.1 THE PHOTOVOLTAIC EFFECT

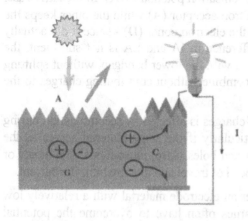

Figure 6. The photovoltaic process:
Absorption of photons (A),
Generation of carriers (G),
Collection of carriers (C).

The production of electrical energy from sunlight is the result of a chain process
(figure 6). Photons are absorbed inside the device (A), carriers are then generated
via exciton dissociation (G), they are finally collected by the electrodes and driven

into the external circuit (C). This figure illustrates the basis of the PV process [75]. A more detailed analysis permits one to understand and eventually improve the many different steps in the PV effect (figure 7). According to [26] these steps are the following:

- Absorption of photons: In most organic devices, only a small portion of the incident light is absorbed for the following reasons: The semiconductor bandgap is too high. A bandgap of 1.1 eV (1100 nm) is required to absorb 77% of the solar radiation on earth, whereas the majority of semiconducting polymers have bandgaps higher than 2.0 eV (600 nm), limiting the possible absorption to about 30%. The organic layer is too thin. The typically low charge carrier and exciton mobilities require layer thicknesses of the order of 100 nm. Fortunately, the absorption coefficient of organic materials is generally much higher than in silicon, so that only about 100 nm is necessary to absorb between 60 and 90%, if a reflective back contact is used.

- Reflection: Reflection losses are probably significant, but are little investigated in these materials. Systematic measurements of photovoltaic materials are desired to provide knowledge of their impact on absorption losses. Anti-reflection coatings, as used in inorganic devices, may then prove useful once other losses such as recombination become less dominant.

- Exciton diffusion: Ideally, all photoexcited excitons should reach a dissociation site. Since such a site may be at the other end of the semiconductor, their diffusion length should be at least equal to the required layer thickness (for sufficient absorption). Otherwise, they recombine and the photons are wasted. Exciton diffusion ranges in polymers and pigments are usually around 10 nm.

- Charge separation: This is known to occur at organic semiconductor/metal interfaces, at impurities such as oxygen or between materials with sufficiently different electron affinities (EA) and ionisation potentials (IA). In the latter case, one material can then act as the electron acceptor (A) while the other keeps the positive charge and is referred to as the electron donor (D) - since it did actually donate the electron to A. If the difference in IA and EA is not sufficient, the exciton may just hop into the material with the lower bandgap, without splitting up its charges. Eventually it will recombine without contributing charges to the photocurrent.

- Charge transport: The transport of charges is affected by recombination during the journey to the electrodes - particularly if the same material serves as the transport medium for both electrons and holes. Also, interactions with atoms or other charges may slow down the speed of travel and thereby limit the current.

- Charge collection: In order to enter an electrode material with a relatively low work function (Al or Ca), the charges often have to overcome the potential barrier of a thin oxide layer. In addition, the metal may have formed a blocking contact with the semiconductor, so that they cannot immediately reach the metal.

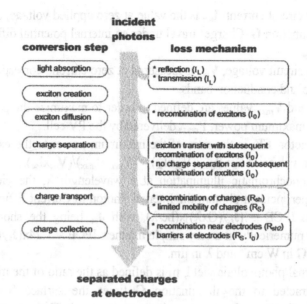

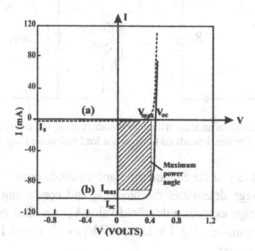

Figure 7. Conversion steps and loss mechanism of light power into electric power (after [26]). Light which is not converted to electricity is converted to heat in the best case, or to damage etc.

3.2 CHARACTERISTICS OF ORGANIC SOLAR CELLS

Figure 8. I-V characteristics of an ideal solar cell in the dark (a) and under illumination (b).

Drawing the current-voltage characteristics of a cell in the dark and under illumination (figure 8), permits one to evaluate most aspects of its photovoltaic performance, as well as its electrical behaviour [76].

- The short circuit current ,I_{sc}, is the value at zero applied voltage, and is a function of the illumination **G**. Charges travel under an internal potential difference typically equal to V_{oc}.

- The open circuit voltage, V_{oc}, is measured at zero current, corresponding to almost flat valence and conduction bands.

- The I_{max} and V_{max} values are defined in order to maximize the power $|I_{max} \times V_{max}|$. This is the maximum power, P_{max}, delivered by the PV cell.

- The fill factor, **FF**, is the ratio of the maximum power to the external short and open circuit values: $FF = P_{max}/(V_{oc} \times I_{sc}) = (V_{max} \times I_{max})/(V_{oc} \times I_{sc})$.

- Under monochromatic illumination at a wavelength λ, the yield of electrons generated per incident photon: the Internal Photon to Current Efficiency (**ICPE**) is defined as: $IPCE = (J_{sc}/(G\lambda)) \times (hc/\lambda)$, with J_{sc} being the short-circuit current density, or numerically in a very simple manner: $IPCE = 1.24 \times (J_{sc}/(G\lambda))$, with J_{sc} in $A.cm^{-2}$, **G** in $W.cm^{-2}$ and λ in μm.

- The external photovoltaic yield, η, is defined as the ratio of the maximum electric power extracted to the illumination **G** times the surface **S** of the module: $\eta = P_{max}/(s \times G)$ (It is often expressed as a percentage). The conversion yield is the key parameter as concerns cell productivity, and it must be evaluated carefully [77].

3.3   EQUIVALENT CIRCUIT

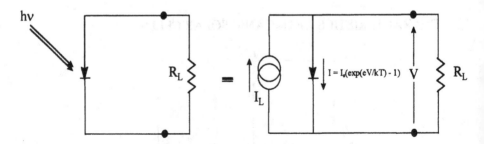

Figure 9. Equivalent circuit of an ideal cell under illumination. I_L is a current source the intensity of which depends on **G**. R_L is the load resistance [78].

The dark characteristics of the cell are the standard diode ones. The sign and value of the applied voltage determines the blocking and conducting ranges. The cell conducts when voltage exceeds a threshold value V_s. An ideal cell can follow the thermionic injection model [78]: $I = I_s(\exp(eV/kT) - 1)$, where I_s is the saturation current under reverse bias.

   Under illumination, the cell can be represented by the equivalent circuit in figure 9. It is described as a current source in parallel with the junction. I_L originates from the charge generated by the illumination. R_L is the resistance of the external circuit. Ideally, the current in the circuit is modelled as: $I = I_s(\exp(eV/kT) - 1) - I_L$.

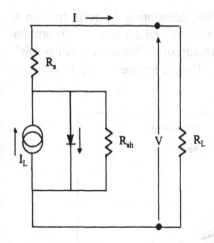

Figure 10. Equivalent scheme of a real PV cell under illumination. R_s and R_{sh} are the series and shunt resistances. R_L is the load resistance of the external circuit.

In real devices, the circuit must be modified to account for serial R_s and shunt R_{sh} resistance losses. An "ideality factor" **n** is also introduced (it is 1 for an ideal diode). The current then becomes: $I\left(1+\dfrac{R_s}{R_{sh}}\right)-\dfrac{V}{R_{sh}}+I_L=I_s\left(\exp\left(\dfrac{e}{nkT}(V-IR_s)\right)-1\right)$.

The equivalent circuit is that of an imperfect current generator, with shunt and series resistances (Figure 10). The series resistance depends on the resistivities of the material and the electrodes, and the metal - organic interfaces at the electrodes. The shunt resistance (several kΩ) corresponds to leaks and shorts in the diode. The slope around zero bias is a measure of the shunt resistance [79].

The relationship between V_{oc} and I_{sc} can be determined when it is assumed that $R_s=0$ and $R_{sh}=\infty$, with $I=0$ and $I_L=I_{sc}$: $V_{oc}=\dfrac{nkT}{e}\ln\left(\dfrac{I_{sc}}{I_s}+1\right)$.

A small shunt resistance will reduce V_{oc}. Additionally, the cell will not deliver any voltage under low illumination **G**. I_{sc} is essentially reduced by the series resistance.

3.4 THE EFFICIENCIES OF ORGANIC SOLAR CELLS

The main criterion for the evaluation of organic solar cells is the energetic conversion efficiency η (as defined above). It is essentially the product of four contributions : $\eta=\phi\cdot A\cdot FF\cdot eV_{max}/h\nu$. The fill factor can be close to unity (~ 0.8), provided the shunt resistance is large (R_{sh} > 25 kΩ) and the series resistance is small (R_s < 50 Ω). The ratio $eV_{max}/h\nu$ between the extracted electron energy (0.5 eV) and the average energy of the absorbed photons (2 eV) easily reaches 1/4 in current organic PV cells. One can expect a photogeneration yield, ϕ, close to 1 at the active junction [80,81,82] (see Section 2.5). In a homogeneous layer, this would require an exciton diffusion length close to the thickness of the diode, which is also possible in some pigments [83]. Organic colorants have an absorption in the visible range greater $\alpha\geq 10^7\,m^{-1}$ over a 100 to 200 nm bandwidth

212

(figure 11). This corresponds to 85% absorption, accounting for reflection on a metallic back electrode in a 100 nm thick layer. Let us take a sunlight absorption value $A = \frac{1}{2}$. We thus get the order of magnitude of a "maximum achievable" photovoltaic conversion efficiency: $\eta \approx 10\%$. This is the necessary level for a realistic technology.

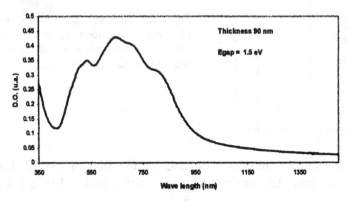

Figure 11. Absorption spectrum of a characteristic photovoltaic pigment. Its gap is rather well optimised. Its absorption coefficient is greater than 8×10^4 cm^{-1} between 500 and 800 nm.

Two parameters require particular engineering control: the **exciton diffusion length** (Section 2.4) and the **charge carrier mobility** (Section 2.3). Both are sizable issues.

- Improvement of the exciton diffusion length has recently permitted the achievement of a 3.6% solar efficiency with a bi-layer molecular cell [84], analogous to the original Tang cell [1]. An exciton diffusion length close to 100 nm is needed in homogeneous solar cell materials. Exciton diffusion is controlled by dipole coupling between molecules. It can be large in "pure" materials (trap free materials) made of "small" molecules exchanging energy through so-called *H* coupling (as opposed to the *J* coupling) [85], for which the luminescence yield vanishes. This is the usual case for small molecules in a crystalline or polycrystalline state, when molecules are stacked parallel to each other. One must notice that it can be larger than in some inorganic materials, because most organic materials behave as direct gap semiconductors.

- The high mobility of carriers (1–3 cm^2/V.s) may explain the exceptional solar efficiency obtained in pentacene crystals [2]. Photo-generated charges must indeed cross the solar cell, from the active junction to the counter electrode, fast enough for the current to be evacuated to the circuit faster than photo-generation brings charges into the device. Otherwise, a space charge field in the cell would screen the internal field, and the photo-generation efficiency would vanish. Under AM1 solar illumination, a 10% efficiency would correspond to a $J = 20$ mA/cm^2 current extracted from the cell. Space charge limited current follows the law: $J_{sc} = (9/8)\varepsilon\mu V^2/L^3$, where V is the working voltage and L the organic semiconductor thickness. We get a lower limit for the mobility of $\mu \geq 2.10^{-4}$ cm^2/V.s. It can be the electron or hole mobility, or both, which must reach this value, depending on the cell structure. The mobility can reach $1 \ cm^2/V.s$ in organic crystals, $10^{-2} \ cm^2/V.s$ in polycrystalline materials and $10^{-4} \ cm^2/V.s$ in amorphous materials. However, it drops easily by 2 orders of magnitude with impurities or defects which act as traps.

3.5 JUNCTIONS

The junction is the place where the exciton dissociates. A monolayer cell will make a Schottky junction with one of the electrodes. A bi-layer cell will preferentially develop a p-n junction at the interface if semiconductors are doped, or a Donor – Acceptor one (figure 3d) if they are intrinsic (undoped). Several examples are given in [86].

A junction will result from the equalization of the chemical potential of the electron (Fermi level in inorganic semiconductors) in the two different materials [86]. The case of a Schottky cell is illustrated in figure 12, for the contact between a metal and a p-type semiconductor. ϕ_m is the ionisation potential (IA) of the metal in a vacuum. ϕ_s is the same for the semiconductor (Fermi level). The nature of the contact depends on the relative height of the Fermi levels ϕ_m and ϕ_s.

- If $\phi_m > \phi_s$, electrons diffuse from the semiconductor to the metal. We thus get a positive accumulation in the semiconductor. No barrier forms at the interface, and the contact is ohmic.

- If $\phi_m < \phi_s$, electrons diffuse from the metal to the semiconductor (holes in the reverse direction). Diffusion stops when the space charge field at the interface is large enough to compensate for the potential difference. Owing to the small carrier density in organic semiconductors, the so-called depletion layer extends in the semiconductor only. This is a rectifying (Schottky) contact.

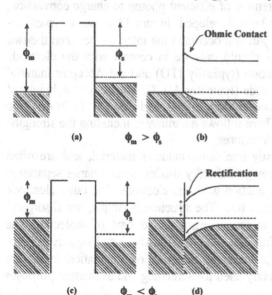

Figure 12. Energy levels of a metal – p-type semiconductor contact. Before and after contact for $\phi_m > \phi_s$. (a, b) and for $\phi_m < \phi_s$ (c, d).

4. Structure and Technology of Organic Solar Cells

4.1 CELL STRUCTURES [6,26,75,86]

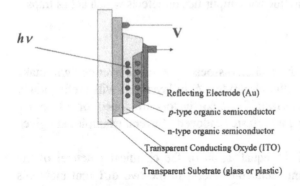

$h\nu$

Reflecting Electrode (Au)

p-type organic semiconductor

n-type organic semiconductor

Transparent Conducting Oxyde (ITO)

Transparent Substrate (glass or plastic)

Figure 13. Typical bi-layer organic solar cell design

Figure 14. Prototype single layer organic solar cells used for materials and technology evaluation. Each unit module is made of six 5x5 mm^2 cells.

In order to meet the specific requirement of efficient photon to charge conversion, different device architectures have been developed. Figure 13 shows a typical bi-layer device (p-n diode). Charge separation occurs at the interface between the two layers. Ideally, the Donor material should only be in contact with the electrode material with the higher work function (typically ITO) and the Acceptor material with the lower work function electrode (typically Al). Of course, as the height of the work function of ITO is between those of Al and Au (Al<ITO<Au), the structure is inverted in figure 13. There follows a summary, including the strengths and weaknesses of the different architectures:

- **Single Layer Cells** consist of only one semiconductor material, and are often referred to as Schottky type devices or Schottky diodes, since charge separation occurs at the rectifying (Schottky) junction with one electrode. The other electrode interface is supposed to be of ohmic nature. The structure is simple, but absorption covering the entire visible range is rare using a single type of molecule. The photoactive region is often very thin and since both positive and negative photo-excited charges may travel through the same material, recombination losses are generally high. Such cells are currently used for screening and evaluation purposes (Figure 14).

- **Double layer cells** benefit from separated charge transport layers that ensure connectivity with the correct electrode and give the separated charge carriers very little chance to recombine with their counterparts. The drawback is the small interface that allows only excitons of a thin layer (thickness = exciton diffusion length + depletion layer) to reach it and become dissociated.
- **Blend cells** exhibit a large interface area, if the molecular mixing occurs on a scale that allows good contact between alike molecules (charge percolation in an **interpenetrated network**) and most excitons to reach the D/A interface. This can usually only be partly achieved, so the defects of the network structure - particularly the connectivity with the correct electrode - represent a technological challenge.

4.2 EMERGING TECHNOLOGIES

In the past, the main difficulty in obtaining large solar conversion efficiencies was related to the small exciton diffusion length of the materials used in organic PV cells [3,87]. In most cases, the useful region of a planar p-n-type solar cell was limited to about 10% of the thickness necessary to absorb a significant proportion of sunlight, where the built-in junction-field is large, as in figure 15 [88]. The efficiency of such a diode is limited to about 1% [1]. This difficulty was recently overcome in a CuPc/C_{60} cell designed with improved electrodes (see Section 4.3), yielding a 3.6% efficiency [84]. This proves that bi-layer cells are good candidates for efficient energy conversion [80]. All the most important bi-layer cells such as the one in figure 15 have proved to be stable for years, which is also an important issue (see Section 4.4).

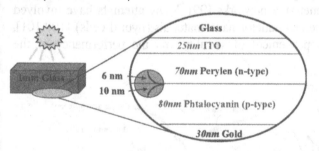

Figure 15. Scheme of a bi-layer p-n-type molecular PV cell, showing the extension of the photoactive region in which excitons dissociate, around the junction (according to [88]).

In the year 2000, the Lucent group at Murray Hill achieved a 4.5% efficiency with a halogen (Br or I_2) doped pentacene monocrystal (formula: ⬡⬡⬡⬡⬡) [89]. The efficiency was even larger (5%) if one considered only the surface of the cell and not the crystal one. The same material deposited as a thin polycrystalline film gave a lower 2.2% efficiency. This breakthrough may result from 4 favourable factors:

(1) pentacene behaves as direct band gap material for absorption, but the exciton decays rapidly to a triplet exciton which has a long lifetime.

(2) The crystal is dense and pure, permitting a large exciton diffusion length.

(3) The hole mobility in pentacene reaches $1\,cm^2/V.s$.

(4) The material was doped, which eases exciton dissociation away from the contact electrodes which otherwise would quench the exciton before charge transfer.

This result opens-up the possibility of building single layer solar cells for efficient PV conversion. However, the absorption cannot cover all the solar spectrum with a single layer of an homogeneous organic material, so that a bi-layer diode remains a potentially attractive device [90]. Additionally, halogen doped pentacene crystals may be highly unstable materials, but this also can be overcome using controlled molecular doping with strong acceptor molecules [91].

In order to circumvent the weak extension of the depletion layer associated with the small diffusion length of excitons which usually greatly limits the potential efficiency of layered organic solar cells, the effective surface of the junction was greatly increased in dye-sensitized solar cells [92] (so called Grätzel cells). The concept was applied successfully to a solid single-layer polymer solar cell using the efficient charge transfer which occurs between C_{60}-derivatives as donors and semiconducting polymers as acceptors [93,94]. Such cells exceed 3% solar efficiency. From a chemical point of view, the advantage is that the materials do not have to be optimised for charge AND exciton transport, because excitons can dissociate everywhere in the mixture, owing to the closeness of the donor and acceptor functions (figure 16). From a topological point of view, the effective surface of the junction is dramatically improved by the interpenetrated network. The difficulty lies in the fabrication of a doubly connected network made of an electron transport material (a conjugated polymer) and a hole transport material (C_{60}) [95]. Attempts have been made to link the donors and acceptors chemically, but experimental results do not yet show any significant progress in this direction [96, 97]. Another actual difficulty lies in the stability of the cells: both structural and photochemical stability [98]. There is still much room for significant progress in the technology of interpenetrated networks [99]. Many attempts have involved replacing C_{60} by inorganic semiconductor nano-materials (hybrid cells) [100,101], as a strategy for further improvement of stability and the performance of the technology.

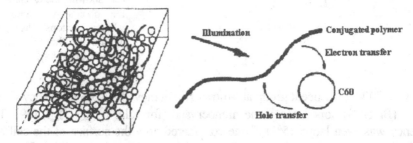

Figure 16. Scheme of an interpenetrating network of a donor polymer and an acceptor C_{60}.fullerene derivative, showing the detail of an ultra-fast charge transfer event (after [73]).

A new route for the improvement of organic solar cell efficiency was demonstrated by the CEA group in Saclay [102,60]. It relies on the orientation of polar chromophores inside the volume of the cell, in order to induce an internal polarization field [103,104]. This field improves exciton dissociation inside the cell as well as the charge carrier mobility. An internal polarization may also help in reducing the potential barriers at the electrodes, thus improving the series resistance. The practical demonstration was made with several blends of polymers incorporating polar chromophores [104,105,106]. A two orders of magnitude improvement was obtained in a blended polythiophene structure (figure 17)

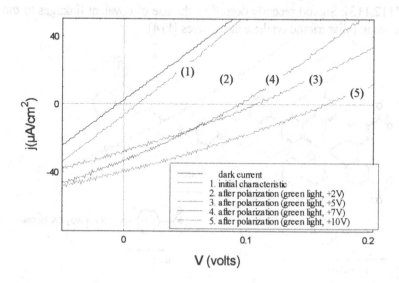

Figure 17. Current density vs. Voltage characteristics recorded in the dark and under 12.5mW/cm^2 red light illumination through ITO for an ITO / P3BT(25%wt) + copolymer MMA-DR1 (50%wt) + PR3072 (25%wt) / Au device, before and after different polarization sequences at room temperature (after [106]).

4.3 ELECTRODES

The organic–inorganic contact at the electrodes is a source of losses due to the series resistance of non-ohmic contacts. A lot of research work has been carried out in relation to light emitting diodes. Current materials used as hole transport layers at the ITO contact are poly(ethylenedioxythiophene) doped with poly(styrene sulfonate) (PEDOT-PSS) [107] (figure 18a) and copper phtalocyanine (CuPc) [108] (figure 18b). The first of these is a wet processed polymer and second is an evaporated molecule. Both are transparent to visible light. They favour hole injection and reduce diffusion of impurities from ITO into the diode. There are also organic materials used as electron injection layers. Bathocupuroine (BCP) [109,110] (figure 18c) can be used to block the excitons, although permitting electron tunnelling from aluminium into the device. Perylene derivatives (figure 5d) can also be used for electron transport at the ITO contact [90].

A recent breakthrough in the concept of electron or hole injection from metallic or semi-metallic electrodes is the use of self-assembled monolayers (SAM) [25] or

LB films [111] of dipolar molecules. In contact with an electrode, anode or cathode, such a monolayer shifts the electrode potential by an amount $\Delta\Phi = e\sigma\mu/\varepsilon$ where e is the electron charge, σ the surface density of dipoles, μ their dipole moment and ε the permittivity of the medium. The injection potential can be adjusted by up to ± 0.6 eV with such molecular monolayers. Some typical molecules used in this respect are shown in figure 18d. These were grafted by ionic or Van der Waals bonds [112,113]. Sigaud recently described the use of covalent linkages to the ITO cathode using polar triethoxysilane derivatives [114].

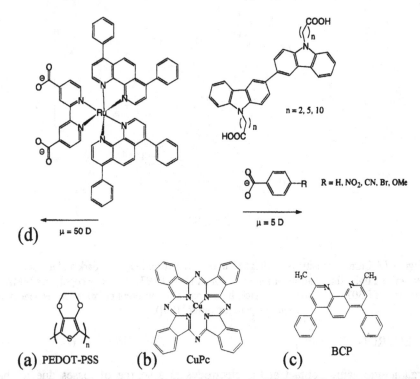

Figure 18. Materials used in contact with electrodes: the PEDOT-PSS polymer (a) and the CuPc molecule (b) as hole transport layers, and the BCP molecule (c) as an exciton blocking and electron transport layer. Polar molecules used as monolayers to adjust the injection potential at electrodes (d).

4.4 DEGRADATION

The degradation issue has made enormous progresses in the recent years in the field of organic electroluminescent devices. Their lifetime now reaches 20000 hours under nominal working conditions. However, a photovoltaic device should be exposed to sunlight for a period longer than 10 years in order to become a competitive alternative to silicon. That is more than 10000 hours of sun. This is not an easy task (figure 19). The causes of degradation have different origins:
- Photochemically, it can be an intrinsic property of the molecules or an extrinsic one (photo-oxidation with oxygen for instance) [63].

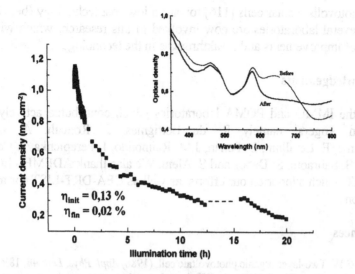

Figure 19. Degradation of an organic solar cell based on a dye-doped polythiophene [76].

- Electrochemically, ionic impurities or water may promote redox reactions at the electrodes.
- Structurally, low T_g materials may reorganise, recrystallise, or diffuse one into another owing to repeated heating and cooling in outdoor conditions.
- The electrodes can react with the molecules, by metal diffusion from ITO into polymers for instance [115].

Improvement of the stability of organic PV cells needs:
the choice of intrinsically stable materials (stable under storage, under sunlight and under reduction or oxidation conditions),
the use of efficient purification methods,
the choice of high T_g materials and structures (stable under outdoor cooling or heating),
a fabrication technique featuring water and oxygen free conditions,
encapsulation of the structures after fabrication,
the fabrication of efficient solar cells, because unconverted light heats and eventually damages the cells (figure 7).

5. Conclusions

We have described several possibilities for fabricating efficient solar cells: single layers, double layers, blended or interpenetrated cells, hybrid cells and oriented cells. All offer room for interesting research in physics, chemistry and technology, and all are possible routes to building an efficient PV technology. Such cells will obviously be soft and possibly bendable or foldable. They will be adapted to the fabrication of solar roofs, but also sails, tents and all kind of plastic outdoor furniture or portable devices. Importantly, plastic solar cells will help orient the future of photovoltaic solar cells [116] toward a low cost technology (below 1€/W module). Several laboratories are now involved in this research, which will speed up the rate of improvements and breakthroughs in the technology.

6. Acknowledgements

We thank the IMMO and POMA laboratories which contributed actively to this research in Angers: namely R. de Bettignies, J. Roncali, A. Gorgues, P. Hudhomme, E. Levillain, L. Perrin, J.M. Raimondo, I. Perepichka, J. Cousseau, M. Cariou, B. Sahraoui, S. Dabos and S. Alem. We also thank ADEME - ECODEV and MENRT which supported our efforts, as well as CEA-DRT-LIST for an active collaboration.

7. References

1. Tang, C.W. Two-layer organic photovoltaic cell, (1986) *Appl. Phys. Lett.* **48**, 183.
2. Schön, J.H., Kloc, Ch., Batlogg, B. Efficient photovoltaic energy conversion in pentacene-based heterojunctions, (2000) *Appl. Phys. Lett.* **77**, 2473.
3. *First European Conference on Organic Solar Cells*, (1998) Cadarache, France, http://iacrs1.unibe.ch/~koenigs/ecos98.htm; and also http://netserv.ipc.uni-linz.ac.at/ICOS02/.
4. Le Barny, P. *et al.* (2000) *C. R. Acad. Sci.* Paris 1, **Série IV**, 493.
5. Kalinowski, J. Electroluminescence in organics, (1999) *J. Phys. D: Appl. Phys.* **32**, R179.
6. Simon, J. and André, J.-J. (1985) *Molecular semiconductors: photoelectrical properties and solar cells*, Springer.
7. Dantas de Morais, T., Chaput, F., Boilot, J.-P., Lahlil, K., Darracq, B. and Lévy, Y. (2000) *C. R. Acad. Sci. Paris* 1 série **IV**, 479.
8. Müllen, K. and Wegner, G. (1998) *Electronic materials: the oligomer approach*, Wiley-VCH, Weinheim.
9. Halim, M., Pillow, J.N.G., Samuel, I.D.W. and Burn, P.L. (1999) *Adv. Mater.* **11**, 371.
10. Kido, J., Hongawa, K., Okuyama, K. and Nagai, K. (1994) *Appl. Phys. Lett.* **64**, 815.
11. Zhang, Z.I., Jiang, X.Y., Xu, S.H. and Nagatomo, T. (1997), in S. Miyata and H.S. Nalwa (eds.), *Organic electroluminescent materials and devices*, Gordon and Breach, Amsterdam, p. 203.
12. Gautier-Thianche, E., Sentein, C., Lorin, A., Denis, C., Raimond, P. and Nunzi, J.M. (1998) *J. Appl. Phys.* **83**, 4236.
13. Cacialli, F. Friend, R.H. Bouche, C.-M., Le Barny, P., Facoetti, H., Soyer, F. and Robin, P. (1998) *J. Appl. Phys.* **83**, 2343.
14. Jiang, X., Register, R.A., Killeen, K.A., Thompson, M.E., Pschenitzka, F. and Sturm, J.C. *Chem. Mater.* **12**, 2542 (2000)
15. Seguy, I., Destruel, P. and Bock, H. (2000) *Synthetic Metals* **111–112**, 15.
16. Mikami, T. and Yanagi, H. (1998) *Appl. Phys. Lett.* **73**, 563.
17. Yanagi, H., Okamoto, S. and Mikami, T. (1997) *Synthetic Metals* **91**, 91.
18. Jandke, M., Strohriegl, P., Gmeiner, J., Bruetting, W. and Schwoerer, M. (1999) *Advanced Materials* **11**, 1518.
19. Appleyard, S.F.J., Day, S.R., Pickford, R.D and Willis,M.R. (2000) *J. Mater. Chem.* **10**, 169.
20. Kim, J., Chitibabu, K.G., Cazeca, M.J., Kim, W., Kumar, J. and Tripathy, S.K (1997), *Mat. Res. Soc. Symp. Proc.* **488**, p. 527.

221

21. Cimrova, V., Remmers, M., Neher, D. and Wegner, G. (1996) *Adv. Mat.* **8**, 146.
22. Wu, A., Fujuwara, T., Jikei, M., Kakimoto, M.-A., Imai, Y., Kubota, T. and Iwamoto, M. (1996) *Thin Solid Films* **284-285**, 901.
23. Tokuhisa, H., Era, M. and Tsutsui, T. (1998) *Appl. Phys. Lett.* **72**, 2639.
24. Arias-Marin, E., Arnault, J.C., Guillon, D., Maillou, T., Le Moigne, J., Geffroy, B. and Nunzi, J.M. (2000) *Langmuir* **16**, 4309.
25. Nuesch, F., Si-Ahmed, L., François, B. and Zuppiroli, L. (1997) *Adv. Mater.* **9**, 222.
26. Petritsch, K. (2000) *Organic solar cell architectures*, PhD thesis, Graz.
27. Papadimitrakopoulos, F., Zhang, X.M. and Higginson, K.A. (1998) *IEEE Proceedings* **4**(1).
28. Wagner, H.J. and Loufty, R.O. and Hsio, C. (1982) *J. Mater. Sci.* **17**, 2780.
29. Dentan, V., Vergnolle, M., Facoetti, H. and Vériot, G. (2000) *C. R. Acad. Sci.* Paris 1 (série IV), 425.
30. Salem, L. (1966) *The molecular orbital theory of conjugated systems*, Benjamin, New York.
31. Nguyen, T.A. (1994) *Introduction à la chimie moléculaire*, Ellipses, Grenoble.
32. Charra, F., Fichou, D., Nunzi, J.M. and Pfeffer, N. (1992) *Chem. Phys. Lett.* **192**, 566.
33. Nunzi, J.M., Pfeffer, N., Charra, F., Nguyen, T.P. and Tran, V.H. (1995) *Nonlin. Opt.* **10**, 273.
34. Kirova, N., Barzovskii, S. and Bishop, A.R. (1999) *Synthetic Metals* **cfc**, 29.
35. Kittel, C. (1972) *Introduction à la physique de l'état solide*, Bordas, Paris.
36. Cojan, C., Agrawal, G.P. and Flytzanis, C. (1977) *Phys. Rev. B* **15**, 909.
37. Su, P.W., Schrieffer, J.R. and Heeger, A.L. (1979) *Phys. Rev. Lett.* **42**, 1698.
38. Moliton, A. (2001) in J.P. Goure (ed.), *Les sources de lumière, traité d'optoélectronique*, Hermes, Paris.
39. Brédas, J.L., Chance, R. R., Silbey, R., Nicolas, G. and Durand, P. (1982) *J. Chem. Phys.* **77**, 371.
40. Leising, G., Tasch, S. and Graupner, W. (1998) in T.A. Skotheim (ed.), *Handbook of conducting polymers*, M. Dekker, Chap. 30,
41. Lange, J. and Bässler, H. (1982) *Phys. Stat. Sol. B* **114**, 561.
42. Schön, J.H., Kloc, C., Dodabalapur, A. and Batlogg, B. (2000) *Science* **289**, 599.
43. Schott, M. (2000) *C. R. Acad. Sci. Paris* 1 (série IV), 381.
44. Emin, D. (1996) in T.A. Skotheim (ed.) *Handbook of conducting polymers*, M. Dekker, Vol. 2, Chap. 26.
45. Gill, W.D. (1976) in J. Mort and D.M. Pai (eds) *Photoconductivity and related phenomena*, Elsevier, p.63.
46. Kepler, R.G., Beeson, P.M., Jacobs, S.J., Anderson, R.A., Sinclair, M.B., Valencia, V.S. and Cahill, P.A. (1995).*Appl. Phys. Lett.* **66**, 3618.
47. Horowitz, G. (1998) *Adv. Mater.* 10, 365.
48. Blom, P.W.M., De Jong, M.J.M. and Vleggaar, J.J.M. (1996) *Appl. Phys. Lett.* **68**, 3308.
49. Martens, H.C.F. Huiberts, J.N. and Blom, P.W.M. (2000) *Appl. Phys. Lett.* **77**, 1852.
50. Reddecker, M., Bradley, D.D.C., Inbasekaran, M. and Woo, E.P. (1999) *Appl. Phys. Lett.* **74**, 1400.
51. Scher, H. (1976) in J. Mort and D.M. Pai (eds.), *Photoconductivity and related phenomena*, Elsevier, p.63.
52. Bussac, M.N. and Zuppiroli, L. (1997) *Phys. Rev. B* **55**, 15587.
53. Schön, J.H., Kloc, C., Haddon, R.C. and Batlogg, B. (2000) *Science* **288**, 656.
54. Schön, J.H., Berg, S., Kloc, C. and Batlogg, B. (2000) *Science* **287**, 1022.
55. Juska, G., Arlauskas, K., Osterbacka, R. and Stubb, H. (2000) *Synth. Metals* **109**, 173.
56. Sirringhaus, H., Brown, P.J., Friend, R.H., Nielsen, M.M., Bechgaard, K., Langeveld-Voss, B.M.W., Spiering, A.J.H., Janssen, R.A.J. and Meijer, E.W. (2000) *Synthetic Metals* **111-112**, 129.

222

57. Schön, J.H., Dodabalapur, A., Bao, Z., Kloc, Ch., Schenker, O. and Batlogg, B. (2001) *Nature* **410**, 189.

58. Malliaras, G.G., Shen, Y., Dunlap, D.H., Murata, H. and Kafafi, Z.H. (2001) *Appl. Phys. Lett* **79**, 2582.

59. Inigo, A.R., Tan, C.H., Fann, W., Huang, Y.-S., Perng, G.-Y. and Chen, S.-A. (2001) *Adv. Mater.* **13**, 504.

60. Sentein, C., Fiorini, C., Lorin, A. and Nunzi, J.M. (1997) *Adv. Mater.* **9**, 809.

61. Murata, H., Malliaras, G.G., Uchida, M., Shen, Y. and Kafafi, Z.H. (2001) *Chem. Phys. Lett* **339**, 161.

62. H. Kageyama, K. Ohnishi, S. Nomura, Y. Shirota, (1997) *Chem. Phys. Lett.* **277**, 137.

63. Turro, J. (1991) *Modern molecular photochemistry*, University Science Books, Mill Valley, Ca.

64. Roncali, J. (1997) *Chem. Rev.* **97**, 173.

65. Mitschke, U. and Bäuerle, P. (2000) *J. Mater. Chem.* **10**, 1471.

66. Martin, R.E., Geneste, F. and Holmes, A.B. (2000) *C. R. Acad. Sci. Paris* 1, (Série IV) 447.

67. « Les Composants Electroniques Organiques », *Etats-Unis Microélectronique* (2001) **24**.

68. http://www.covion.com/

69. http://www.syntec-synthon.com/

70. http://www.adsdyes.com/

71. http://www.hwsands.com/

72. Godovsky, D., Chen, L., Pettersson, L., Inganas, O., Andersson, M.R. and Hummelen, J.C. (2000) The use of combinatorial materials development for polymer solar cells, *Adv. Mater. Opt. Electron.* **10**, 47.

73. Sariciftci, N.S. (1999) Polymeric photovoltaic materials, *Current Opinions in Solid State and Materials Science* **4**, 373.

74. Law, K.Y. (1993) Organic photoconductive materials: recent trends and developments, *Chem. Rev.* **93**, 449.

75. Chamberlain, G.A. (1983) Organic solar cells: a review, *Solar Cells* **8** 47.

76. Sicot, L. (1999) *Etude et réalisation de cellules photovoltaïques en polymère*, PhD thesis, Orsay.

77. Rostalski, J. and Meissner, D. (2000) Monochromatic versus solar efficiencies of organic solar cells, *Sol. Energy Mat. Sol. Cells* **61**, 87.

78. Sze, S.M. (1981) Physics of semiconductor devices, J. Wiley.

79. Ricaud, A. (1997) Photopiles solaires, Presses polytechniques et universitaires romandes.

80. Tsuzuki, T., Shirota, Y., Rostalski, J. and Meissner, D. (2000) The effect of fullerene doping on photoelectric conversion using titanyl phthalocyanine and a perylene pigment, *Sol. Energy Mat. Sol. Cells* **61**, 1.

81. Fromherz, T., Padinger, F., Gebeyehu, D., Brabec, C., Hummelen, J.C. and Sariciftci, N.S. (2000) Comparison of photovoltaic devices containing various blends of polymer and fullerene derivatives, *Sol. Energy Mat. Sol. Cells* **63**, 61.

82. Schmidt-Mende, L., Fechtenkötter, A., Müllen, K., Moons, E., Friend, R.H. and MacKenzie, J.D. (2001) Self-organized discotic liquid crystals for high-efficiency organic photovoltaics, *Science* **293**, 1119.

83. Stuebinger, T. and Bruetting, W. (2001) Exciton diffusion and optical interference in organic donor–acceptor photovoltaic cells, *J. Appl. Phys.* **90**, 3623.

84. Peumans P. and Forrest, S.R. Very-high-efficiency double-heterostructure copper phthalocyanine/C_{60} photovoltaic cells, (2001) *Appl. Phys. Lett.* **79**, 126.

85. Pope, M. and Swenberg, E. (1982) Electronic processes in organic crystals, Clarendon Press, Oxford.

86. Wöhrle, D. and Meissner, D. (1991) *Adv. Mat.* **3**, 129.

223

87. Meissner, D. (1999) Plastic solar cell, *Photon* 2; and also: European organic solar cell initiative, EUREC projects, http://www.eurec.be/htm/projects/euroscilinkshtm.htm
88. Rostalski, J. and Meissner, D. (2000) Photocurrent spectroscopy for the investigation of charge carrier generation and transport mechanisms in organic p/n-junction solar cells, *Sol. Energy Mat. Sol. Cells* 63, 37.
89. Schön, J.H., Kloc, Ch., Bucher, E. and Batlogg, B. (2000) Efficient organic photovoltaic diodes based on doped pentacene, *Nature* 403, 408.
90. Sicot, L., Geffroy, B., Lorin, A., Raimond, P., Sentein, C. and Nunzi, J.M. (2001) Photovoltaic properties of Schottky and p–n type solar cells based on polythiophene, *J. Appl. Phys.* 90, 1047.
91. Pfeiffer, M., Beyer, A., Fritz, T. and Leo, K. (1998) Controlled doping of phthalocyanine layers by cosublimation with acceptor molecules: A systematic Seebeck and conductivity study, *Appl. Phys. Lett.* 73, 3202.
92. O'Regan, B. and Grätzel, M. (1991) *Nature* 353, 737.
93. Sariciftci, N.S., Smilowitz, L., Heeger, A.J. and Wudl, F. (1992) *Science* 258, 1474.
94. Brabec, C.J., Sariciftci, N.S. and Hummelen, J.C. (2001) Plastic solar cells, *Adv. Funct. Mater.* 11, 15.
95. Liu, J., Shi, Y. and Yang, Y. (2001) Solvation-induced morphology effects on the performance of polymer-based photovoltaic devices, *Adv. Funct. Mater.* 11, 420.
96. Eckert, J.-F., Nicoud, J.-F., Nierengarten, J.-F., Liu, S.-G., Echegoyen, L., Barigelletti, F., Armaroli, N., Ouali, L., Krasnikov, V. and Hadziioannou, G. (2000) Fullerene - Oligophenylenevinylene hybrids: Synthesis, electronic properties, and incorporation in photovoltaic devices, *J. Am. Chem. Soc.* 122, 7467.
97. Angeles Herranz, M. and Martin, N. (1999) A new building block for Diels-Alder reactions in p-extended tetrathiafulvalenes: synthesis of a novel electroactive C_{60}-based dyad, *Organ. Lett.* 1, 2005.
98. Neugebauer, H., Brabec, C., Hummelen, J.C. and Sariciftci, N.S. (2000) Stability and photodegradation mechanisms of conjugated polymer/fullerene plastic solar cells, *Sol. Energy Mat. Sol. Cells* 61, 35.
99. Miller, J.S. (2001) Interpenetrating lattices-materials of the future, *Adv. Mater.* 13, 525.
100. Arango, A.C., Johnson, L.R., Bliznyuk, V.N., Schlesinger, Z., Carter, S.A. and Hörhold, H.-H. (2000) Efficient titanium oxide/conjugated polymer photovoltaics for solar energy conversion, *Adv. Mater.* 12, 1689.
101. Peng, X., Manna, L., Yang, W., Wickham, J., Scher, E., Kadavanich, A. and Alivisatos, A.P. (2000) Shape control of CdSe nanocrystals, *Nature* 404, 59.
102. Sentein, C., Fiorini, C., Lorin, A. and Nunzi, J.M. (1997) *Dispositif semiconducteur en polymère comportant au moins une fonction redresseuse et procédé de fabrication d'un tel dispositif*, European Patent.
103. Sentein, C., Fiorini, C., Lorin, A., Sicot, L. and Nunzi, J.-M. (1998) Study of orientation induced molecular rectification in polymer films, *Optical Materials* 9, 316.
104. Sicot, L., Fiorini, C., Lorin, A., Raimond, P., Sentein, C. and Nunzi, J.-M. (2000) Improvement of the photovoltaic properties of polythiophene-based cells, *Sol. Energy Mat. Sol. Cells* 63, 49.
105. Sentein, C., Fiorini, C., Lorin, A., Nunzi, J.-M., Raimond, P. and Sicot, L. (1999) Poling induced improvement of organic-polymer device efficiency, *Synth. Metals* 102, 989-90.
106. Nunzi, J.-M., Sentein, C., Fiorini, C., Lorin, A. and Raimond, P. (2001) Oriented polymer photovoltaic cells, *SPIE Proceedings* 4108, 41.
107. Kim, J.S., Granstrom, M., Friend, R.H., Johansson, N., Salaneck, W.R., Daik, R., Feast, W.J. and Cacialli, F. (1998) *J. Appl. Phys.* 84, 6859.
108. VanSlyke, S.A., Chen, C. H. and Tang, C.W. (1996) *Appl. Phys. Lett.* 69, 2160.
109. O'Brien D.F., Baldo, M. A., Thompson, M.E. and Forrest,S.R. (1999) *Appl. Phys. Lett.* 74, 442.

110. Peumans, P., Bulovic´, V. and Forrest, S.R. (2000) *Appl. Phys. Lett.* **76**, 2650.
111. Young-Eun Kim, Heuk Park, and Jang-Joo Kim, (1996) *Appl. Phys. Lett.* **69**, 599.
112. Appleyard, S.F.J. and Willis, M.R. (1998) *Opt. Mater.* **9**, 120.
113. Appleyard, S.F.J., Day, S.R., Pickford, R.D. and Willis, M.R. (2000) *J. Mater. Chem.* **10**, 169.
114. Sigaud, P. (2001) *PhD Thesis*, Ecole Polytechnique, Palaiseau, France.
115. Gautier, E., Lorin, A., Nunzi, J.-M., Schalchli, A., Benattar, J.J. and Vital, D. (1996) Electrode interface effects on ITO/Polymer/Metal light emitting diodes, *Appl. Phys. Lett.* **69**, 1071.
116. Goetzberger, A. and Hebling, C. (2000) Photovoltaic materials, past, present, future, *Sol. Energy Mat. Sol. Cells* **62**, 1.

PHOTOVOLTAIC APPLICATIONS

M. PALFY
Solart-System Ltd.
20 Gulyas Street BudapestXI. Hungary 1112
E-mail: solartsy@elender.hu; www.elender.hu/~solartsy

1. Introduction

When the silicon solar cell appeared on the scene, it was expected to find a large market in terrestrial applications. Workers at Bell Telephone Laboratories originally foresaw a splendid future for the device as a terrestrial solar energy converter, and they sought to prove this capability by installing a solar cell array on a telephone pole in Georgia, to power a repeater amplifier. This array fulfilled its function satisfactorily for over a year, with bird droppings being the only problem encountered. However, based on original installation costs, the cost of the energy generated was not competitive with conventional power.

In early planning for the space program, decision makers apparently considered solar cells to be an insignificant and unreliable curiosity, and decided to equip US spacecraft with chemical batteries which were able to provide power for only a few weeks. Eventually, the Vanguard I satellite was the first to be equipped with solar cells, in 1958 [1].

There is no longer any question concerning the use of solar cells in spacecraft, and in many different terrestrial applications.

2. General Survey

The world photovoltaic (PV) module production is now more than 280 MegaWatt peak (MWp) per year. The annual average growth in world production is about 20%, but Japan had 80% of this growth from 1997 to 1998 [2].

The annual world photovoltaic module production can be seen in Figure 1, and world photovoltaic cell and module shipments for 1998 are presented in Figure 2. The cumulative values of world photovoltaic cell and module production are shown in the Figure 3. These values are very close to the total photovoltaic capacity installed.

The average price of photovoltaic modules is below 4$/Wp, but amorphous silicon modules in higher quantities (some ten kWp or higher) can be purchased at a price of about 2.5$/Wp.

J.M. Marshall and D. Dimova-Malinovska (eds.),
Photovoltaic and Photoactive Materials - Properties, Technology and Applications, 225–238.

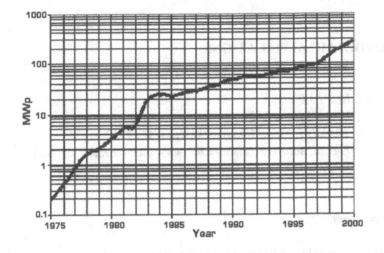

Figure 1. Annual world photovoltaic module production.

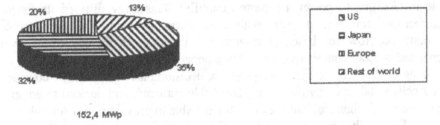

Figure 2. World photovoltaic cell and module shipments in 1998.

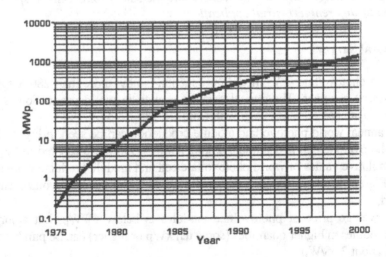

Figure 3. Cumulative values of the world photovoltaic cell and module production.

The market sectors are very different. There have been rapid growths in consumer product applications, and in grid-connected systems. However, few applications are associated with so-called traditional electricity production, i.e. central power generation. The world photovoltaic module shipments are demonstrated, by 1998 market sectors, in the Figure 4.

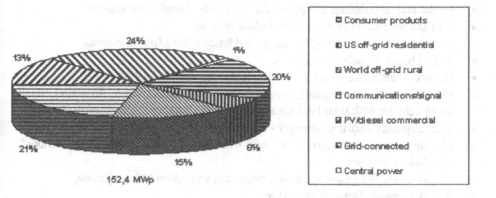

□ Consumer products

▣ US off-grid residential

▣ World off-grid rural

▤ Communications/signal

▨ PV/diesel commercial

▨ Grid-connected

□ Central power

Figure 4. World photovoltaic module shipments, by market sector, in 1998.

3. Application Fields

The application fields of photovoltaic solar cells and modules are now very wide ranging. Without any sequence of importance, the main fields of the application are as follows [3]:

- Remote areas: Off-grid residential, rural, holiday and recreation houses, ranches, tourist facilities. *Stand alone systems.*
- Remote areas: Off-grid settlements. *Local grid systems.*
- Communications: Microwave repeaters, TV relay stations, rural telephony systems, public telephones, wireless telephones, broadcasting, motorway emergency telephony systems. *Stand alone systems.*
- Water supply: Drinking water and water supply for daily use. Irrigation. Drainage. *Stand alone systems.*
- Cathodic corrosion protection: Gas, oil and water pipe lines, and other underground objects. *Stand alone systems.*
- Grid-connected systems: Central power (power stations) above 100 kWp.
- Grid-connected systems: Local decentralized systems, 1 kWp – 1 MWp. Interactive grid-connection. (excess energy demand is supplied *by* the grid; surplus energy is supplied *to* the grid).
- Consumer products: Calculators, clocks, watches, toys, art objects, mobiles, radio and TV sets, electronics goods, garden lights, street number boards, car battery conditioners, PR goods (key cases, ball point pens, wrist watches, mobile and/or lighting goods etc.) *Stand alone systems.*

- Security systems: Protection of remote objects. Protection of residential, rural, holiday and recreation houses, ranches and tourist facilities. *Stand alone systems.*
- Ventilation. Off work cars, and buses. Cockpits of other vehicles, engines. Caravans, measuring vehicles, stations. Incubators, breeders. Farm products, fruit and grain dryers. *Stand alone systems.*
- Electric fences. Wild and farm produce protection in agriculture, forestry and zoos. *Stand alone systems*
- Feeder and air bubbling equipment for fish ponds. *Stand alone systems.*
- Bird and wild alarm systems. *Stand alone systems.*
- Navigational lights: Port lights, buoys and bacons. *Stand alone systems.*
- Sailing boats. *Stand alone systems.*
- Gliders. *Stand alone systems.*
- Electric vehicles: Racing cars. (yearly races in Switzerland, USA, Australia). Small urban vehicles with solar back up and/or solar "fuel stations". *Stand alone systems.*
- Solar airplanes. (Solar Challenger - 3 kWp) *Stand alone systems.*
- Electrically powered boats: Boats with solar back up and/or solar "fuel stations". *Stand alone systems.*
- Caravans, dwelling vans and boats, measuring vans. *Stand alone systems.*
- Railway signals. *Stand alone systems.*
- Road signals. *Stand alone systems.*
- Public lights. *Stand alone systems.*
- Do-it-yourself kits. *Stand alone systems.*
- Measuring stations: Meteorological measuring and monitoring systems. Soil moisture, rain gauge, temperature measurement and telecommunication systems in agriculture. Water level measurement and telecommunication systems. Level indicators in large storage tanks. *Stand alone systems.*
- Drinking water facilities: Chlorinating and osmotic desalination equipment. *Stand alone systems.*
- Solar thermal facilities: Fluid circulation in collectors, swimming pools. *Stand alone systems.*
- Air conditioning. *Stand alone systems.*
- Decorative facades and shading, roofs. *Stand alone or grid-connected systems.*
- Hydrogen production: Energy storage, technological applications and local welding. *Stand alone systems.*
- Refrigeration: Stable and mobile systems. (camel, donkey and horseback, vaccine cooling) *Stand alone systems.*
- Mobile power stations: Commissioning, maintenance in roads, remote places, deserts. Catastrophe and emergency power stations. *Stand alone systems.*
- Sensors: Visible/infra red radiation, dimmers, auxiliary flash lamps, light barriers, electrical arcs, isolation, district heating, fire alarms etc. *Cells.*
- Military equipment: Sensors, measuring equipment, telecommunications, personal facilities. *Stand alone systems.*

The preceding list is almost certainly incomplete, but the wide range of photovoltaic applications is hopefully clear.

4. Photovoltaic Equipment

Photovoltaic equipment generally has the following main parts:

- Solar module(s)
- Battery(s)
- Charge controller(s)
- Inverter(s)
- Junction box(es)
- Supporting mechanical structure

Other than the solar modules, the remaining main parts are summarized as the balance of system (BOS). The general layout of a photovoltaic system is demonstrated in Figure 5.

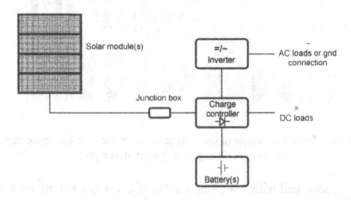

Figure 5. General layout of a photovoltaic system.

If the electricity generated by solar modules is consumed immediately, there is no need for batteries. Grid-connected systems also generally have no batteries, because the electricity is fed directly to the local grid. Also, in many cases, there are only DC loads, so that there is no need for an inverter (i.e. an electronic device which converts DC voltages to AC ones).

The solar modules are fixed on mechanical structures capable of withstanding harsh environmental conditions. In the northern hemisphere, solar modules face to the south, or close to it. The tilt angle is generally set to an optimal value, although a seasonal setting would in some cases be advantageous. On vehicles, a horizontal setting is preferable. To generate additional electricity, sun trackers can be used, but this is an economical consideration.

Solar modules can be connected in series and/or in parallel, in the junction box. Lightning arresters or overvoltage protection can also be located here.

The charge controller is an electronic device that controls the charge state of the batteries. It generally has a temperature sensor for monitoring the battery temperature. Thus, it is practical to locate it in the battery room.

5. Dimensioning

As seen in Sections 2 and 3, most applications are in stand-alone systems. From the point of view of dimensioning, the most important parts of stand-alone systems or autonomous power supplies are the solar modules and the batteries. For dimensioning, it is practical to demonstrate the local solar irradiance data in cumulative form, as in Figure 6.

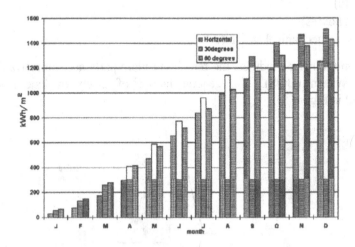

Figure 6. Monthly variations of the average values of the solar irradiance energy, at different tilt angles, in Budapest, Hungary.

Consider a solar cell with a working current of 0.8 A at 1 kW/m² solar irradiance level. The short-circuit current of the cell is proportional to the irradiance. We can also make a good estimate if we take the working current to be proportional to the irradiance. From this data, the charge that the cell produces over the year can be calculated, as in Figure 7.

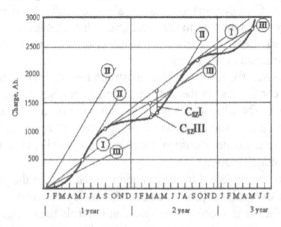

Figure 7. The cumulative charge delivered by a solar cell (working current: 0.8 A at 1 kW/m²) produced at a 30 degree tilt angle in Budapest, Hungary, over several years.

This solar cell can produce an average charge of 1200 Ah, at a tilt angle of 30°, in Budapest, over one year. In the ideal case (no loss) and with a continuous demand, the daily load can be 0.14 A (3.25 Ah/day). This case is represented by line **I**. If we are going to load continuously at 0.14 A, then a storage device - i.e. a battery - is needed. If this is fully charged on 1st January of Year 1, then the demand surplus (above the solar cell production) is covered by the battery. It will only become fully re-charged at the end of May. From June to October, the surplus solar charge will be lost, or the load can be raised to 0.23A (5.43 Ah/day), as represented by line **II**. From October, the demand surplus is once more covered by the battery, which will only become fully re-charged in the next September. The most rapid discharge rate for the battery is in March, and its value is $C_{szI} = 300$ Ah. This value is the minimum needed for seasonal storage.

In normal (non-ideal) practical circumstances, the load can be approximately 70% of that for the ideal case, i.e. 0.1 A (2.27 Ah/day), as indicated by line **III**. The minimal required battery capacity is then $C_{szIII} = 200$ Ah.

If the required current load is larger then that which the solar cell can deliver, then it becomes necessary to connect cells in parallel. The number of cells required, n_p, is:

$$n_p = \frac{Q_L}{Q_M} \tag{1}$$

where Q_l is the load, and the charge of the solar cell is Q_m. One can take 70% charge losses from the cell into consideration, as mentioned above.

The number of solar cells to be connected in series is determined by the voltage level demanded. In crystalline silicon solar modules, 30-36 cells must be connected in series to yield a nominal 12 V D.C. For higher voltages, the number of series-mounted modules must increase proportionately.

In calculating the required battery capacity, it is practical to take into account the local conditions of temperature and cloud cover. The required battery capacity, C, can be expressed as:

$$C = (C_{sz} + C_A) K \tag{2}$$

where C_{sz} is the minimal seasonal storage capacity. The autonomy capacity is $C_A = D \times Q_l$, where D is the number of cloudy days per month, Q_l is the load, and K is a correction factor for temperature (K = 1 above 0°C, and 3 at -10°C).

A suggestion for the number of cloudy days per month is made in Table 1.

TABLE 1. Suggested number of cloudy days per month.

Latitude [L°]	Cloudy days [D]
0° - 30°	5-10
30° - 40°	10-15
40° - 50°	15-20
50° - 60°	20-25
> 60°	25

The number of batteries to be connected in series is determined by the required voltage level. The nominal voltage of a lead-acid cell is 2 V, while that of a NiCd cell is 1.2 V. It is common to connect 6 cells in series in a lead acid battery. For higher voltages, the number of series components must increase appropriately.

6. Autonomous Power Supplies

Stand-alone systems or autonomous power supplies are generally employed where no electrical grid is available. The extension of the existing grid and the electricity supplied can both be costed, allowing the real cost of the electricity to be calculated. The choice of an autonomous power supply or a grid extension is generally an economical decision. Such a cost analysis is presented in Figure 8, in which the electricity demand is plotted against the required length of the grid extension. The curve is very close to linear. If the crossing point falls below the curve, then a solar autonomous power supply is more economical. For higher demands, a grid extension is indicated.

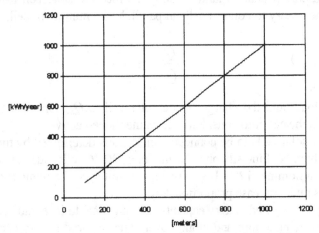

Figure 8. Break-even point for extension of an existing grid or use of an autonomous power supply, in Hungary.

The following data have been taken into consideration in the calculation:
- 20000 Euro/km grid extension cost
- Euro/kWh electricity tariff
- 4 Euro/Wp solar module cost
- 2.5 Euro/Wp battery cost
- 1Euro/Wp electronic parts cost
- Euro/Wp inverter cost
- 1.5 Euro/Wp mechanical structure cost
- 30 years life span
- 1300 kWh/m^2 yearly solar irradiance on a horizontal surface

The data, except the solar irradiance, are very similar throughout Europe. The higher the local yearly irradiance, the steeper the curves in Figure 8.

Various solar autonomous power supplies installed in Hungary will now be illustrated.

The first Hungarian solar autonomous power supply was installed in Transdanubia in 1975.

Figure 9. Wireless telephony transmitter station

Technical data:

Application: Wireless telephony system
Solar cells: VKI pilot cells in 6 arrays; 24 modules per array
Total solar array area: 6 m², on roof
Nominal power: 200 Wp
Batteries: Lead-acid, 24 V / 1000 Ah in the cellar
Alignment: To south, 50° fixed tilt angle
Nominal voltage: 24V DC

Figure 10. Telecommunications autonomous power supply

Technical data:

Application: Telecommunications system
Solar cells: VKI pilot cells in 4 x SM 12/1 modules
Total solar module area: 1 m²
Nominal power: 64 Wp
Battery: Lead-acid Gel 4 x 240Ah/6V in box
Alignment: Optional, tilt angle: 40-60°, changeable
Nominal voltage: 12/24V DC

Figure 11. Solar power supplies for telecommunications and water pumping

Technical data:

Application: Microwave repeater and water pumping
Solar cells: VKI SC 3 cells in 48 x SM 12/1 modules
Total solar module area: 11.3 m²
Nominal power: 768 Wp
Battery: Lead-acid Gel 12 x 1200Ah/2V
Alignment: Optional, 60° fixed tilt angle
Nominal voltage: 24/96V DC

Figure 12. Solar power supplies for cathodic corrosion protection of pipelines

Technical data:

Application: Cathodic corrosion protection
Solar cells: VKI SC 3 cells in 30 x SM 6/1 modules
Total solar module area: 3.75 m^2
Nominal power: 240 Wp
Battery: Lead-acid Gel 3 x 1600Ah/2V in box
Alignment: Optional, tilt angle: 10-80°
Nominal voltage: 6V DC

The first rural telephony system was installed in Transdanubia in 1992.

Figure 13. Solar power supply for a rural telephony system.

Technical data:

Application: Rural telephony system
Solar cells: Pannonglas 12 x SM 12/2.5 modules
Total solar module area: 5.3 m^2
Nominal power: 490 Wp
Battery: Lead-acid Gel 8 x 150Ah/12V in container
Alignment: To south, 60° fixed tilt angle
Nominal voltage: 24V DC

Figure 14. Solar burglar alarm system.

Technical data:

Application: Burglar alarm
Solar cells: 1 x SA2-12 amorphous silicon module, wall mounted
Nominal power: 2.5 Wp
Battery: Lead-acid Gel 7 Ah/12V, in the house
Alignment: To NE, at a fixed 90° tilt angle
Nominal voltage: 12V DC

Figure 15. Solar burglar alarm and signaling system.

Technical data:

Application: Burglar alarm and signaling
Solar cells: 1 x VLX 32 silicon module mounted on pole.
Nominal power: 32 Wp
Battery: Solar lead-acid 100 Ah/12V in the house
Alignment: To south, at a fixed 65° tilt angle
Nominal voltage: 12V DC

Figure 16. Solar education units, technical data as below.

Left Hand Unit: (Multipurpose Educational Solar Power Supply - MESPS)
Application: Solar PV education;
Solar cells: 4xSM2160 module
Total solar module area: 2 m²
Nominal power: 240 Wp
Battery: Lead-acid Gel 12x300 Ah/2V in box;
Alignment: Optional, tilt angle: 15-60°
Nominal voltage: 24V DC

Centre (Rear) Unit:
Application: Solar drying and PV education;
Total solar module area: 0.4 m²;
Alignment: Optional, 40° fixed tilt angle

Solar cells: 2xS20 modules
Nominal power: 40 Wp
Nominal voltage: 12V DC

Right Hand Unit: (Multipurpose Educational Solar Thermal Unit - MESTU)
Application: Education, Solar thermal collector. Alignment: Optional, tilt angle: 15-60°

Various multipurpose educational systems produced by Solart-System Ltd. will now be described in more detail.

6.1. MULTIPURPOSE EDUCATIONAL SOLAR POWER SUPPLY

The Multipurpose Educational Solar Power Supply (MESPS) (Figure 16 (left-hand unit) and Figure 17), is an excellent tool for studying the electrical utilization of solar energy at different latitudes. It is both a simple study aid device, and a solar power supply for other devices like radios, TV sets, videos and PCs. It is also suitable for lighting, water pumping, irrigation etc.

236

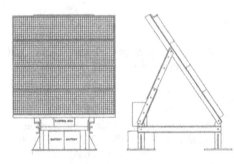

Figure 17. Multipurpose solar power supply. Technical data (see also Figure 16):

Load voltage: 22.2-27.4 V; Maximum load current: 10 A; Timed load: 6 programs per week; Peak power output: 240 W (4 x 60 W); RS232 PC interface. Options: DC-AC inverters, Monitoring systems, DC or AC instruments; For additional electricity, the modular construction allows multiple units to be formed.

The approx. 2 m² solar modules are mounted on an adjustable aluminium support, with holes for setting the tilt angle. The support has two stay plates, suitable for ground-mounting the MESPS. The control box, consisting of the control electronics, the program switch, the junctions and the measuring terminals, is normally fixed on the rear side. The batteries lie on the rear stay plate, providing a heavy load for fixing the equipment. Both the control box and the batteries may also be located separately.

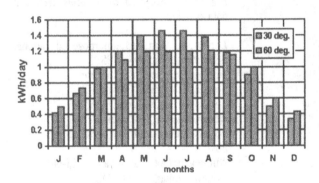

Figure 18. MESPS solar gains expected at different tilt angles, facing due south in Budapest.

6.2. MULTI-PURPOSE EDUCATIONAL SOLAR THERMAL UNIT

The Multi-purpose Educational Solar Thermal Unit (MESTU) (Figure 16, right-hand unit, and Figure 19 below) is an excellent tool for studying the thermal utilization of solar energy at different latitudes. It is not only a simply study aid device, but also a solar hot water production system for schools, washing, etc.

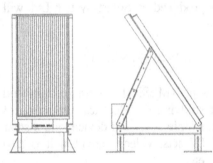

Figure 19. Multipurpose educational solar thermal unit. Main technical data:
Maximum operating pressure: 3 bar; Flow rate: 70-90 litres/hr.; Storage tank: 150 litres; Nominal voltage: 230 V AC, 50 Hz; Auxiliary heating: Electrical, 230 V, 1.8 kW.
Options: DC control electronics, DC circulating pump, AC/DC water pump, Monitoring systems, DC and AC instruments, Meteorological measuring and monitoring station. The modular construction allows multiple units to be employed, for more hot water.

The ~ 2 m^2 solar collector is mounted on an adjustable aluminum support, with holes for setting the tilt angle. Details are similar to those for the MESPS (Section 6.1). The Control Box, consisting of the control electronics, junctions and measuring terminals is normally fixed on the rear side. The hot water storage tank, circulating pump and pressure equalizer are located separately.

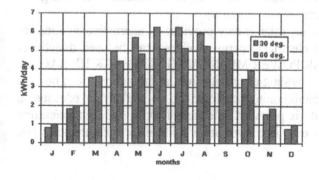

Figure 20. Solar gains expected at different tilt angles, facing due south in Budapest.

6.3. SOLAR POWER MEASURING AND BACK UP BAG

Figure 21. Solar Power Measuring and Backup Bag. Main technical data:

Energy converter: silicon solar cells (can be set up separately; tilt angle adjustable);

Maximum output power: 100 W; Energy storage: battery; Nominal Voltages: 4.5-9, 12 and 24VDC, 220VAC 50Hz; Charge control: electronic; Measuring ranges: 0.1mV-1000VDC; 0.1mV-750VAC; 0.1Ω-40MΩ; 0.1μA-20ADC; 0.1μA-20AAC; 0.1nF-400μF; 10μH-400mH, 4MHz; - 40°C-1200°C.

C-MOS output: 1-10-100kHz; dual LCD display (analog-digital); max./min. storage; true RMS; RS 232 PC interface; PC power socket, 6 x ⌀ 2.1 mm, 4.5-9, 12 and 24 (+/-)VDC.

Dimensions: 460x335x150 mm; mass 10 kg; portable.

For on-site measurements and commissioning, electricity is generally needed. In many cases the energy demand is not high, because only small items of equipment are needed, e.g. measuring instruments, PCs (laptops, notebooks), date acquisition systems etc. However, the required forms of electricity differ. The equipment is often situated in the open air, and far from the public electricity network.

In educational studies of renewable energy, field tests and other demonstrations, electricity is similarly needed in most cases. For various measurements, electrical instruments and electronic devices for data processing, telecommunication, etc. are needed. When such field tests are conducted far from the public electricity network, the electricity is generally provided by small a power station, e.g. batteries or a petrol/diesel generator. However, the transportation and continuous operation of these small power stations can be difficult. Portable measuring instruments such as data acquisition systems and PCs generally have their own energy sources, but their operation is limited. Thus, for longer tests, it becomes necessary to change or recharging these energy sources.

For field tests in the open air, solar energy provides an alternative solution for electricity requirements. The electrical energy produced can be used directly or stored in batteries. The SOLAR POWER MEASURING AND BACK UP BAG is a small portable power station providing electricity from solar energy. It can be used in field tests or system commissioning/development, as well as in educational demonstrations. The electricity is provided in different forms and voltage/current ranges. Furthermore it has is own instrumentation suitable for measurements and for interfacing with a PC.

6.4. SOLAR EDUCATIONAL MODULE

The Solar Educational Module SEM is a complex tool for the study of solar energy conversion. The SEM combines the MESPS (Section 6.1), the MESTU (Section 6.2) and the SPMBB (Section 6.3). Taking into consideration the conditions and demands of education in secondary and high schools, we have adopted the following main design and construction criteria:

- uniform modular structure
- easy mounting, for single and multiple units
- different fixtures, for outdoor applications
- suitability for both measurement and monitoring
- safe, accident free structure
- easy operation
- professional solar technology look
- outdoor construction, with a minimum life span of 10 years
- durability, for frequent adjustment and measurement

The SPMBB, MESTU and MESPS also have operating and technical descriptions, and study guides. Thus, addition of teaching materials yields a complete system.

Options: DC-AC inverters; Monitoring systems; Weather monitoring; Alternative software (e.g. Sunarch); Other DC and AC instruments.

7. References

1. Wolf, M. (1972) Historical development of solar cells, *Proc. 25th Power Sources Symp.* 120-124.
2. Maycock, P. (1999) Photovoltaic technology, *Renewable Energy World* 2 No. 5, 73-76.
3. Pálfy, M. (1999) Photovoltaic power supplies I-IV *Elektrotechnika* 92, 215-218, 324-326, 395-399, 423-426.

PAST, ACTUAL AND FUTURE EU-FUNDED RESEARCH, TECHNOLOGICAL DEVELOPMENT AND DEMONSTRATION ACTIONS IN THE FIELD OF PHOTOVOLTAICS

THIERRY LANGLOIS d'ESTAINTOT
European Commission
B-1049 Brussels, Belgium

Abstract

After a brief review of the European Union institutional structure, the Research Directorate and its Fifth Framework programme are described. A focus on renewable energies is then given, with particular attention to photovoltaics. R&D activities in PV have been openly supported since 1980, and the Sixth Framework Programme (2002-2006) is expected to continue this trend.

1. Institutions of the European Union

The European Union is built on an institutional system which is the only one of its kind in the world.

The Member States delegate sovereignty for certain matters to independent institutions, which represent the interests of the Union as a whole, its member countries and its citizens. The Commission traditionally upholds the interests of the Union as a whole, while each national government is represented within the Council, and the European Parliament is directly elected by citizens. Democracy and the rule of law are therefore the cornerstones of the structure.

This "institutional triangle" is flanked by two other institutions: the Court of Justice and the Court of Auditors. A further five bodies make the system complete.

1.1. THE EUROPEAN PARLIAMENT

Elected every five years by direct universal suffrage, the European Parliament is the expression of the democratic will of the Union's 374 million citizens. Brought together within pan-European political groups, the major political parties operating in the Member States are represented.

Parliament has three essential functions:

It shares with the Council the power to legislate, i.e. to adopt European laws (directives, regulations, and decisions). Its involvement in the legislative process helps to guarantee the democratic legitimacy of the texts adopted.

J.M. Marshall and D. Dimova-Malinovska (eds.),
Photovoltaic and Photoactive Materials - Properties, Technology and Applications, 239–252.
© 2002 *Kluwer Academic Publishers.*

It shares budgetary authority with the Council, and can therefore influence EU spending. At the end of the procedure, it adopts the budget in its entirety.

It exercises democratic supervision over the Commission. It approves the nomination of Commissioners, and has the right to censure the Commission. It also exercises political supervision over all the institutions.

1.2. THE COUNCIL OF THE EUROPEAN UNION

The Council is the EU's main decision-making body. It is the embodiment of the Member States, the representatives of which it brings together regularly at ministerial level.

According to the matters on the agenda, the Council meets in different compositions: foreign affairs, finance, education, telecommunications, etc.

The Council has a number of key responsibilities:

It is the Union's legislative body. For a wide range of EU issues, it exercises that legislative power in co-decision with the European Parliament;

It co-ordinates the broad economic policies of the Member States;

It concludes, on behalf of the EU, international agreements with one or more States or international organisations;

It shares budgetary authority with Parliament;

It takes the decisions necessary for framing and implementing the common foreign and security policy, on the basis of general guidelines established by the European Council;

It co-ordinates the activities of Member States and adopts measures in the field of police and judicial co-operation in criminal matters.

1.3. THE EUROPEAN COMMISSION

The European Commission embodies and upholds the general interest of the Union. The President and Members of the Commission are appointed by the Member States after they have been approved by the European Parliament.

The Commission is the driving force in the Union's institutional system:

It has the right to initiate draft legislation and therefore presents legislative proposals to Parliament and the Council;

As the Union's executive body, it is responsible for implementing the European legislation (directives, regulations, and decisions), budget and programmes adopted by Parliament and the Council;

It acts as guardian of the Treaties and, together with the Court of Justice, ensures that Community law is properly applied;

It represents the Union on the international stage and negotiates international agreements, chiefly in the field of trade and co-operation.

1.4. THE COURT OF JUSTICE

The Court of Justice ensures that Community law is uniformly interpreted and effectively applied. It has jurisdiction in disputes involving Member States, EU institutions, businesses and individuals. A Court of First Instance has been attached to it since 1989.

1.5. THE COURT OF AUDITORS

The Court of Auditors checks that all the Union's revenue has been received and all its expenditure incurred in a lawful and regular manner, and that financial management of the EU budget has been sound.

1.6. THE EUROPEAN CENTRAL BANK

The European Central Bank frames and implements European monetary policy; it conducts foreign exchange operations and ensures the smooth operation of payment systems.

1.7. THE ECONOMIC AND SOCIAL COMMITTEE

The European Economic and Social Committee represents the views and interests of organised civil society vis-à-vis the Commission, the Council and the European Parliament. The Committee has to be consulted on matters relating to economic and social policy. It may also issue opinions on its own initiative on other matters, which it considers to be important.

1.8. THE COMMITTEE OF THE REGIONS

The Committee of the Regions ensures that regional and local identities and prerogatives are respected. It has to be consulted on matters concerning regional policy, the environment and education. It is composed of representatives of regional and local authorities.

1.9. THE EUROPEAN INVESTMENT BANK

The European Investment Bank (EIB) is the European Union's financial institution. It finances investment projects, which contribute to the balanced development of the Union.

1.10. THE EUROPEAN OMBUDSMAN

All individuals or entities (institutions or businesses) resident in the Union can apply to the European Ombudsman if they consider that they have been harmed by an act of "maladministration" by an EU institution or body.

2. List of the European Commission's Directorates-General and Services

GENERALSERVICES
- European Anti-Fraud Office
- Eurostat
- Press and Communication Publications Office
- Secretariat General

POLICIES
- Agriculture
- Competition
- Economic and Financial Affairs
- Education and Culture
- Employment and Social Affairs
- Energy and Transport
- Enterprise
- Environment
- Fisheries
- Health and Consumer Protection
- Information Society
- Internal Market
- Joint Research Centre
- Justice and Home Affairs
- Regional Policy
- **Research**
- Taxation and Customs Union

EXTERNALRELATIONS
- Development
- Enlargement
- EuropeAid - Co-operation Office External Relations
- Humanitarian Aid Office - ECHO
- Trade

INTERNALSERVICES
- Budget
- Financial Control
- Group of Policy Advisers
- Internal Audit Service
- Joint Interpreting and Conference Service
- Legal Service
- Personnel and Administration
- Translation Service

3. Directorate-General Research

The Directorate General's mission is evolving as work on the European Research Area (ERA) continues. It can be summarised as follows:

- to develop the European Union's policy in the field of research and technological development and thereby contribute to the international competitiveness of European industry;
- to co-ordinate European research activities with those carried out at the level of the Member States;
- to support the Union's policies in other fields such as environment, health, energy, regional development etc.;
- to promote a better understanding of the role of science in modern societies and stimulate a public debate about research-related issues at European level.

One of the instruments used for the implementation of this policy is the multi-annual Framework Programme that helps to organise and financially support co-operation between universities, research centres and industries - including small and medium sized enterprises. The current Fifth Framework Programme covers the period 1998 - 2002 and has a total budget of close to € 15 billion.

In carrying out the various tasks, the Directorate General works closely with other Commission departments such as the Joint Research Centre, which also falls under the responsibility of Commissioner Busquin, the Directorates General for the Information Society, Energy and Transport, the Environment, Enterprise, etc.

4. Fifth Framework Programme of the European Community for Research, Technological Development and Demonstration Activities (1998 - 2002)

4.1. INTRODUCTION

The Fifth Framework Programme (FP5) sets out the priorities for the European Union's research, technological development and demonstration (RTD) activities for the period 1998-2002. These priorities have been identified on the basis of a set of common criteria reflecting the major concerns of increasing industrial competitiveness and the quality of life for European citizens.

The Fifth RTD Framework Programme differs considerably from its predecessors. It has been conceived to help solve problems and to respond to major socio-economic challenges facing Europe. To maximise its impact, it focuses on a limited number of research areas combining technological, industrial, economic, social and cultural aspects. Management procedures will be streamlined with an emphasis on simplifying procedures and systematically involving key players in research.

A major innovation of the Fifth Framework Programme is the concept of "Key actions". Implemented within each of the four thematic programmes. "Key actions" will mobilise the wide range of scientific and technological disciplines - both fundamental and applied - required to address a specific problem so as to overcome barriers that may exist, not only between disciplines but also between the programmes and the organisations concerned.

The Fifth Framework Programme has two distinct parts: the European Community (EC) framework programme covering research, technological development and demonstration activities; and the Euratom framework programme covering research and training activities in the nuclear sector.

A budget of 14 960 million euros has been agreed for the period up to the year 2002. Of this, 13 700 million euros is foreseen for the implementation of the European Community section of Fifth Framework Programme, and 1 260 million euros have been allocated to the Euratom programme.

4.2 SCHEMATIC OVERVIEW

FIRST ACTIVITY			
Quality of life and management of living resources EUR 2 413 m	User-friendly information society EUR 3 600 m	Competitive and sustainable growth EUR 2 705 m	Energy, environment and sustainable development EC - EUR 2 125 m EURATOM - EUR 979 m
Food, nutrition and health Control of infectious diseases The "cell factory" Environment and health Sustainable agriculture, fisheries and forestry and integrated development of rural areas including mountain areas The ageing population and disabilities	Systems and services for the citizen New methods of work and electronic commerce Multimedia content and tools Essential technologies and infrastructures	Innovative products, processes, organisation Sustainable mobility and intermodality Land transport and marine technologies New perspectives for aeronautics	Sustainable management and quality of water Global change, climate and biodiversity Sustainable marine ecosystems The city of tomorrow and cultural heritage Cleaner energy systems, including renewables Economic and efficient energy for a competitive Europe Controlled thermonuclear fusion and nuclear fission
Research and technological development activities of a generic nature Support for research infrastructures			
SECOND ACTIVITY			
Confirming the international role of Community research (EUR 475 m)			
THIRD ACTIVITY			
Promotion of innovation and encouragement of participation of SMEs (EUR 363 m)			
FOURTH ACTIVITY			
Improving human research potential and the socio-economic knowledge base (EUR 1 280 m)			
Improving the socio-economic knowledge base (EUR 165 m)			
Joint Research Centre			
EUR 1 020 m (EC and Euratom)			

5. The EU and Energy Research

5.1. WHY ENERGY RESEARCH?

Energy research is essential to support the development of a more sustainable energy policy. It ranges from the nurturing of new, cleaner production technologies based on different energy sources to the demonstration of innovative energy conservation techniques.

The challenge for energy research is to reconcile conflicting pressures. These are a growing energy demand and the need to reduce the environmental impact of energy production and use while, at the same time, seeking to improve:

- security and diversity of supply
- economic competitiveness, and
- social benefit.

Energy is vital to modern society. It is a key component of citizens' quality of life and an important factor in economic competitiveness and employment. However, there is a price to pay. Energy production and use can, in some cases, degrade the environment, and the growing recognition of this has catapulted energy policy to the top of the political agenda worldwide.

The key objectives for energy research are:

- promoting sustainable development: with substantial research, technological development (RTD) and demonstration in the energy field;
- ensuring security and diversity of energy supply: by making use of the knowledge and technologies developed by energy research programmes;
- improving industrial competitiveness: by helping to reduce energy demand and costs and generating technical innovations;
- enhancing economic and social cohesion: via advances in understanding how new energy technologies affect and are taken up by society, and how they lead to improved societal well-being, such as the creation of new jobs or the supply of energy to remote communities.

5.2. NON-NUCLEAR ENERGY RESEARCH

The European Union (EU) is committed to realising a European Research Area (ERA) in the field of energy research. In this context, the European Commission (EC) both develops non-nuclear energy research policy and implements the EU's multi-annual research Framework Programmes.

Non-nuclear energy (NNE) research addresses the EU policy objectives of achieving more sustainable energy systems and services. It is of great strategic importance to the EU in both the short and the longer term.

5.2.1. *Research Topics*

The scope of non-nuclear energy research is very wide, covering the:
- development of cleaner energy systems, including renewable energies,
- economical and efficient use of energy, and
- socio-economic aspects of energy.

5.2.2. *Energy Production from Renewable Sources*

Renewable energies can help diversify energy supply with little adverse environmental impact. Renewable energy sources tap naturally occurring energy flows to produce electricity, heat and fuel. Such resources are often produced on a stand-alone use close to their point of consumption.

These renewable sources are:
- Wind - harnessing the naturally occurring energy of the wind to generate electricity, both onshore and offshore.
- Solar thermal - concentrating the energy of the sun to generate electricity or provide hot water.
- Photovoltaics - using semi-conductor materials to capture the energy in sunlight and to convert it directly into electricity.
- Biomass - converting organic matter such as wood, plants and agricultural waste to provide heat, produce fuel and generate electricity.
- Geothermal - using steam and hot water generated by heat from the earth's core to produce electricity and provide heating.
- Water - exploiting the power of falling water (hydropower) and that from the sea (wave and tidal power) to generate electricity.
- Integration of renewable energy sources and distributed generation - overcoming both the technical and non-technical problems associated with integrating new distributed energy resources into energy systems.

6. The EU and PV Technology - Key Questions and Answers about PV

6.1. WHAT ARE THE KEY ADVANTAGES OF PV TECHNOLOGY?

PV cells are modular and light, have no moving parts, have no direct impact on the environment and require only minimal maintenance. In theory, they are available anywhere on the planet, can be installed and operated in areas of difficult access and are an easy route to a power source for the developing world. Furthermore, they have a long life and durability as well as low operating costs.

6.2. WHAT IS THE FUTURE POTENTIAL OF PV TECHNOLOGY?

Photovoltaic electricity generation has a low energy density per hectare, and its efficiency does not depend on the size of a photovoltaic plant, making it suitable for a wide range of applications. Namely, photovoltaic systems can be used in small, de-centralised plants as well as in larger, central power plants; they can be built in sizes from cm^2 up to km^2, and can be deployed in many locations.

6.3. WHAT DOES IT OFFER THE EU SPECIFICALLY?

In the long-term, the EU objective is a sustainable energy supply based on Renewable Energy Sources (RES). PV could contribute substantially to the total EU energy consumption and can be easily integrated into current electricity systems.

6.4. HOW DOES IT CONTRIBUTE TO THE ACHIEVEMENT OF EU OBJECTIVES?

In the EU White Paper on Renewable Energy Sources, photovoltaic energy forms part of an EU action in order to double the share of RES from 6% in 1998 to 12% in 2010. In this paper, it is estimated that by 2010 the installed PV capacity will have increased to 3GWp from today's 250MWp.

The EU initiated a "Campaign for Take-Off" for renewable energy sources including the construction of 500,000 grid-connected, roof and façade PV systems and an export initiative for 500,000 PV village systems, to kick-start decentralised electrification in developing countries.

7. A Look into the Future

7.1. WHAT FUTURE APPLICATIONS OF PV ARE ANTICIPATED?

As PV energy is area dependent (providing approximately 1GWhr/hectare/year), it would fit well into a concept of decentralised electricity production.

7.2. WHERE MIGHT WE BE WITH THIS TECHNOLOGY IN THE SHORT, MEDIUM AND LONG TERMS?

A new concept for local energy production is envisaged for the future:
- Centralised large-scale electricity production where small and medium size producers will be connected to electricity grids.
- Development of stand-alone energy systems producing electricity and heat for single houses or apartments with different components such as PV, fuel cells etc.
- The development of "mini-grids" for a limited area should be investigated. In such grids small energy systems such as PV systems and fuel cells in individual houses would be connected together to supply energy. They would be supplemented by conditioning and storage systems, and could serve as an interface between the main grid and single houses.

7.3. WHAT ARE THE RESEARCH AND TECHNOLOGICAL DEVELOPMENT REQUIREMENTS?

For large-scale market introduction of PV power, reduced investment cost, easy integration and long-term supply reliability are key issues.

Further Pre-market R&D efforts should be carried out in three main areas.

- Low cost PV integration into buildings.
- Intelligent and extremely low-cost control systems.
- Research on electricity storage and making new technologies suitable for PV use.

7.4. HOW CAN WE DEVELOP OF A STRONGER MARKET FOR PV TECHNOLOGY?

In addition to technological development, market development is a prerequisite for further growth. Namely,

- Setting the legal, fiscal and political framework: a legal framework to define the conditions for accession to the electricity grid, as well as technical standards for the quality, reliability and safety of the grid. Fiscal measures, such as subsidies, exemption from VAT or Eco-taxes, to support the introduction of renewable energies and to allow renewable energy technologies to compete in the market.
- Marketing of PV added value. The added value of photovoltaics, in terms of quality, reliability and service delivered require an improved common marketing effort.
- Building a 21st century image. The creation of an image for photovoltaics of being high-tech but green, readily available almost everywhere, capable of being easily integrated into new or existing structures, and adapted for entirely new applications could greatly contribute to the growth of the PV market.

7.5. WHY AREN'T MORE PV MODULES IN REGULAR USE?

A major bottleneck is the fact that photovoltaic solar energy conversion is the most expensive form of renewable energy today. The cost is still an order of magnitude too high, when compared with conventional electricity production.

7.6. WHY DO WE NEED RTD?

PV has the potential to cover the world total energy demand. A major contribution to a reduction of CO_2 emissions and sustainable energy production can only be expected if the cost can be considerably reduced. Therefore, RTD is to play a key role in achieving this objective.

7.7. WHAT RTD IS NECESSARY FOR PV?

Material and manufacturing costs are the two major factors that influence the price of PV cells. Even though silicon is the second most abundant material on earth, the silicon used for PV cells must be very pure. In addition, the manufacture of PV cells at present is labour- and capital-intensive, although methods of automation have been undertaken. How quickly PV becomes cost-effective depends on whether research solves these material and production problems.

A further major problem that needs to be solved is the intermittent character of PV electricity production. Thus, electricity supply has to be adapted to demand by storage systems or combined systems, in which PV is used together with small power systems, such as fuel cells.

7.8. WHY IS EU SUPPORT NECESSARY?

The competitiveness of EU industry is a major issue in this rapidly increasing market. Currently, EU industry is lagging behind the US, where the PV industry exports to the EU. Thus, increased EU RTD support, as well as new policies on RES penetration, are pre-requisites in order for the EU to bridge the gap.

The EU added value for PV is high, as there is a high level of PV RTD activity in most member states. Co-operation between the EU programme and the programmes of the member states would assist in achieving a critical mass and providing a complementary approach. Furthermore, standardisation of the technology will require a common EU approach.

8. Bottlenecks and Barriers

8.1. WHAT ARE THE ISSUES THAT NEED TO BE ADDRESSED?

Reduction of PV costs, expected through increased automation in manufacturing, production of new types of cells and an increase in the lifetime and efficiency of the complete PV system need to be addressed.

8.2. WHAT ARE THE MAJOR TECHNICAL BARRIERS TO BE OVERCOME?

The main barriers to be overcome are:
- High Cost: RTD should be aimed at the reduction of both module and system costs.
- Technology limitations of crystalline cells: Alternative production processes, reducing the processing steps and handling requirements to improve the manufacturing line and to make better use of expensive equipment.

- Acceptability of thin-film technologies: Research in order to overcome their relatively low efficiency, toxicity in production and disposal, and to increase their expected lifetime.
- Storing the electricity: Research into new electricity storage technologies to be adapted for photovoltaic use. Reliability, maintenance, ease of use in stand-alone applications and processes, which both convert and store electricity, need to be addressed.

8.3. WHAT ARE THE MAJOR NON-TECHNICAL BARRIERS TO BE OVERCOME?

The liberalisation of the energy markets, as increased competition will mean that energy companies will be reluctant to invest in new, risky, sustainable technologies.

Social acceptance of the technology is also a major concern for wide dissemination of PV.

8.4. WHAT ARE THE RTD PRIORITIES IN THIS AREA?

The main long-term objective is a module cost of 0.5€/Wp. To that end, RTD should be focused on:
- Crystalline silicon cells and modules;
- Thin film cells and modules;
- Advanced system technologies

In the future, more emphasis should be placed on the socio-economic aspects and benefits to be derived by increased use of photovoltaic electricity.

8.5. HOW CAN THE COST OF PV BE REDUCED?

Increased production capacity. In the EU White Paper on Renewable Energy Sources, the PV system cost is expected to decrease to below 3€/Wp by 2005, which results in a PV electricity cost of around 0.15€/kWhr.

Further options for cost reductions are:
- an increase in the lifetime of PV systems from 20 years to 30 years;
- greater use of higher yearly insulation in the southern EU countries,
- an increase in levels of efficiency by between 30% and 50% in the long-term.
- In densely populated areas, south facing roofs, facades of buildings and possibly windows should be used for PV electricity production. PV elements may be designed so as to replace tiles, windows, and walls.

9. Overview of Current Programme Activities

The main areas of PV research funded under the current EU programme are the following:

- Low cost and high quality silicon feedstock
- Optimisation of crystalline silicon process technologies, with particular emphasis on the cost and efficiency of wafer cell production
- Thin film technologies: highly efficient mass production, understanding of material limitations aiming at reducing costs
- Innovative PV concepts for PV cells and modules which have a potential for large cost reductions (such as tandem and concentrator cells, new materials, etc.)
- RTD on cost reduction for other new and innovative components and systems.

The PV projects funded by the EU can be divided into the following categories: projects dealing with PV cells and modules, those dealing with PV systems, projects about buildings incorporating PV and lastly a category consisting of other projects. Figure 1 illustrates the distribution of funding between the four different categories.

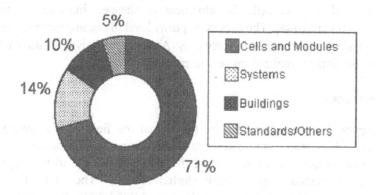

Figure 1. RTD components of EC funding for PV projects under the ENERGIE programme (1998-2002) (total EC funding: 43 Mio€)

9.1. WHAT GOES ON IN EACH OF THE SUB-AREAS OF THE TECHNOLOGY?

9.1.1. *Crystalline Silicon Cells*

Cells produced on the basis of crystalline silicon today constitute about 85% of the world's module shipments, and further improvements both in device and manufacturing technology are needed.

The problem of materials supply has to be solved by refined production methods, improved cell efficiency and the development of technologies leading to a doubling of the wafer area produced from the same amount of feedstock.

The goal is to obtain high efficiency at the lowest possible cost by:

- Fine-tuning of the existing high efficiency structures
- Research for more efficient light trapping schemes and simpler cell structures
- Carrier generation processes in which low-energy photons create charged carriers.
- Cost reduction
- Recycling of chemicals and the reduction of chemical waste production
- Recycling of complete modules and cells after their lifetime

9.1.2. *Thin Film Cells*

Thin film technologies show good potential for cost reduction. However, it is imperative that RTD efforts be devoted to increasing their efficiency and acceptability

9.1.3. *Advanced PV Concepts*

RTD is also focused on the exploration of advanced concepts.

Organic and Polymer cells: Development of cheaper thin-film organic and polymer cells and modules. The necessary purity level of organic materials is small and large-scale production is relatively easy. As concerns polymers, their inherent processing advantages make them very attractive.

10. Conclusions

Strong support has been provided by the EU in the field of photovoltaics. In particular, the European Commission has clearly indicated interest since 1980. Its Fifth Framework programme confirmed the effort of the EU in this topic, and many new EC-funded projects have started recently. The Sixth Framework Programme is now being prepared for the period 2002-2006. It is expected that support for PV will be maintained or even increased in view of economic and strategic interests proved by high worldwide growth.

EVALUATION OF THE GAP STATE DISTRIBUTION IN a-Si:H BY SCLC MEASUREMENTS

A. ERAY and G. NOBILE[*]
*Hacettepe University, Faculty of Eng. Dept. of Physics Eng.,
Beytepe, Ankara-Turkey, and*
[*]ENEA Research Center, Loc. Granatello, I-80055, Portici (NA), Italy

1. Introduction

The states in the mobility gap of amorphous hydrogenated silicon play an important role in determining its optical and electrical properties, and it is essential to know their density and distribution in the gap. Space Charge Limited Current (SCLC) is a classic and a well-known method for determining this density of states [1-4]. In SCLC measurements, the superlinear I-V dark characteristics due to injected charges from the ohmic contacts are measured. When the injection is large enough to displace significantly the quasi Fermi level from its equilibrium position, the current start to rise super-linearly. The magnitude of the current and the shape of the curve depend on the density of states just above the equilibrium Fermi level, so this portion of the curve can be used to determine this quantity.

We consider an n^+-i-n^+ structure, which is frequently used for the measurement of SCLC in a-Si:H. Since undoped a-Si:H can be slightly n-type and the contacts in a n-i-n structure are blocking for holes, the SCL current is predominantly by electrons. Electrons injected into the i-layer are captured by localized states in the mobility gap, creating a space charge. Consequently, for n-i-n samples, the electron injection yields a shift of the electron quasi-Fermi level towards the conduction band edge, so the SCLC characteristic is an indicator of the defect density at and above the Fermi level. In the literature, there are several methods for the analyzing the J-V characteristics to obtain an energy profile for the gap states [1-4].

In this study, the analysis of the J-V characteristics of good quality a-Si:H n^+-i-n^+ structures (having a deep dangling bond density of 8×10^{15} $cm^{-3}eV^{-1}$) has been performed, as a function of temperature. In order to obtain the J-V characteristics of the a-Si:H n-i-n samples in the space charge region, a computer controlled I-V measuring system and a temperature controller have been designed and constructed [5]. The den Boer analysis has been used to obtain the defect density distribution in a limited part of the band gap of the sample. We emphasize that in a good quality sample, even if it is thin, the den Boer analysis gives correct information about the density of states.

J.M. Marshall and D. Dimova-Malinovska (eds.),
Photovoltaic and Photoactive Materials - Properties, Technology and Applications, 253–256.
© 2002 *Kluwer Academic Publishers.*

2. Experimental Procedure

Device-quality a-Si:H n-i-n samples grown on stainless steel substrates were provided by ENEA Research Center, Portici, Italy. The structure of the samples is as follows:

$$\text{Stainless-steel / Ag / ZnO / n}^+ \text{ a-Si:H / i a-Si:H / n}^+ \text{ a-Si:H / Ag.}$$

The doped n^+ and intrinsic layers were 300 and 3000 Å thick, respectively. There must be metal contacts on the n^+ layers to make electrical measurements in n-i-n structures. For that purpose, silver electrodes were used. SCLC measurements were carried out with a PC controlled I-V measuring set-up, designed and constructed at Hacettepe University, Physics Engineering Dept. [5]. We measured the I-V characteristics of n-i-n samples in the dark, in the voltage range 100 mV to 2V. The temperature during measurements of the J-V characteristics on the computerized system was kept in the range of ± 0.2 K. All measurements were carried out in a vacuum of 1×10^{-3} torr.

3. Results and Discussion

The temperature dependence of the J-V characteristics for n^+-i-n^+ structure is shown in Figure 1. All the curves show the same general behaviour: an ohmic region at lower voltages, followed by a supra-linear increase of the current. The arrows in the Figure show the transition from the ohmic to the space charge region. The transition point moves to higher voltages with increasing temperature, which is consistent with an increase in the intrinsic electron concentration.

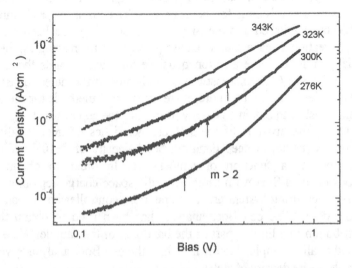

Figure 1. The temperature dependence of the J-V characteristics for an n^+-i-n^+ structure. m denotes the power in the relation $J \sim V^m$.

To obtain the thermal activation energy of the dark conductivity in the ohmic region and space charge regions, the temperature dependent dark conductivity formula $\sigma_D = \sigma_0\exp(E_a/kT)$ was used, and the activation energies were determined from the $\ln\sigma_D$ against $1/T$ plots. By applying that procedure for different applied voltages, the voltage dependency of the activation energy was obtained, as in Figure 2. The activation energies obtained in the space charge regimes decrease as the applied bias increases, representing a Fermi level shift towards the conduction band. The activation energy of an n^+-i-n^+ structure reflects the effects of the n^+ contacts (band bending) and the low density of states above mid-gap, which places E_F close to the conduction band. Without any such bending, the activation energy would be determined by E_F in the bulk a-Si:H, which in high quality materials is near mid-gap (in the range 0.75-0.8 eV). This is in agreement with numerical calculations, where the bulk values are obtained only for thickness greater than 4 μm [6,7].

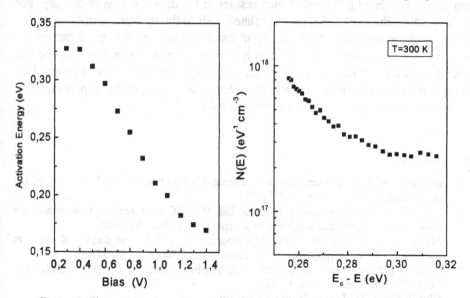

Figure 2. The voltage dependency of the activation energy, as obtained from the temperature dependent dark conductivity.

Figure 3. Energy profile of the gap states extracted from the J-V characteristic

From the measured J-V characteristics at 300 K, the shift of the quasi-Fermi level was calculated for each voltage increment. For each such shift, the density of gap states was then evaluated, using the den Boer procedure. The sweep of the quasi Fermi level (E_c-E) was determined by subtracting these quasi-Fermi level shift values from the starting position of the quasi-Fermi Level shift, as obtained from the ohmic part of the J-V characteristics. In Figure 3, the density of gap states is depicted against the values of (E_c-E). At first sight, these DOS values appear much higher than those from published data. However, these states are in the upper part of the gap, owing to the low activation energy value of these thin samples.

Since the DOS is not thickness dependent in good quality samples [6], the experimental DOS value obtained by the den Boer approach should be interpreted as the correct value. In that model, depending on the value of activation energy, the starting point of the quasi-Fermi level shifts will change. Thus, the DOS values obtained in thin samples having much lower activation energy values reflects information about different states which reside in the upper part of gap.

4. Conclusions

Den Boer's analysis of n^+-i-n^+ samples with different thicknesses gives information about states in different regions of the gap. For thin samples, the density of states obtained by analyzing the J-V data yields the overlapping region of mid gap states and conduction band tail states, which resides in the upper portion of the gap. For thick samples, the explored region is related with to the mid-gap density of states. With a high enough bias, the Fermi level can be swept into the upper part of the gap. We emphasise that in a good quality sample, even if it is a thin one, the den Boer approach to SCLC gives correct information about the density of states. This information comes from the states in a limited upper region of the gap, because of the low activation energy featured by thin samples.

5.   References

1.   den Boer, W. (1981) Determination of midgap density of states in a-Si:H using SCLC measurements, *J. Phys. (Paris)*, **42**, C4-451.
2.   Nespurek, S. and Sworakowski, J. (1980) Use of SCLC measurements to determine the properties of enegetic distributions of bulk traps, *J. Appl. Phys.*, **51**, 2098.
3.   Mackenzie, K.D., LeComber, P.G. and Spear, W.E. (1982) The density of states in amorphous silicon determined by SCLC measurements, *Phil. Mag*, **46**, 377-389.
4.   Cech, V. (1997) Determination of the bulk density of states in a-Si:H by steady state SCLC, *Solid State Electronics*, **41**, 81-86.
5.   Okat, T. (2001) Ms Thesis, Hacettepe University.
6.   Smith, J.H. and Fonash, S.J. (1992) Assessments of density of states extraction from the SCLC measurements: a numerical simulation, *J. Appl. Phys.*, **72**, 5305-5310.
7.   Molenbroek, E.C., Van Der Werf, C.H.M., Feenstra, K.F., Rubinelli, F. and Schropp, E.I. (1997) SCLC in nin devices incorporating glow-discharge and hot-wire deposited a-Si:H, *MRS Symp. Proc.*, **467**, 717-722.

CREATION AND ANNEALING KINETICS OF LIGHT INDUCED METASTABLE DEFECTS IN a-Si$_{1-x}$C$_x$:H

A.O. KODOLBAŞ and Ö. ÖKTÜ

Hacettepe University, Department of Physics Engineering
TR-06532 Beytepe, Ankara, Turkey

Abstract

We have used the Constant Photocurrent Method and steady-state photoconductivity measurements to investigate the creation and annealing kinetics of light induced metastable defects, and their effect on photocarrier lifetimes, in a set of good quality a-Si$_{1-x}$C$_x$:H alloys (x ≤ 0.11) at room temperature. The annealing activation energy distribution for the alloys was deduced using the method proposed by Hata and Wagner [1]. A narrow Gaussian distribution of annealing activation energies, peaking at about 1 eV, accounts for the observed annealing behaviour for the unalloyed sample. For the alloys, the peak positions of the Gaussian distributions shift to higher energies, and their half-widths decrease with increasing carbon content. The relationship between the inverse mobility-lifetime product and the light induced metastable defect density during the creation and annealing cycles were also investigated for these alloys.

1. Introduction

Despite the large number of studies [2-3] and proposed microscopic models [4-5] for the origin of metastable degradation in a-Si:H and related alloys, a detailed mechanism is still missing. Newer kinetic measurements emerge, invalidating the proposed microscopic models. However, detailed kinetic measurements are fairly limited, especially on hydrogenated amorphous silicon-carbon alloys. A kinetic investigation of light induced degradation in silicon-carbon alloys can provide new information on the variations of the degradation behaviour with composition. It can also give hints about the variation of the phenomenon with the properties of a-Si:H itself. In this study, we have utilized CPM and steady state photoconductivity measurements for a set of a-Si$_{1-x}$C$_x$:H samples (x ≤ 0.11), to investigate the creation and annealing kinetics of light induced defects. The concept of a distribution of annealing activation energies is used to describe the observed annealing of the defects. The results are discussed within the framework of the weak-bond breaking [4] and the hydrogen-collision models [5].

257

J.M. Marshall and D. Dimova-Malinovska (eds.),
Photovoltaic and Photoactive Materials - Properties, Technology and Applications, 257–260.
© 2002 *Kluwer Academic Publishers.*

2. Experimental Details

a-Si$_{1-x}$C$_x$:H samples with x ≤ 0.11 were provided by the Dept. of Interface Physics, Utrecht University, Netherlands. Selected physical parameters are given in Table 1. Our experimental procedure has been described previously [6]. IR filtered light of intensity 200 mW/cm^2 from a quartz halogen lamp was used for the metastable defect creation. All CPM spectra for different light exposures or annealing steps were taken at the same constant photocurrent, optimised for the most damaged state of individual samples. The relative changes in the light induced defect density after each creation and annealing step were determined from room temperature CPM measurements [6]. A 660 nm LED was used to measure the photocurrent at a constant photon flux of 3x10^{13} photons /cm^{2}s.

TABLE 1: Selected physical parameters of a-Si$_{1-x}$C$_x$:H samples used in the experiments. x, d, E$_g^T$, E$_a$, E$_0$, N$_{d0}$ are carbon content, film thickness, Tauc optical gap, dark conductivity activation energy, Urbach parameter, and defect density in the annealed state, respectively.

Sample	x	d(nm)	E$_g^T$ (eV)	E$_a$ (eV)	E$_0$ (meV)	N$_{d0}$(cm^{-3})
1	0	1836	1.73	0.51	46	5.65x10^{15}
2	0.05	991	1.83	0.88	70	7.51x10^{16}
3	0.11	1123	1.95	0.94	75	1.52x10^{17}

3. Results and Discussion

The relative increase in the light induced defect density during successive illumination steps up to 10 h is presented in Figure 1. The measured slopes agree with the generally observed t$^{1/3}$ dependence of the light-induced defect density [2-6]. This result also provides an independent confirmation of the intrinsic nature of light induced metastable defects in a-Si:H [2,4]. Despite the constant defect creation rate, the defect creation efficiency (CE) decreases with carbon content. The insert in Figure 1 illustrates this, by showing how the relative magnitude of the induced defect density after 10h illumination varies with the Tauc gap. As the carbon content increases, photocarriers will be trapped deeper in the wider band tail states. This reduces the probability of creation of light induced defects in the presence of the higher density of deep defect states. Thus, the creation efficiency decreases [2].

   The annealing kinetics of the light induced defects were studied using our prior procedure [6], for specimens illuminated for 10h at 300K. The annealing temperatures were T$_A$ = 373K, 393K, 413K and 433K. The decrease in the normalized light-induced defect density at these annealing temperatures is plotted in Figure 2. The light induced metastable defects start annealing at higher temperatures in alloys of higher carbon content. Also, the high temperature annealing characteristics for the three samples were quite different. To quantify this, the concept of a distribution of annealing activation energies is proposed [1]. Our calculation procedure was as in [6]. The density of the remaining light-induced defects, after annealing at a temperature T$_A$ for a time t$_A$, can be calculated as:

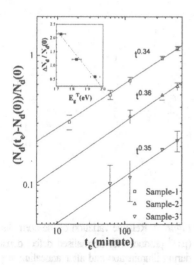

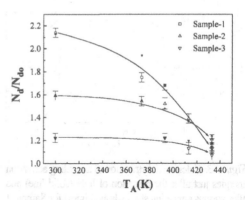

Figure 1. Relative increase in light induced defect density during 10h illumination. Insert: the variation of CE with the Tauc gap.

Figure 2. Annealing of the normalized density of defects, as a function of annealing temperature for the alloys.

$$N_d(T_A, t) = N_{do} + \int_0^\infty \Delta N_d(E_A, T_{ill}, t = 0) \exp(-t/(\tau_0 \exp(E_A/kT))) dE_A. \quad (1)$$

Here, $\Delta N_d(E_A, T_{ill}, t=0)$ is the calculated annealing activation energy distribution function at the degradation temperature just after the illumination is stopped, and N_{do} is the density of stable defects. Very good matches to the experimental values in Figure 2 were obtained by taking a Gaussian distribution of annealing activation energies, as in Figure 3. The peak value of the calculated distribution of annealing activation energies shifts from 1.0 to 1.1 eV with increasing carbon content. Also, the annealing behaviour of Sample 3 can almost be described with a single activation energy of 1.1 eV. The results suggest that non-radiative recombination of energetically-distant photocarriers trapped in the band tails [4] may supply the energy necessary for the metastable complex $M(Si-H)_2$, formed upon the collision of two mobile Si-H/DB [5], to relax to a more stable configuration, and thus shift the peak of the calculated distribution of annealing energies to 1.1 eV.

There is no simple relationship between light induced defect density and associated photoconductivity, as predicted by weak-bond breaking model [4]. The degradation and annealing paths are different, and one often observes a hysteresis-like relationship between the two [2,6]. Our results, in Figure 4, are in agreement with those studies for Samples 1 and 2. Defects created at earlier times during illumination degrade the normalised inverse $\mu\tau$-product deduced from photoconductivity measurements more strongly, and they anneal out more easily than those created at later times [4,6]. Conversely, a linear relationship is emerging for Sample 3. The difference in the recombination rates results from different distributions of defect annealing activation energies [6]. The observed hysteresis-like relationship starts to become linear when the created defects have the same E_A values as those in Figure 3(c).

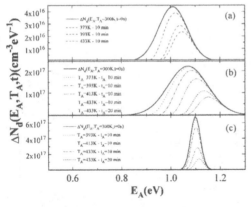
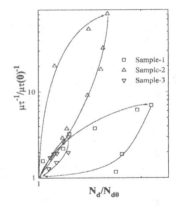

Figure 3. Calculated distribution of annealing activation energies just after the cassation of light (solid line) and after various annealing steps (dashed lines) for Sample 1 (a), Sample 2 (b) and Sample 3 (c).

Figure 4. Relative variation of the normalised $(\mu\tau)^{-1}$ product with normalised defect density during illumination and after annealing steps. The lines are a guide to the eye.

4. Conclusions

The defect creation rate is found to be insensitive to carbon content in a-$Si_{1-x}C_x$:H ($x \leq 0.11$) alloys. This result is in agreement with both of the proposed microscopic models proposed to explain the observed changes [4-5], and indicates the intrinsic nature of these defects in a-Si:H. On the other hand, a decrease in the defect creation efficiency, together with a shift in the peak value of the calculated distribution of annealing activation energies, suggest the possible role of non-radiative recombination of energetically distant photocarriers [2, 4]. The hysteresis-like relationship between the light induced defect density and the resultant photoconductivity is linked to the distribution of defect annealing activation energies. The relationship approaches linearity for the highest carbon content alloy.

5. Acknowledgements

The authors thank J. Bezemer (Utrecht University) for providing the samples used. The work is supported by Hacettepe University Research Grant 01.01.602.007.

6. References

1. Hata, N. and Wagner, S. (1992) A comprehensive defect model for amorphous silicon, *Appl. Phys. Lett.* **72(7)**, 2857-2872.
2. Fritzsche, H. (1997) Search for explaining the Staebler-Wronski effect, *MRS Symp. Proc.* **467**, 19-31.
3. Stutzmann, M. (1997) Microscopic aspects of the Staebler-Wronski effect, *MRS Symp. Proc.* **467**, 37-48.
4. Stutzmann, M., Jackson, W.B. and Tsai, C.C. (1985) Light-induced metastable defects in hydrogenated amorphous silicon: a systematic study, *Phys. Rev. B* **32(1)**, 23-47.
5. Branz, H.M. (1999) Hydrogen collision model: Quantitative description of metastability in amorphous silicon, *Phys. Rev. B* **59(8)**, 5498-5512.
6. Kodolbaş, A.O., Eray, A. and Öktü, Ö. (2001) Effect of light-induced metastable defects on photocarrier lifetime, *Solar Energy Materials and Solar Cells* **69(4)**, 325-337.

POTENTIAL PV MATERIALS BASED ON InN POLYCRYSTALLINE FILMS: GROWTH, STRUCTURAL AND OPTICAL PROPERTIES

V.Ya. MALAKHOV

Institute for Problems of Materials Science, National Academy of Sciences of the Ukraine, 3, Krzhizhanovsky St., 03142 Kiev, Ukraine

Abstract

The fabrication details, plus the basic structural and optical properties of low temperature plasma enhanced reactionary sputtered (LTPERS) indium nitride thin films are presented. RHEED and AFM studies of surface morphology were performed. Optical absorption and reflectance spectra of textured films were taken, to reproduce accurately the dielectric function and to determine the optical effective mass of electrons (0.11) and the direct band gap energy (2.03 eV). Some TO (485 cm^{-1}) and LO (585 cm^{-1}) phonon features of InN polycrystalline films in the NIR and Raman spectra are observed and discussed. The attractive possibilities of InN layers are discussed, based on a model calculation of the InN/Si tandem system for potential applications in PV devices including high efficiency solar cells.

1. Introduction

Over the last ten years, indium nitride has become the focus of growing interest. It is a III-V semiconductor with a direct band gap of about 2.0 eV, and therefore has potential applications in photonic devices such as LEDs, lasers, color displays and especially high efficiency solar cells [1]. The physical properties of InN films obtained by different methods have been studied in numerous works [2-7]. However optical and electric parameters such as the dielectric and optical constants, energy gap, effective masses of carriers and phonon wavenumbers await more accurate definition. A lack of single crystal samples explains the situation regarding the above data. No structural or thermal properies of InN have been studied for epitaxial films on lattice-matched substrates. As reported [3], the TO and LO phonon features observed in the transmittance spectra of reactively sputtered InN thin films are close to those determined from the reflectance of a $Ga_{1-x}In_xN$ solid solution [4]. However, Kwon *et al.* [5] did not reveal a LO phonon peak in the Raman spectra of InN monocrystalline films. For this reason, the present paper presents structural and optical data for InN polycrystalline films synthesized previously and recently by LTPERS [6-7]. It also offers potential possibilities for using InN/Si heterojunctions in PV devices, including high efficiency solar cells.

261

J.M. Marshall and D. Dimova-Malinovska (eds.),
Photovoltaic and Photoactive Materials - Properties, Technology and Applications, 261–264.
© 2002 *Kluwer Academic Publishers.*

2. Experimental Details

Because the dissociation temperature of InN films is about 650°C [2], a low temperature growth technique is required. In our case, LTPERS equipment was used. Intensive Ti-wire evaporation, for gettering, was used throughout the deposition, to reduce oxygen contamination inside the reactor and in the growing films. High quality smooth surfaced Si, quartz, fluorite and compound ceramics were used as substrates. During film growth, the substrate temperature was about 350°C, due to intensive ion bombardment of the top electrode (anode) during sputtering. Film thicknesses varied in the range 100 - 2000 nm. To determine the chemical composition of the sputtered InN films, an Auger spectrometer (JAMP-10) was used. The surface morphology and the microstructure of a cross-section of the films were investigated using a Philips SEM5V scanning electron microscope, plus standard AFM equipment. The crystalline structure parameters of the sputtered films were determined using an X-ray diffractometer (DRON-3) employing Cu(kα) radiation, and also an electronograph (EG-100a). Reflectance and transmittance measurements in the visible and NIR regions (25000-200 cm^{-1}) were carried out using a Bruker IFS66 Fourier transform spectrometer (FTIR) and a Carl Zeiss M40 grating spectrometer, respectively. A Raman spectrometer (Dilor XY), equipped with a microscope, was used to study the phonon spectra of the nitride films.

3. Results and Discussion

X-ray diffraction patterns of α-InN (wurtzite) layers deposited on ceramic substrates demonstrate a very strong InN (002) diffraction peak. This suggests a textured crystalline structure of the films, with the c axis perpendicular to the plane of substrate. The same result was obtained from the RHEED pattern of InN films on ceramics (Figure 1). An AFM image of the natural surface morphology of films on polished ceramics is shown in Figure 2. These results, plus an Auger investigation, show that no outsider phases except the InN one were present in the films. An off-stoichiometric In/N ratio, with an abundance of In, was caused by nitrogen vacancies. A perceptible oxygen concentration was revealed in the films, probably leading to amorphous indium oxide partially forming (bound oxygen).

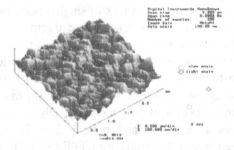

Figure 1. RHEED pattern from an InN layer deposited on a ceramic

Figure 2. AFM image of the natural surface of an InN film synthesized on a ceramic

Moreover, other unactive oxygen molecules can be incorporated both in voids between the InN grains or/and sites on film's surface.

Further transmittance and reflectance spectra in the visible range were taken, to determine the band gap energy of InN thin films on CaF_2 (fluorite) substrates. The absorption coefficient, α, of the InN layer and yielded $\mathbf{E_g} \cong 2.03$ eV, which is very close to earlier values [6-7]. In addition, UV and visible reflectance spectra have been used to reproduce accurately the dielectric function of wurtzite InN, for assignments of peak structures to interband transitions (1.5-12.0 eV) as well as to determine the dielectric constant (9.3) and refractive index (>3.0) [6]. Also, the electron plasma resonance energy (0.6 eV), damping constant (0.18 eV) and optical effective mass of electrons (0.11) were calculated from visible reflectance spectra, confirming previous values [6]. Moreover, the PDS method was used to explore some peculiarities in the free carrier absorption spectra for InN polycrystalline films deposited on different substrates. In order to obtain the necessary information about phonon features of the InN films, and to determine precisely some optical parameters, the Drude-Lorentz formalism was used for dielectric function modelling to reflectance spectra from both the InN film surface and a bare fluorite substrate. Figure 3 shows good agreement between experimental data and the calculated curve. Also, in our opinion, the reflectance peaks at 485 and 590 cm^{-1} are respectively connected with TO and LO modes of vibration of indium nitride [3-5]. An identical result was obtained from a study of Raman spectra for InN textured films (>1 μm thick) on ceramics (Figure 4). The broadening of the peak at 485 cm^{-1} in the Raman spectra of nitride films is close to that observed in the reflectance spectra. However, because of the imperfect crystalline structure of our samples, we can observe only two apparent optical phonon modes: E_1(TO) at 485 cm^{-1} and A_1(LO) at 585 cm^{-1}. For this reason, it was also difficult to find the main LO phonon mode at 694 cm^{-1}, as predicted by Osamura et al. [4].

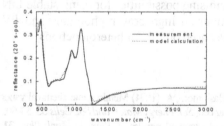

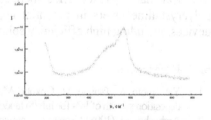

Figure 3. NIR reflectance spectra of an InN polycrystalline layer on CaF$_2$

Figure 4. Typical Raman spectra of InN textured films deposited on ceramics

4. Theoretical and Practical Considerations

In order to estimate the potential possibilities of InN films for solar cell fabrication, both theoretical and practical considerations were used. To achieve an optimum power conversion efficiency η for solar cells based on an InN n-layer (emitter) and a p-Si (base), large values of α(E), minority carrier lifetime τ, diffusion length L (at least $\alpha L > 3$ for front side illumination) and a small surface recombination velocity S have to be combined. This can be achieved with a compromise of optimum band gaps E_g (~1.0 and 2.0 eV) for an efficient tandem system consisting of several stacked single cells between a large I_{sc} or V_{oc} [1].

Summarizing, a solar cell base material should exhibit an appropriate energy gap and a strong absorption ($\alpha_{InN} > 10^4$ cm^{-1}) adjusted to the solar spectrum, plus a dopability by carriers featuring high mobilities and long lifetimes. Another important advantage of an InN/Si heterojunction in future solar cells is a protective function, including protection from radiation using the absorptive InN layer as a top coating. The existing problem of heteroepitaxy of InN films on Si will be resolved by research into an appropriate (e.g. AlN) buffer layer on an InN-Si interface, to match the lattice periods and improve the heterojunction operating parameters. The fabrication of high quality heterojunctions with InN/Si, and its further characterization, are also of great importance for future high efficiency solar cells.

5. Conclusions

Polycrystalline α-InN (wurtzite) thin films synthesized by LTPERS were textured, with a preferable <002> orientation. A perceptible oxygen concentration was identified in the films, probably leading to the partial formation of amorphous indium oxide (bound oxygen). Other inactive oxygen molecules may be incorporated in voids between InN grains or surface sites. A direct energy gap of 2.03 eV for InN polycrystalline films was confirmed, as well as a value of 0.11 for the electron effective mass, as obtained earlier by the author using identical InN film samples. Moreover, the E_1(TO) at 485 cm^{-1} and A_1(LO) at 585 cm^{-1} phonon features observed from Raman and NIR reflectance spectra are in a good agreement with those obtained by other authors. A comparison of the data obtained now with those obtained 25 years ago for identical samples shows no differences, and thus demonstrates the long term stability of this compound in respect of its optical and electrical characteristics. Thus, there are promising possibilities for using stable InN polycrystalline layers in potential applications in high power photonics and PV devices, including high efficiency solar cells based on InN/Si heterojunctions.

6. References

1. Yamamoto, A., Tsujino, M., Ohkubo, M. and Hashimoto, A. (1994) Metalorganic chemical vapor deposition growth of InN for InN/Si tandem solar cell, *Sol. Energy Mater. and Sol. Cells* **35**, 53-60.
2. Ambacher, O. (1998) Growth and applications of group III nitrides, *J. Phys. D: Appl. Phys.* **31**, 2653-2710.
3. Tansley, T., Egan, R. and Horrigan, E. (1988) Properties of sputtered nitride semiconductors, *Thin Solid Films*, **164**, 441-448.
4. Osamura, K., Naka, S. and Murakami, Y. (1975) Preparation and optical properties of Ga$_{1-x}$ In$_x$N thin films, *J. Appl. Phys.* **46**, 3432-3437.
5. Kwon, H.-J. and Lee, Y.-H. (1996) Raman spectra of InN monocrystalline films, *Appl. Phys. Lett.* **69**, 937-940.
6. Tyagai, V., Evstigneev, A., Krasiko, A., Andreeva, A. and Malakhov, V. (1977) Optical properties of indium nitride films, *J. Sov. Phys. Semicond. (USA)* **11**, 1257-1261.
7. Malakhov, V. (1999) Growth and optical characterization of InN thin films synthesized by LTPERS, *Proc. Euromat'99* **9**, 75-79.

LIGHT SOAKING EFFECT IN a-Si:H BASED n-i-p AND p-i-n SOLAR CELLS

G. NOBILE and M. MORANA*
*ENEA Research Center, Loc. Granatello, 80055 Portici (NA), Italy, and
* University of Cagliari, Department of Electronic Engineering.*

Abstract

Hydrogenated amorphous silicon solar cells have been realized in both a p-i-n configuration on Corning glass substrates, and a n-i-p configuration on stainless steel substrates. The performance degradation of the two kinds of cells under solar illumination has been examined over a 140 hour period. During degradation, the two devices have been kept under load in the maximum power condition that is normally used in a solar plant. The performance of the Corning glass deposited device exhibited a higher rate of degradation than did that of the other cell. A discussion of the possible reasons for this behaviour is given.

1. Introduction

The Staebler-Wronski effect [1] in a-Si:H based thin film solar cells represents a major obstacle to the practical utilization of this kind of material for the power conversion of solar energy. This effect depends on the structure of the cell. When the p^+ layer is deposited first, as in a substrate/p-i-n structure, it has been observed that a slight contamination of the intrinsic layer by boron is beneficial in terms of the conversion efficiency. In this case, above all in single-chamber reactors, the B_2H_6 trapped on the walls of the reactor during deposition of the p-layer can be released during i-layer deposition. The consequent slight contamination by boron electrically compensates the undoped amorphous silicon, which is normally slightly n-type, leading to a less defective material and to higher efficiency solar cells. In contrast, if the n^+ layer is deposited first, as in the case of a substrate/n-i-p structure, the slight contamination of the i-layer by phosphorus can cause an increase of the concentration of defects, which should lead to a lower efficiency. This contamination effect is nullified in modern multichamber reactors, where there is no cross-contamination between single layers. Nowadays, in this so-called "reversed structure", efficiencies equivalent to the "standard structure" can be obtained [2].

The "reversed structure" presents further advantages relative to the "standard" one. The possibility of illuminating the cell from the top permits a wider choice of substrates, flexible or otherwise, such as stainless steel or plastic, widening the possible fields of application and reducing the costs.

J.M. Marshall and D. Dimova-Malinovska (eds.),
Photovoltaic and Photoactive Materials - Properties, Technology and Applications, 265–268.
© 2002 *Kluwer Academic Publishers.*

The effect of dopants on the Staebler-Wronski effect in the p-i-n and n-i-p configurations is not yet fully understood. In some cases, a lower rate of degradation has been observed in cells using a microcrystalline p-layer.

In this work, we examined the degradation behaviour of both kinds of structure. A comparison of the results allows some insight into the degradation mechanisms that can be involved when using different deposition procedures.

2. Experimental Details

Single junction a-Si:H solar cells were realized in both the p-i-n and the n-i-p configuration. The "normal" p-i-n structure was deposited on commercial Asahi "U" TCO coated Corning glass. The reactor utilised for device preparation was a large area Nextral ND-400 single chamber PECVD system. The device area was 1 cm². The starting efficiency of this solar cell (annealed state) was about 10%. Further fabrication details have been reported elsewhere [3].

The "reversed structure" n-i-p solar cell produced for this work was realised in a low cost PECVD system: model ORD-1, Energy Conversion Devices Inc. Fabrication details have been reported elsewhere [4]. This device was deposited on a thin, flexible stainless steel substrate previously coated with a back reflector/scatterer (Ag/ZnO). The single n-i-p device was contacted with indium tin oxide (900 Å) and an Ag gridline. The device area was 1 cm². Details of the properties of the RF-sputtered indium tin oxide are reported elsewhere [5].

The two cells were first brought to the "annealed state" by heating in vacuum at 150 °C for 1 hour. They were then exposed to simulated solar radiation (100 mW/cm²) using an ORIEL 81173 solar simulator, for a period of 140 hours. The cell temperature under irradiation was held at 25 °C. To simulate requirements during normal use in a solar plant, in which the cells should operate continuously at maximum power, they were connected to an electronic load. At fixed time intervals, the J-V characteristics of cells under illumination were recorded, and the data were used to set the new load current corresponding to maximum power conditions. Also, the degradation process was paused at fixed times, and a new quantum yield spectrum was recorded in the wavelength range 350 - 800 nm.

3. Results and Discussion

Figure 1 summarises the results of the first degradation experiment performed on the p-i-n cell deposited on Corning glass. This cell had an original efficiency in the annealed state of 9.96%, with a fill factor of 70%. After 140 hours of light soaking, all electrical parameters suffered some degradation. In particular, the efficiency decreased to 7%, and the short circuit current fell from 17 mA/cm² to 15.8 mA/cm². As can be seen from Figure 1, the main cause of the reduced efficiency was the decrease in fill factor, due to a strong increase in both the series and the shunt resistance. Figure 2 shows the quantum efficiency in the "annealed" and "final" degraded states (after 140 hours illumination), and also at two intermediate intervals (after 20 and 68 hours light soaking). This type of cell exhibits a strong degradation even at the p-i interface, since the quantum yield (QY) drops by 31% at 350 nm, at the end of the degradation cycle. The largest degradation occurs in the first cell layers, with the QY decreasing by about 40% at 450 nm. At longer wavelengths, the uniform absorption gives a fairly uniform reduction in quantum efficiency (~ 11%) across the whole wavelength range up to 800 nm. These data indicate that for this type of cell, the interface layer, as well as the part of the cell which may suffer from boron contamination, are strongly sensitive to degradation by illumination.

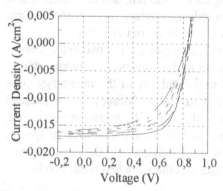

Figure 1. J-V characteristic for the p-i-n cell deposited on Corning glass, for different degradation times from the "annealed" state to 140 hours of illumination.

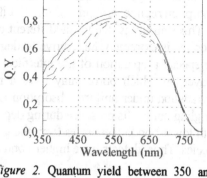

Figure 2. Quantum yield between 350 and 800 nm for the p-i-n cell on Corning glass, for different degradation times from the "annealed" state to 140 hours of illumination.

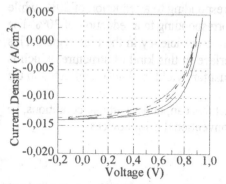

Figure 3. J-V characteristic for the n-i-p cell deposited on stainless steel, at different degradation times from the "annealed" state to 140 hours of illumination.

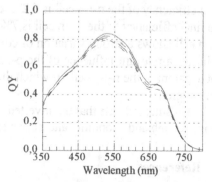

Figure 4. Quantum yield between 350 and 800 nm for the n-i-p cell on stainless steel, at different degradation times from the "annealed" state to 140 hours of illumination.

Figure 3 shows J-V curves for the n-i-p solar cell deposited on stainless steel. This cell had an "annealed" efficiency of 7.7%, and a fill factor of 62.3%. The lower initial efficiency of this cell is due to shadowing by gridlines and to limited light trapping. Light enters the cell in the normal direction and is only scattered, to some extent, at the back of the device. The limited scattering efficiency is shown in Figure 4 by the presence of interference fringes. These effects lead to a reduced current density of 13.8 mA/cm^2. Note that the fill factor is also slightly lower for this type of cell, due to the presence of substrate defects that cause micro-shorts within the cell, and to the thin ITO top contact, giving a higher series resistance.

After 140 hours of illumination, the efficiency of the n-i-p cell decreased to 6.17%, while the short circuit current decreased only to 13.3 mA/cm^2. Figure 4 provides information on the degradation behaviour for the n-i-p cell, extending from the "annealed" state to the final degraded state. In this case, the p-i interface is almost stable, since the variation of the quantum efficiency at 350 nm is less than 5%. It remains below 5% throughout the range 350 - 500 nm, and thus deep within the intrinsic layer. The quantum efficiency is reduced slightly more in the uniform absorption regime, around 600 nm, but it never falls by more than 10%.

This stability is due to the different deposition recipes for the two cells. The p-i interface is extremely insensitive to degradation in the "reversed" structure, due to the special preparation of the interface, and to the stability of the microcrystalline p-layer, in which boron may have a reduced influence. Moreover, the reduced degradation under uniform absorption is due to the role of hydrogen in passivating dangling bonds. Its presence during deposition contributes to the removal of species with lower bonding energies from the growing film, leaving only the most compact species. Thus, defects have higher bond energies and there are less weak bonds.

4. Conclusions

We have shown that the light soaking effect is more evident in the p-i-n cell. The final efficiency of the n-i-p cell is 6.2%, corresponding to a reduction of 20%; while the final efficiency of the p-i-n cell is 7%, corresponding to a reduction of 30%. The p-i-n cell shows a strong variation of conversion efficiency in the blue region. This indicates a relevant influence of the p-i interface in this kind of structure, which is not apparent in the n-i-p cell, in which the quantum yield is almost unaffected in the same region.

The results indicate that the inverted structures can, under suitable conditions, be an interesting and economic alternative to conventional p-i-n devices.

5. References

1. Staebler, D.L. and Wronski, C.R. (1977) Reversible conductivity changes in discharge produced amorphous silicon, *Appl. Phys. Lett.* **31**, 292-294.
2. Guha, S., Yang, J., Banerjee, A., Glatfelter, T., Hoffman, K., Ovshinsky, S.R., Izu, M., Ovshinsky, H.C. and Deng, X. (1994) Amorphous silicon alloy photovoltaic technology: from R&D to production, *Proc. MRS Spring Meeting* **336**, 645-655.
3. Privato, C., Avagliano, S., Conte, G., Di Domenico, D., Mangiapane, P., Nobile, G. and Rubino, A. (1994) Optimization of p-SiC:H, i and n-Si:H layers in a large area PECVD reactor, *Proc. 12th EPVSEC*, 370-373.
4. Nobile, G. and Morana, M. (in press) Light soaking effect in a-Si:H based n-i-p and p-i-n solar cells, *Solar Energy Materials and Solar Cells*.
5. Thilakan, P., Terzini, E., Nobile, G., Loreti, S., Minarini, C., Polichetti, T. and Sasikala, G. (1999) Large area device quality indium-tin-oxide thin films by magnetron sputtering, *Solid State Phenomena* **67-68**, 255-260.

ACCELERATED AGEING TESTS OF SOLAR CELLS AND ENCAPSULATIONS

V. ŠÁLY, M. RUŽINSKÝ and P. REDI*
Slovak University of Technology, Faculty of Electrical Engineering and Information Technology, Ilkovičova 3, SK-81219 Bratislava, Slovak Rep.
* *Universita degli Studi di Firenze, Dipartimento di Ingegneria Elettronica, Via S. Marta 3, I-50139 Firenze, Italia.*

1. Introduction

Today's power PV module is a complete, enclosed package of solar cells, interconnects, power leads and a transparent cover or optical concentrator, depending on the type of module. The estimated lifetime of cells used as solar converters is ~ 25 years. Encapsulation materials must be highly transparent, resistant against thermal and UV oxidation (degradation) at low and high temperatures, resistant against humidity, have good mechanical and electrical properties (resistance), and be thermally consistent with the cell. At present, the most viable materials for industrial use seem to be ethylene vinyl acetate (EVA), ethylene methyl acrylate (EMA), poly/n/butyl acrylate (P-n-BA) and aliphatic polyether urethane (PU). EVA and EMA are dry films for vacuum-bag lamination at temperatures up to 150°C. P-a-BA and PU are liquid casting systems. Both the cells themselves and the encapsulation of the photovoltaic panels undergo degradation during the device lifetime. As a particular panel degradation mechanism, electrochemical corrosion of the cell metallization leads to encapsulant deterioration, reducing the insulation resistance and especially its electrical stability [1]. The voltage between two electrified cells or between a cell and the grounded metallic frame may result in dissolution of the cell metallization into the surrounding encapsulant, which then becomes insufficiently resistive to prevent electrical breakdown. Humidity is very effective in promoting this process. Also, discoloration effects and a browning of the encapsulant can reduce the efficiency, because of reduced light transmittance [2].

In order to avoid failure of the electrical equipment, and to estimate the real lifetime, accelerated ageing tests are often done under extreme conditions (elevated temperature, electric field, UV radiation). As to the electric insulation, present photovoltaic modules can easily withstand voltages up to 5,000 V, and this limit can be extended to more than 10,000 V with simple and inexpensive improvements [3].

2. Ageing of Electrical Insulation

Insulating (encapsulating) materials for photovoltaics operate under different conditions (temperature, irradiation) from the usual room values. Oxidation is the

J.M. Marshall and D. Dimova-Malinovska (eds.),
Photovoltaic and Photoactive Materials - Properties, Technology and Applications, 269–272.
© 2002 *Kluwer Academic Publishers.*

most effective process influencing organic matter, generally used as an encapsulant. An elevated temperature accelerates the influence of oxygen, and results in a deterioration of the mechanical and electrical properties of the insulator.

In some simple cases the lifetime τ at temperature ϑ can be expressed as:

$$\tau = \tau_0 \exp(-b\vartheta) \qquad (1)$$

where τ_0 is an ideal lifetime at $\vartheta = 0$, and b is a constant. This is Montsinger's law. It implies that the process of ageing is active at all finite temperatures, and that the lifetime of an insulating material decreases with increasing temperature.

3. Experimental Procedure

Tests were made in laboratory conditions on small experimental PV modules, in which 13 polycrystalline silicon solar cells (Kyocera) in series were covered in polycarbonate and protected by RTV silicon rubber (Wacker), and on small commercially produced MSX-01 OEM Solarex modules, laminated with EVA. Some samples were biased by a dc voltage of 1500 V. The insulation resistance, R_{ins}, of the encapsulant, and the electrical parameters under standard irradiation (AM1.5 spectrum, 100 mW/cm^2, room temperature) and in the dark were investigated. The parallel equivalent capacitance and resistance of the cell structure in the frequency range 40-500 kHz were also measured. A complex impedance diagram was used to estimate the influence of accelerated ageing on the electrical properties of the solar cell structures.

A solar simulator (ORIEL model 6722), a standard halogen lamp, a radiometer (Oriel) and an impedance tester (HIOKI) were used for the measurements. The insulating properties of the encapsulant were also measured using a Statron Teralog 6202 operating at 1000 V dc. After initial characterisation, the samples were stored in a climatic chamber at a temperature of 80 °C and a relative humidity of 70 %. An electrical voltage of 1500 V dc, limited in current to 3 mA, was applied to stress the samples. After each cycle of accelerated ageing (30 days), the samples were visually checked and re-measured.

Impedance spectroscopy is often used to characterize dielectric or semiconductive structures. The data are presented as a plot of the real versus the imaginary part of these functions. A solar cell p-n structure can be simply described via a connection of discrete capacitors and resistors in series and/or in parallel. Analysis of the circuit complex impedance and comparison between its real and imaginary parts allows an estimation of the values of the main circuit components. A simple equivalent circuit can be created by parallel connection of capacitance and resistance (shunt resistance R_{sh}) connected to a series resistance, R_s [4]. Usually, $R_{sh} \gg R_s$ and the Nyquist diagram is a circular arc with its centre on or below the real axis, and with a radius $R_{sh}/2$. In the solar cell structure, R_{sh} controls current leakage, and the demand is for R_{sh} to be as large as possible.

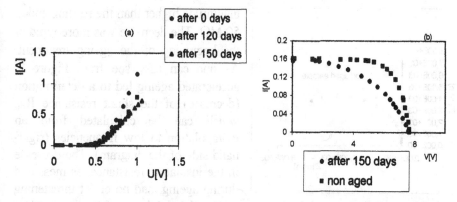

Figure 1. An example of the influence of ageing on the current-voltage characteristics in the dark (a) and under illumination (b).

4. Results and Discussion

After five ageing cycles (totalling 150 days), degradation of the photovoltaic parameters and visual damage to the Ag metallization and encapsulant were observed (as compared to similar non-stressed cells/modules). In both cases - silicon rubber and EVA - almost complete and uniform yellowing of the encapsulant appeared. On the cells covered with EVA, the characteristic metallic brightness of the cell metallization, including the output leads, disappeared after ageing. In the case of the module with silicon rubber as the encapsulant, the brightness of the metallization was comparable with that of non-aged samples. The forward biased current-voltage characteristics were measured at room temperature, in the dark. Figure 1a illustrates the effect of ageing time. The slopes of the curves decrease with the ageing time. The value of the ideality factor, n, in the equation:

$$I = I_L - I_0 \left[\exp\frac{qU}{nkT} - 1 \right] \qquad (2)$$

usually varies between 1 and 2. The higher value reveals a higher recombination current via the solar cell structure, causing a reduction in the slope of the I-V characteristic. An example of current-voltage curves under standard illumination, measured at the beginning and end of accelerated ageing, is shown in Figure 1b. The deterioration leads mainly to a decrease in the fill factor from 0.72 at the beginning to 0.51 after 150 days. Practically, the value of the open-circuit voltage does not change. On the other hand, a small decrease in the short-circuit current can be observed. This can be either the result of an increased recombination current or the result of reduced transmittance caused by the yellowing of the encapsulant. We expect both these effects to be active. The insulation resistance, R_{ins}, of the encapsulant decreased relative to the starting values, but was still a few orders of

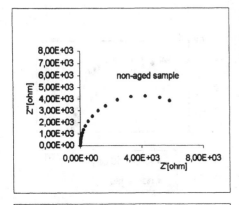

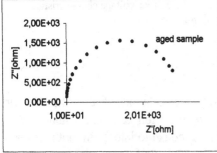

Figure 2. Complex impedance diagram of a non-aged sample, and one after 150 days of accelerated ageing

magnitude higher than the recommended 50 MΩ. The decrease was more rapid at the beginning of the ageing treatment. As one can easy see from Figure 2, accelerated ageing led to a deterioration (decrease) of the shunt resistance R_{sh}, which can be calculated from an extrapolation to low frequencies (right-hand side of the diagram). The decrease in the insulation resistance, as measured during ageing, had no effect threatening electrical breakdown of the encapsulant – silicon rubber or EVA. Electrically, no significant influence of a dc bias stress of 1500 V on the cell and encapsulant parameters was observed, after the 150-day period.

It is not possible to predict the lifetime of the investigated samples using the present results. For this, measurements would have to be performed at different temperatures, and critical values for the cell parameters would be needed. However, some possibilities for visualising the effects of ageing have been demonstrated.

5. Acknowledgements

The work is based on research activities between the Slovak University of Technology Bratislava, the University of Florence and SEI (Sistemi Energetici Integrati) Prato (the industrial partner). It has been also supported by VEGA – Scientific Agency of the Ministry of Education of the Slovak Republic, under project No. 1/7619/20.

6. References

1. Mon, G.R., Orehotsky, J., Ross, G.R. and Whitla, G. (1984) Predicting electrochemical breakdown in terrestrial photovoltaic modules. *Proc. 17th IEEE Photovoltaic Specialists Conf.*, Kissimmee, USA, pp 682-692.
2. Pern, F.J. (1994) Factors that affect the EVA encapsulant discoloration rate upon accelerated exposure *Proc. 1st WCPEC*, Waikoloa, Hawaii, USA, pp 897-900.
3. Redi, P. (1991) Considerations about the design of PV modules for central power plants, *Proc. 10th European Photovoltaic Solar Energy Conf.*, Lisbon, Portugal, pp 959-961.
4. Pellegrino, M., Nardelli, G. and Sarno, A. (1997) An indoor technique for assessing the degradation of PV modules, *Proc. 14th European Photovoltaic Solar Energy Conf.*, Barcelona, Spain, P1B15.

DIFFERENTIAL SPECTRAL RESPONSIVITY OF μc-Si:H SOLAR CELLS

M. SENDOVA-VASSILEVA, O. ANGELOV, ST. KANEV and
D. DIMOVA-MALINOVSKA
Central Laboratory for Solar Energy and New Energy Sources,
Bulgarian Academy of Sciences, 72 Tzarigradsko Chaussee,
1784, Sofia, Bulgaria.

1. Introduction

It is very important to be able to characterise the spectral response of solar cells in conditions close to the ones in which they actually work, i.e. under visible light illumination. For this reason, a system for measuring the differential spectral responsivity (DSR) has been built in our laboratories. This technique was first developed and described in the literature [1,2].

The DSR of a set of a-Si:H and μc-Si:H solar cells deposited by VHF PECVD has been studied. The shape of the spectrum depends on the type of cell - microcrystalline or amorphous, the thickness of the intrinsic layer, the presence of texture on the ZnO:Al transparent contact, and other parameters. The results are discussed, taking into account the carrier generation, recombination and extraction mechanisms in these devices.

2. Experimental Details

The solar cells examined were single junction thin film *p-i-n* cells deposited on glass. One of them was an a-Si:H solar cell with an Asahi (U) front contact, an a-SiC:H *p*-layer (10 nm), an a-Si:H *i*-layer (530 nm), an a-Si:H *n*-layer (20 nm) and a Ag back contact. It was deposited by PECVD at 13.56 MHz. The μc-Si:H solar cells had a ZnO:Al etched or smooth front contact, a μc-Si:H *p*-layer(20 nm), a μc-Si:H *i*-layer (different thicknesses between 500 and 2200 nm), an a-Si:H *n*-layer (20 nm) and a ZnO:Al/Ag back contact. The cells were deposited by PECVD, using a 5% silane concentration in H_2 and a frequency of 94.7 MHz.

A diagram of the equipment built to measure the DSR is presented in Figure 1a. The sample was illuminated by a halogen lamp, and at the same time by monochromatic light modulated by a chopper. The spectrum of the DSR generated by the monochromatic light was recorded, using a lock-in amplifier and an electrometer controlled by a computer. The wavelength range of the background

J.M. Marshall and D. Dimova-Malinovska (eds.),
Photovoltaic and Photoactive Materials - Properties, Technology and Applications, 273–276.

illumination could be changed using optical filters. Measurements were performed under short-circuit conditions, with a current through the halogen lamp (9) of 3.5 or 4.7 A. The latter light intensity generated in the solar cells a short circuit current (I_{sc}) equal to that obtained under AM 1.5 illumination.

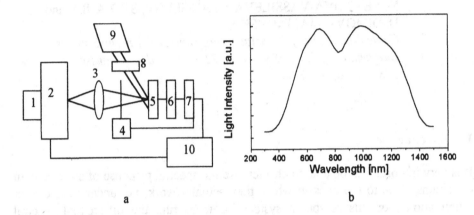

a   b

Figure 1. (a) Experimental set-up: 1 - halogen lamp (30W), 2 - 1/4 meter monochromator DK240, 3 - lens, 4 – chopper, 5 - sample, 6 - current to voltage converter, 7 - lock-in amplifier, 8 - removable optical filter, 9 - halogen lamp (50W) for background illumination, 10 - computer. (b) Spectral distribution of the light intensity coming from the monochromator.

The spectral distribution of the light intensity coming from the monochromator, using a 30W halogen lamp, a 1200 g/mm grating and a 1.5 mm slit was measured with a pyroelectric non-selective detector (Hamamatsu P2613-06). This curve (Figure 1b) was used to correct the measured DSR signal for the spectral distribution of light coming from the monochromator.

3. Results and Discussion

Figure 2a demonstrates that for an *i*-layer thickness of 500 nm, the µc-Si:H cell shows a better spectral response than the a-Si:H one, in both the short and the long wavelength region. In the range above 700 nm, this is due to the higher absorption coefficient of µc-Si:H at these wavelengths. For wavelengths below 550 nm, the results suggest less efficient charge carrier extraction in the amorphous material [3].

In Figure 2b, µc-Si:H solar cells differing only in the thickness of the intrinsic layer are compared. The difference at long wavelengths is entirely due to the better utilisation of weakly absorbed long wavelength light in the thicker devices. The much better spectral sensitivity of the cell with the thinnest *i*-layer, in the region below 550 nm, must be due to the better extraction efficiency of this device, for carriers generated near the surface of the cell by strongly-absorbed light [3,4].

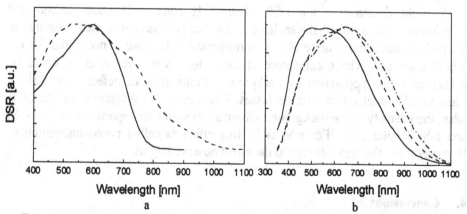

Figure 2. (a) DSR of an a-Si:H solar cell (solid line) and a μc-Si:H solar cell (dashed line) with approximately equal *i*-layer thicknesses (~500 nm). (b) Dependence of the DSR of *p-i-n* μc-Si:H solar cells on the thickness of the μc-Si:H *i*-layer, in cells with a textured front ZnO:Al contact: solid line – 0.5 μm *i*-layer, dashed line – 1 μm, dash dot line – 2.2 μm. The curves are normalized.

A wet chemical etching process (0.5% HCl) is used to obtain surface roughness in the ZnO:Al front contact, deposited by magnetron sputtering. The use of a textured ZnO:Al front contact leads to a significant improvement of the μc-Si:H solar cell sensitivity in the long wavelength region, where the absorption of the *i*-layer material is low, as shown in Figure 3a. This effect is due to "light trapping" which causes the incident light to be reflected several times, and thus to pass through the material more than twice, generating additional charge carrier pairs [3,5].

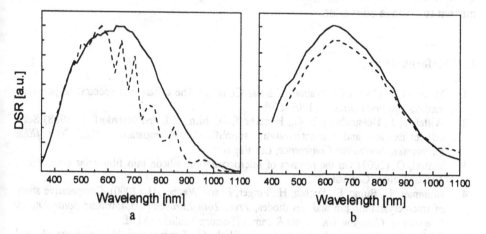

Figure 3. (a) Improvement of the DSR of *p-i-n* μc-Si:H solar cells (*i*-layer thickness 1μm), due to the wet chemical etching of the front ZnO:Al contact: solid line – etched ZnO:Al, dashed line – smooth ZnO:Al. The curves are normalized. (b) Change in the intensity of the DSR of a *p-i-n* μc-Si:H solar cell (*i*-layer thickness 2.2 μm) when background illumination with a 50W halogen lamp (3.5A) is used: solid line - measurement in the dark, dashed line - measurement under background illumination.

A slight decrease in the DSR of μc-Si:H solar cells under background illumination with a 50W halogen lamp (3.5A), in comparison to measurements in the dark, is observed (Figure 3b). A corresponding decrease is not found for the a-Si:H solar cell. The exact nature of this effect is not yet clear. It could be connected to the population by background illumination of defect states, which leads to additional recombination losses. Changes in the magnitude of the DSR when the intensity of the background illumination varies are reported in [1] for c-Si and a-Si:H solar cells. For c-Si cells, this effect is called photo-augmentation. However it has the opposite sign to the effect observed by us.

4. Conclusions

Differential spectral responsivity measurements give useful information on the performance of solar cells. The technique has been applied for the characterisation of a-Si:H and μc-Si:H cells. The dependence of the DSR on the type and thickness of the intrinsic layer, background illumination and front contact texture has been demonstrated and discussed.

5. Acknowledgements

The solar cells were prepared at the Institute for Photovoltaics, Forschungzentrum Jülich, D-52425 Jülich, Germany, and kindly provided by A. Lambertz and Dr. F. Finger. We are very grateful to Dr. G. Popkirov for the construction of the current to voltage converter.

6. References

1. Metzdorf, J. (1987) Calibration of Solar Cells. 1 : The differential spectral responsivity method, *Applied Optics* **26**, 1701-1708.
2. Wittchen, J., Holstenberg, H.-C., Hünerhoff, D., Min., Z.J. and Metzdorf, J. (1988) Solar cell calibration and characterization: simplified DSR apparatus, *Proc. 20th IEEE Photovolaic Specialists Conference*, Las Vegas.
3. Vetterl, O. (2001) On the physics of microcrystalline silicon thin film solar cells, *Ph.D. Thesis*, Juelich, Germany.
4. Brammer, T., Bunte, E., Stiebig, H., Finger, F. and Wagner, H. (2000), Comparative study of microcrystalline pin and nip diodes, *Proc. 16th European Photovoltaic Solar Energy Conference, Glasgow, UK*, James & James (Science Publishers) Ltd.
5. Stiebig, H., Brammer, T., Repmann, T., Kluth, O., Senoussaoui, N., Lambertz, A. and Wagner, H. (2000) Light scattering in microcrystalline silicon thin film solar cells, *Proc. 16th European Photovoltaic Solar Energy Conference, Glasgow, UK*, James & James (Science Publishers) Ltd.

TEMPERATURE AND COMPOSITIONAL DEPENDENCE OF RAMAN SCATTERING AND PHOTOLUMINESCENCE EMISSION IN $Cu_xGa_ySe_2$ THIN FILMS

C. XUE, D. PAPADIMITRIOU, Y.S. RAPTIS, N. ESSER[1],
W. RICHTER[1], S. SIEBENTRITT[2] and M.CH. LUX-STEINER[2]

National Technical University, Dept of Physics, 15780 Athens, Greece.
[1]Institut für Festkörperphysik, TU-Berlin, 10623 Berlin, Germany.
[2]Hahn-Meitner Institut, Glienickerstr. 100, 14109 Berlin, Germany.

Abstract

Raman scattering and photoluminescence emission of $Cu_xGa_ySe_2$ thin films grown by MOCVD and by PVD were studied at room and low temperatures in dependence of composition. The simultaneous broadening of both Raman- and PL-bands with increasing Ga-content, and the simultaneous intensity enhancement of defect related Raman modes and PL-bands observed in Ga-rich samples with decreasing temperature are discussed.

1. Introduction

Chalcopyrite compounds are studied for their applications in photovoltaics. Recently, conversion coefficients up to 18% were reported [1] for $Cu(In,Ga)(S,Se)_2$ based solar cells, while less-developed $CuGaSe_2$ based cells operate at a limit of 9,3% [2], so far.

In this work, the structural and emission properties of $Cu_xGa_ySe_2$ (CGS) layers grown by MOCVD on GaAs (100) and by PVD on Mo-glass substrates are studied at room and low temperatures down to 20 K, as a function of composition. All the studied samples were grown at the Hahn-Meitner-Institute in Berlin. According to the SEM-images of $CuGaSe_2$ samples of different compositions shown in Figure 1, Ga-rich samples exhibit structural modifications dependent on the Ga-content, while Cu-rich samples consist of stoichiometric $CuGaSe_2$ and Cu_xSe platelet crystallites covering the surface (with a density dependent on the Cu excess).

Raman and photoluminescence (PL) spectra were excited by the 514.5 nm line of an Ar^+-laser, and recorded from the same sample spot with a double grating monochromator/cooled PMT detection system.

J.M. Marshall and D. Dimova-Malinovska (eds.),
Photovoltaic and Photoactive Materials - Properties, Technology and Applications, 277–280.

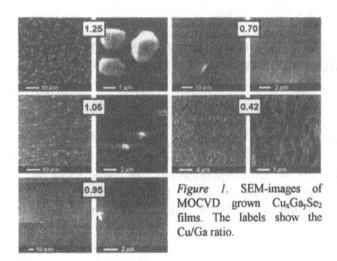

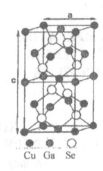

Figure 1. SEM-images of MOCVD grown $Cu_xGa_ySe_2$ films. The labels show the Cu/Ga ratio.

Figure 2. The crystal structure of $CuGaSe_2$.

2. Experimental Results

RAMAN SPECTROSCOPY: The crystal structure of ternary chalcopyrites belongs to space group D_{2D}^{12} (8 atoms per primitive unit cell), as in Figure 2. The spectrum of $CuGaSe_2$ consists of 22 Raman active modes [3]. Figure 3 shows the temperature dependence of the experimentally observed modes, for three different compositions (Cu-rich, stoichiometric, and Ga-rich) of $CuGaSe_2$ epitaxial layers (MOCVD) and polycrystalline films (PVD). The most intense modes, A_1 at 185 cm^{-1} (breathing-mode) and E_1 at 276 cm^{-1} (stretching-mode), were used for structural characterization. With increasing Ga content, the frequency of the two modes is almost unchanged, while their intensities decrease and the band widths increase. This indicates increasing disorder in Ga-rich samples, in agreement with the SEM images. With increasing temperature, both modes decrease in intensity, shift to lower frequencies and exhibit band broadening due to anharmonicity effects. At lower temperatures, additional modes emerge at 194, 197, 235 and 242 cm^{-1}. Those at 194 and 197 cm^{-1} are defect related. The intensity ratio of "defect-modes" to breathing mode shows a very strong temperature dependence in Ga-rich samples, but is almost temperature independent in stoichiometric and Cu-rich samples (Figure 5). This behaviour is interpreted, in agreement with the SEM-images, as due to disorder-enhanced anharmonicity with increasing Ga content.

PL-SPECTROSCOPY: Figure 4 shows the temperature and compositional dependence of PL spectra in the range 1.4-2.1 eV. PL-spectra of the stoichiometric and the (nominally) "Cu-rich" samples have three edge emission bands at E_a=2.02, E_b=1.81 and E_c=1.71 eV originating from valence band splitting (Figure 6). These bands are absent from the spectra of Ga-rich samples. The PL-bands due to donor-acceptor-pair transitions (DAP) in Cu-rich samples consist of two strong DA-emission peaks at 1.66 and 1.63 eV, accompanied by free-to-bound excitonic peaks (FB), and one weak peak at 1.60 eV, which has been identified [4] as a

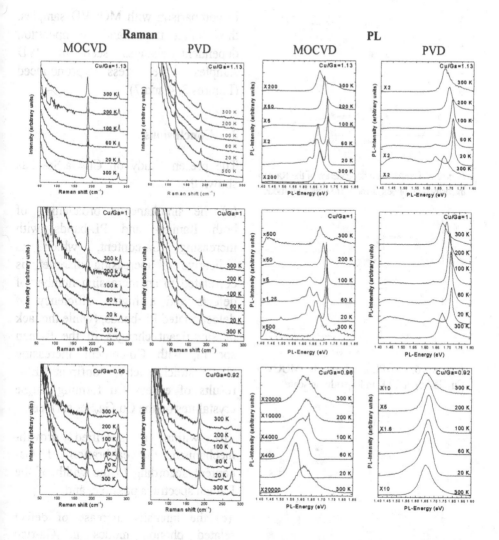

Figure 3. Temperature and compositional dependence of Raman spectra of MOCVD and PVD grown Cu$_x$Ga$_y$Se$_2$ films.

Figure 4. Temperature and compositional dependence of PL-spectra of MOCVD and PVD grown Cu$_x$Ga$_y$Se$_2$ films.

phonon-replica (Figure 6). The PL spectra of Ga-rich samples show only an asymmetric broad band in the range 1.52-1.60 eV (Figure 4) associated [5] with so called quasi donor-acceptor-pair transitions (QDAP). With increasing temperature, the PL-bands exhibit, independently of composition, an intensity decrease, a red energy shift, and band-broadening. With increasing Ga-content, the PL-emission exhibits, independently of temperature, an intensity enhancement, a red energy shift and band broadening (Figure 7). The PL-energy decrease is attributed [4,5] to potential fluctuations leading to band bending. The PL-band broadening is associated with increasing disorder and potential fluctuations.

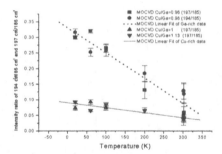

Figure 5. Temperature dependence of the Raman intensity ratio of defect-modes to breathing-mode.

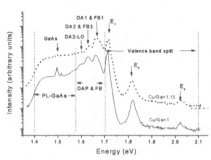

Figure 6. PL-spectra of MOCVD $Cu_xGa_ySe_2$ at 20 K in logarithmic scale.

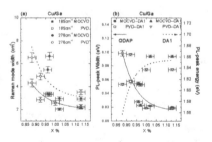

Figure 7. Compositional dependence of Raman bandwidth and of PL-energy and width for $Cu_xGa_ySe_2$ at 20 K (the line is a guide to the eye).

In comparison with MOCVD samples, the temperature and composition dependent changes seen in PVD samples are less pronounced (Figures 3, 4 and 7).

3. Conclusions

The present study on $Cu_xGa_ySe_2$ has shown that:

(a) the simultaneous broadening of both Raman- and PL-bands with increasing Ga-content, which is indicative of increasing disorder, is correlated to an increasing number of defects and a higher efficiency of defect related PL-bands, (while the lack of significant changes in the Raman spectra, with Cu-content increasing above stoichiometry, confirms SEM-results of excess Cu forming Cu_xSe crystallites on the $Cu_xGa_ySe_2$ surface),

(b) within a specified growth-mechanism, the calibration of Ga-content according to the width of the Raman spectra is possible, and

(c) the intensity increase of defect related phonon modes in Ga-rich samples at low temperatures may be associated with the intensity enhancement of the defect related PL.

4. References

1. Contreras, M., Egaas, B., Ramanathan, K., Hiltner, J., Swartzlander, A., Hasoon, F. and Noufi, R. (1999) *Prog. Photovoltaics*, **7**, 311.
2. Nadenau, V. *et al.* (1997), in "CuGaSe2 based thin film solar cells with improved performance", *Proc. 14th PV Solar Energy Conference*, Barcelona.
3. Tanino, H., Maeda, T., Fujikake, H., Nakanishi, H., Endo, S. and Irie, T. (1992), *Phys. Rev. B* **45**, 13323.
4. Bauknecht, A. (1999), Ph.D. thesis, Freie Universität Berlin.
5. Bauknecht, A., Siebentritt, S., Albert, J. and Lux-Steiner, M. Ch. (2001), *J. Appl. Phys.* **89**, 4391.

ENHANCEMENT OF THE PHOTOVOLTAIC EFFICIENCY OF $Ge_{0.2}Si_{0.8}$/Si PHOTODIODES

M.M. POCIASK, T. KĄKOL, E.M. SHEREGII, M.A. POCIASK and
G.M. TOMAKA
*Institute of Physics, University of Rzeszów, Rejtana 16A, 35-310
Rzeszów, Poland*

1. Introduction

In the search for higher photovoltaic (PV) conversion efficiencies, multiple band gap concepts have received increased theoretical attention in the last few years. Many dual band gap systems are based on the semiconductor couple Ge_xSi_{1-x}/Si. However, this is not the optimum theoretical pairing for maximum system efficiency, because of both economic and technological criteria. The Si cell may work at a moderate concentration of infrared light which would be unconverted by GaAs (energy lower than about 1.5 eV). However, by changing the basic solar cell material from Si to a Ge_xSi_{1-x} alloy, with the least possible modification, solar technology can fully profit from present knowledge and future improvements [1-4].

The purpose of the present work is to examine the possibilities of heterostructure Ge_xSi_{1-x}/Si materials. We explore, by theoretical modelling, how to construct Si-like solar cells showing enhanced infrared efficiencies.

Polycrystalline SiGe specimens were obtained covering the full range $0 < x < 1$, and were thoroughly characterised using optical absorption. Ge_xSi_{1-x} is usually obtained as very thin (strained) or thicker (unstrained) epitaxial layers forming heterojunctions with the Si or Ge substrate. The band gap energies at 300 K were calculated and experimentally checked. The absorption coefficient of Ge_xSi_{1-x} for infrared photon energies (hv<1.5eV) is $\alpha = 10^1\text{-}10^2$ cm^{-1}, for a 1.1eV photon energy. Increasing the Ge content in Ge_xSi_{1-x} decreases the energy gap, E_g, from 1.12 eV for pure Si to 0.66 eV for pure Ge. There is a strong decrease in the E_g curve at $x = 0.85$.

Unstrained crystalline Ge_xSi_{1-x} alloys have been shown to be powerful candidate materials for infrared photovoltaic conversion. The use of Ge_xSi_{1-x} solar cells instead of Si ones in GaAs/Si systems should improve their efficiency by up to 2 percentage points, i.e. from 29 to 31% for $x = 0.5$, and 30% for $x = 0.15$ [2].

2. Solar Cell Efficiency Simulation

Let us analyse the theoretical efficiency of barrier photocells [5,6]:

$$\eta_m = (kT/E_g)(1-R)\beta f(z)\psi(x_i), \tag{1}$$

281

J.M. Marshall and D. Dimova-Malinovska (eds.),
Photovoltaic and Photoactive Materials - Properties, Technology and Applications, 281–284.
© 2002 *Kluwer Academic Publishers.*

where R is the reflectivity coefficient, $\beta = (1 + (sd/D_p))^{-1}$ is the surface recombination coefficient, d is the distance from the illuminated surface, D_p is the hole diffusion coefficient, s is the surface recombination velocity, and

$$f(z) = \frac{y_m^2}{z(1 + z - y_m)}, \quad 0 \le y_m \le z, \quad z = \frac{i_l}{i_s}, \quad y = \frac{i}{i_s}. \tag{2}$$

Here, y_m is the root of the equation $y_m = (1 + z - y_m)\ln(1 + z - y_m)$, which is a condition for the power maximum in the outside circle: i_l is the flux-induced current component, i_s is the diode saturation current, i is the total current in the photovoltaic cell and

$$\psi(x_1) = x_1 \frac{\int_{x_1}^{\infty} \frac{x^2 dx}{e^x - 1}}{\int_0^{\infty} \frac{x^3 dx}{e^x - 1}}, \qquad x_1 = \frac{E_g}{kT_1}, \qquad x = \frac{\hbar\omega}{kT_1}, \tag{3}$$

where $\psi(x_1)$ is a function describing the relationship between the generation of carriers and the power of the electromagnetic irradiation. Because solar irradiation is close to the full black body irradiation, the Planck formula for $T = T_1 = 6000$ K (temperature of the solar surface) was used. The form of the $\psi(x_1)$ function is presented in Figure 1.

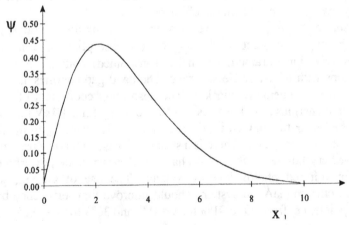

Figure 1. The $\psi(x_1)$ function.

Analysing expressions (1) and (2) and Figure 1, we can observe that a barrier photocell can give a maximum 44 % efficiency when all external and internal energy losses are neglected. The function $\psi(x_1)$ has its maximum at about $E_g = x_1 \cdot kT_1 = 1.1$ eV, i.e. at the Si band gap.

3. Theoretical Considerations on Heterojunction Efficiency

If we could compose a heterojunction with Si and Si-like band gap materials, we could theoretically increase the efficiency of the system.

3.1. ENERGETIC MODEL OF A CELL WITH A Ge_xSi_{1-x} /Si HETEROJUNCTION

Let us analyse the energetic model of a photoelectric cell with a Ge_xSi_{1-x}/Si heterojunction. The n-type semiconductor (Si, with energy gap $E_g(Si)$) is the illuminated side of the heterojunction, while the p-type one ((Ge_xSi_{1-x}) with energy gap $E_g(Ge_xSi_{1-x}) < E_g(Si)$) is the rear side. All photons with energy $\hbar\omega < E_g(Si)$ are absorbed very weakly in the n-area, and the barrier arising for holes counteracts their flow to the irradiated surface. Photons with energy $\hbar\omega > E_g(Si)$ are strongly absorbed in the n-area, and the p-n junction is not accessible for them. Photons with energy $\hbar\omega < E_g(Ge_xSi_{1-x})$ do not generate electron-hole pairs in the junction region. Only in the photon energy range $E_g(Si)$ to $E_g(Ge_xSi_{1-x})$ is the photocell sensitive.

Si-like material with E_g less then 1.1 eV can be chosen as a heterojunction solar cell material, but the point is to choose material with optimal optical and electrical parameters. As we see, GeSi is a good candidate.

3.2. EFFICIENCY FOR Ge_xSi_{1-x} /Si SYSTEM SIMULATION

Now let us consider $\psi(x_1)$ for Ge_xSi_{1-x} as a function of the germanium component (Figure 2). It should be noted that with increasing Ge content in a Ge_xSi_{1-x} alloy, the function $\psi(x_1)$ decreases. Concluding, the efficiency of the Ge_xSi_{1-x}/Si system is:

$$\eta_{Ge_xSi_{1-x}/Si} = \eta\big|_{Si} + \eta\big|_{Ge_xSi_{1-x}} = A\,\psi(x_1)\big|_{Si} + B\,\psi(x_1)\big|_{Ge_xSi_{1-x}}. \qquad (4)$$

For our simulations, we have chosen (neglecting energy losses):

$$A = \frac{kT}{E_{g\,Si}}(1 - R_{Si})\beta_{Si} f_{Si}(z) = 1, \qquad\qquad z\big|_{Si} = 10^4,$$

$$B = \frac{kT}{E_{g\,Ge_xSi_{1-x}}}(1 - R_{Ge_xSi_{1-x}})\beta_{Ge_xSi_{1-x}} f_{Ge_xSi_{1-x}}(z) = \frac{1}{p}, \qquad z\big|_{Ge_xSi_{1-x}} = 10^2,$$

$$\frac{1}{p} = \frac{f_{Ge_xSi_{1-x}}(z)}{f_{Si}(z)} = 0.4.$$

The results of the modelling of $\eta_{Ge_xSi_{1-x}/Si}$ (see Equation (4)) for $0 < x < 0.8$ are presented in Figure 3.

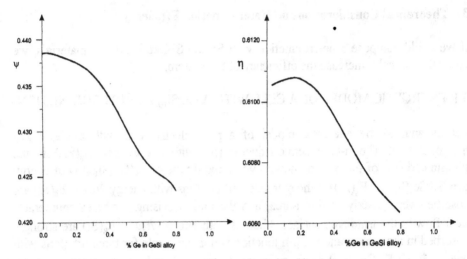

Figure 2. The $\psi(x_1)\big|_{Ge_xSi_{1-x}}$ function versus Ge composition, for $0 < x < 0.8$.

Figure 3. The efficiency, $\eta_{Ge_xSi_{1-x}/Si}$, versus Ge composition, for $0 < x < 0.8$.

4. Conclusions

For Ge concentrations in the range $x = 0.1$ to 0.2, the $\psi(x_1)$ function has its maximum, and consequently the efficiency $\eta_{Ge_xSi_{1-x}/Si}$ is at its peak value. For Ge contents less than 15%, the increase in the production costs of solar cells is not significant. If the experimental efficiency of such a heterojunction reaches almost one half (30%) of the theoretical value, it becomes a very promising device for the future development of solar cell technology.

5. References

1. Luque, A. and Araujo, G.L (1990) *Physical Limitations to Photovoltaic Energy Conversion*, Adam Hilger, Bristol.
2. Ruiz, J.M., Casado, J. and Luque, A. (1994) Assessment of crystalline Ge_xSi_{1-x} infrared solar cells for dual bandgap PV concepts, *Proc. 12th European Photovoltaic Solar Energy Conference*, Amsterdam, The Netherlands, 572-574.
3. Noel, J.P., Rowell, N.L., Houghton, D.C. and Perovic, D.D. (1990) Intense photoluminescence between 1.3 and 1.8 μm from strained $Si_{1-x}Ge_x$ alloys, *Appl. Phys. Lett.* **57**, 1037-1040.
4. Pearsall, T.P. (1988) The physics and applications of Ge-Si strained layer superlattices, *Rev. Solid State Science* **1**, 505-513.
5. Tauc, J. (1966) *Photoelectric and Thermoelectric Phenomena Semiconductors*, PWN, Warsaw (in Polish).
6. Boncz-Brujewicz, W.L. and Kałasznikow, S.G. (1985) *Semiconductor Physics*, PWN, Warsaw (in Polish).

CHARGE CARRIER TRANSPORT AND PHOTOVOLTAGE IN LAYERS BASED ON POLY(3,3`-PHTHALIDYLIDENE-4,4`-BIPHENYLILENE)

A.R. TAMEEV, A.A. KOZLOV, A.V. VANNIKOV and A.N. LACHINOV*

A.Frumkin Institute of Electrochemistry, Russian Academy of Sciences Leninsky Pr., 31, Moscow 117071, Russia, and
***Institute of Molecule and Crystal Physics, Russian Academy of Sciences, Prospekt Oktyabrya 151, Ufa 450075, Russia*

Abstract

Poly(3,3`-phthalidylidene-4,4`-biphenylilene) was investigated, for use as an electron-hole-transport layer in photovoltaic cells. The drift mobility of charge carriers decreased by one to two orders of magnitude when magnesium phthalocyanine dye molecules at 5 wt.% were dispersed into the polymer. The photovoltaic properties were studied for polymer-dye structures sandwiched between ITO and Al electrodes.

1. Introduction

In the search for organic compounds for possible use as active materials in molecular devices, a promising polymer poly(3,3`-phthalidylidene-4,4`-biphenylilene), hereinafter denoted as PPB (Figure 1), has been investigated [1]. The use of conjugated organic polymeric materials in molecular-based devices offers the possibility of combining, in a single material, the electrical properties of semiconductors with the flexibility, light weight and processability of typical polymers. In thin films, PPB has exhibited a dielectric-to-metal transition under the action of pressure, electric field or heat. The reason for this transition is not yet clear, and awaits a more detailed investigation.

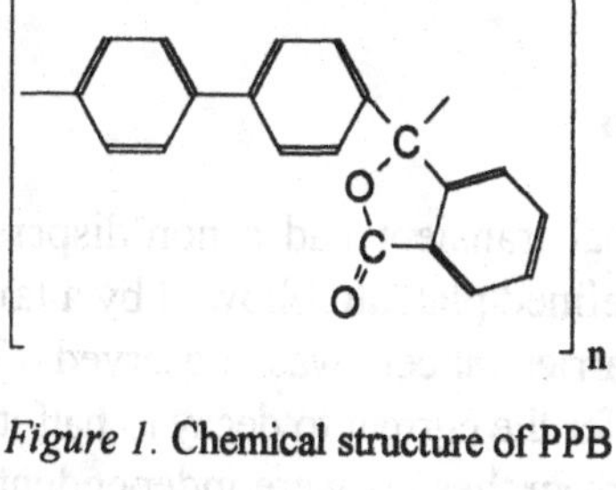

Figure 1. Chemical structure of PPB

285

J.M. Marshall and D. Dimova-Malinovska (eds.),
Photovoltaic and Photoactive Materials - Properties, Technology and Applications, 285–288.
© 2002 *Kluwer Academic Publishers.*

Photoactive phthalocyanine dyes can be used to sensitize the PPB to visible-wavelength light, and to produce photocurrents or photovoltages. They have attractive properties: (1) they are chemically very stable; (2) they have highly absorbing chromophores within the solar spectrum; (3) they often exhibit semiconducting behaviour and produce a rectifying interface [2]. Also, a wide variety of metal phthalocyanines have been prepared. We chose magnesium phthalocyanine (MgPc), allowing us to cast layers of the PPB doped with MgPc from a cyclohexanon solution. In this paper, we present a study of photovoltaic (PV) currents in layered structures based on PPB and MgPc, as well as the charge carrier mobility in PPB.

2. Experimental Details

The charge carrier drift mobility was measured by a conventional time-of-flight (TOF) technique [3,4]. The current transients were recorded with a Tektronix model 340A digital oscilloscope. The transit time is related to the drift mobility as $\mu = d/(F\,t_T)$, where F is the applied electric field and d the thickness of the charge transport layer (CTL). Sandwich specimens were prepared for the TOF experiment as follows: a CTL was first cast onto the aluminum substrate from a solution of PPB, or a mixture of PPB and MgPc (PPB:MgPc, the concentration of MgPc was 5 wt.%), in cyclohexanon. The layer was dried overnight at room temperature. The layer thickness, which varied between 3 and 10 μm, was determined using a micro-interferometer (MII-4). A 0.2 μm thick charge generation layer of MgPc was deposited on top of the CTL. Finally, a semitransparent aluminium electrode was deposited on top of the generation layer. Both this layer and the top electrode were deposited by thermal evaporation in a vacuum of 10^{-5} Torr. PPB with a molecular weight of 50,000-80,000 was used. The MgPc was from Fluka.

PV currents were measured in layered structures of ITO/PPB/MgPc/Al and ITO/PPB:MgPc/Al with a thickness of 0.8 μm, using a filtered tungsten-halogen lamp (3400 K black body spectrum) and a digital electrometer (Belvar V7-57/2). Both the PPB/MgPc and the PPB:MgPc structure had a similar optical density of 0.4, at the long-wave absorption band maximum of MgPc. The absorption spectra were recorded with a diode ray spectrophotometer (Ocean Optics, PC2000).

3. Results and Discussion

For the PPB films, a typical transient had a non-dispersive profile with a short initial spike, then a well-defined plateau followed by a tail, i.e. normal transport of the generated charge carrier sheet was observed. Transit times, t_T, were determined from the time for the current to decay to half its plateau value [5,6]. The drift velocities calculated from these t_T were independent of the sample thickness, within experimental error. The transport of both holes and electrons was observed, and a Poole-Frenkel plot of the field dependence of the mobility is presented in Figure 2. In the PPB:MgPc system, the electron and hole mobilities are smaller by one and one-two orders of magnitude, respectively, in comparison to PPB. This is because the MgPc molecules serve as charge carrier traps. Such an influence of the dispersed dye molecules on the hole mobility is associated with the lower ionization potential of MgPc (IP = 4.95 eV [2]) relative to PPB (IP = 6.1 eV [7]). In the case of electron transport, it is associated with the higher electron affinity of MgPc (EA = 3.34 eV [2]), compared to PPB (EA = 2.15 eV [8]).

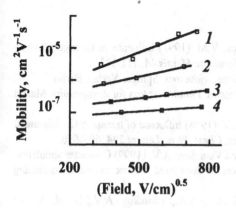

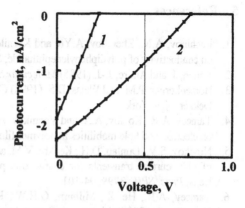

Figure 2. Hole (*1,3*) and electron (*2,4*) drift mobility in PPB (*1* and *2*) and in MgPc doped PPB (*3* and *4*).

Figure 3. *I-V* curves in the 4th quadrant for diodes based on a MgPc (5 wt.%) doped PPB single layer (*1*) and a double layer PPB/PcMg structure (*2*) under irradiation at 4 mW/cm² through the ITO electrode.

Figure 3 shows the current-voltage (*I-V*) curves of the photocurrent in ITO/PPB:MgPc/Al (*1*) and ITO/PPB/MgPc/Al (*2*) structures. In the polymer-dye system, charge carriers (or excitons) are known to transfer from photoactive dye molecules to polymer. The concentration of charge carriers photo-injected into (or generated in) the PPB will be higher in structure (*1*) than in structure (*2*), because the PPB/MgPc interface has a larger area in system (*1*) and both structures have a similar light absorption. On the other hand, the PV current is defined by diffusion of charge carriers through the bulk, and the diffusion coefficient is known to be proportional to the mobility, in accordance with the Einstein relationship. A superposition of the free charge generation and transport processes leads to a short-circuit current in system (*1*), with a magnitude comparable to that in (*2*) (Figure 3). The difference in the magnitudes of the photo-emf (open-circuit voltage) may be associated with the importance of the potential barrier (Schottky-type) on the Al/MgPc contact [2] in the cells studied. This contact is present only in structure (*2*), so this cell demonstrates a higher photo-emf than (*1*).

In conclusion, we have demonstrated that the drift mobility of charge carriers in PPB decreases by one or two orders of magnitude when MgPc molecules (5 wt.%) are dispersed into the polymer. This correlates well with data from photovoltaic current measurements in appropriate layered structures.

4. Acknowledgments

This work was supported in part by the Russian Fund for Basic Research (Project 00-03-33144) and the International Science and Technology Centre (PP 2207).

5. References

1. Lachinov, A.N., Zherebov, A.Yu. and Kornilov, V.M. (1991) Influence of uniaxial pressure on conductivity of polydiphenylenephthalide, *Synthetic Metals* **44**, 111-113.
2. Simon, J. and Andre, J.-J. (1985) *Molecular Semiconductors*, Spinger-Verlag, Berlin.
3. Borsenberger, P.M. and Weiss, D.S. (1998) *Organic Photoreceptors for Xerography*, Marcel Dekker, New York.
4. Tameev, A.R., Kozlov, A.A. and Vannikov, A.V. (1998) Influence of transport site alignment on electron and hole mobilities in polymer films, *Chem. Phys. Letters* **294**, 605-610.
5. Novikov, S.V., Dunlap, D.H., Kenkre, V.M. and Vannikov, A.V. (1999) Computer simulation of photocurrent transients for charge transport in disordered organic materials containing traps, *Proc. SPIE*, **3799**, 94-101.
6. Tameev, A.R., He, Z., Milburn, G.H.W., Kozlov, A.A., Vannikov, A.V., Danel, A. and Tomasik, P. (2000) Electron drift mobility in pyrazolo[3,4-b]quinoline doped polystyrene layers, *Appl. Phys. Letters* **77**, 322-324.
7. Zykov, B.G., Baydin, V.N., Bayrina, Z.Sh., Timoshenko, M.M., Lachinov, A.N. and Zolotukhin, M.G. (1992) Valence electronic structure of phtalide-based polymers, *J. Electron Spectroscopy and Related Phenomena* **61**, 123-129.
8. Wu, C.R., Johansson, N., Lachinov A.N., Stafstrom, S., Kugler T., Rasmusson J. and Salaneck W.R. (1994) The chemical and electronic structure of the conjugated polymer poly(3,3'-phthalidyliden-4,4'-biphenylilene), *Synthetic Metals*, **67**, 125-128.

DIAGNOSTICS OF LARGE-AREA SOLAR CELL HOMOGENEITY BY LOCAL IRRADIATION

V. BENDA
CTU Prague, Department of Electrotechnology,
Technicka 2, 166 27 Praha 6, Czech Republic.

1. Introduction

A low volume concentration of recombination centres is very important for the fabrication of high-efficiency crystalline silicon solar cells. The fast and non-destructive diagnostics of solar cells can give information about both the input material and the quality of the technological operations. The diagnostic method using local irradiation, which is described in the present paper, can be used to investigate the influence of technology on the homogeneity of solar cells, and consequently can help to increase their efficiency and reliability.

2. Description of the Method

The equivalent circuit shown in Figure 1 can approximate the V-A characteristic of a solar cell. The characteristic of an illuminated cell of area A can then be expressed as [1]:

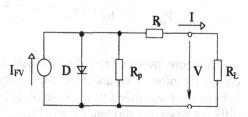

Figure 1. An equivalent circuit of a solar cell

$$I = AJ_{FV} - AJ_{01}\left[\exp\left(e\frac{V - R_s I}{kT}\right) - 1\right] - AJ_{02}\left[\exp\left(e\frac{V - R_s I}{2kT}\right) - 1\right] - \frac{V - R_s I}{R_p},\quad (1)$$

where

$$J_{01} = n_i^2 e\left(\frac{D_n}{L_n}\frac{1}{p_{p0}} + \frac{D_p}{L_p}\frac{1}{n_{n0}}\right)\quad (2)$$

represents the diffusion components of the p-n junction reverse current, and

$$J_{02} = \frac{en_i d}{\tau_{sc}}\quad (3)$$

is the generation-recombination component of the p-n junction reverse current.

J.M. Marshall and D. Dimova-Malinovska (eds.),
Photovoltaic and Photoactive Materials - Properties, Technology and Applications, 289–292.
© 2002 *Kluwer Academic Publishers.*

J_{FV} is the density of current generated within the volume of a cell structure (shown schematically in Figure 2) of thickness H by incident light:

$$J_{FV} = \int_0^\infty J_{FV}(\lambda)d\lambda, \qquad (4)$$

where

$$J_{FV}(\lambda) = e\int_0^H G(\lambda)dx - e\int_0^H \frac{\Delta n}{\tau}dx - J_{sr}(0) - J_{sr}(H) \qquad (5)$$

is the current density generated by light of the wavelength λ. The first term on the right hand side represents the current density generated by the incident light. The second term represents the recombination in the bulk of the device, and $J_{sr}(0)$ and $J_{sr}(H)$ represent surface recombination. It follows from Equation (5) that an increase in the rate of recombination results in a reduced cell efficiency. Non-uniformity in either the generation or the recombination rate over the area of the solar cell results in a non-uniform distribution of J_{FV}, J_{01} and J_{02}.

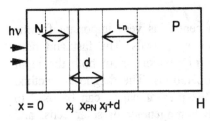

x = 0 x_j x_{PN} x_j+d   H

Figure 2. The structure of a solar cell

The open circuit condition is represented by $I = 0$ in Equation (1). Assuming a high R_p, one can obtain an open circuit voltage:

$$V_{OC} = \frac{2kT}{e}\ln\left(\frac{-J_{02} + \sqrt{J_{02}^2 + 4J_{01}(J_{02} + J_{01} + J_{FV})}}{2J_{01}}\right). \qquad (6)$$

Therefore, by measuring the distribution of V_{OC} over the area of a solar cell, we can obtain information about the cell quality. From detailed simulations of solar cell behaviour, it was found that the use of different wavelengths of incident light gives different types of important information about non-uniformity in recombination.

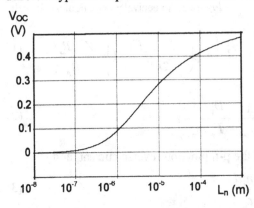

Figure 3. Simulated dependence of V_{OC} on diffusion length L_n, for a solar cell illuminated with a laser diode: $\lambda = 870$ nm.

If infrared light is applied, the absorption length is relatively long (about 30 μm for $\lambda = 900$ nm and 20 μm at 870 nm). The influence of the local centre concentration on the carrier lifetime in the bulk (p-type base) of the cell dominates, and the measured distribution of the open circuit voltage, V_{OC}, corresponds to the distribution of recombination centres in the bulk. The influences of the distribution of recombination centres in the highly doped diffusion layer, and of surface recombination, are negligible. The simulated dependence of V_{OC} on the electron diffusion length L_n in the p-base of the cell structure, under illumination by an 870 nm laser diode, is shown in Figure 3. In this manner, local infrared irradiation gives important information about the starting material and the fabrication process quality.

Figure 4. Schematic diagram of the measuring equipment

If visible light is applied, the absorption length is much shorter (about 3μm for $\lambda =$ 635 nm, and 1 μm at 500 nm), and excess carriers are generated in the close vicinity of the p-n junction. Recombination in the highly doped region (relatively very homogeneous, because Auger recombination is dominant) and surface recombination have a much greater influence than in the case of incident infrared light, and the influence of bulk recombination is much lower.

Information on the positions and extents of local defects in large-area solar cells can be obtained [2] using local irradiation by a laser diode of suitable wavelength. The cell is placed on an X-Y translation table, and separate areas between the busbars are locally illuminated by the laser diode. The basic arrangement is shown in Figure 4. Because only a dc voltage is measured, the measurement is very quick and reliable. It can easily be automated and used for in-process checking.

3. Results and Discussion

In our experiments, standard single-crystal Si cells of area 102x102 mm^2 were used, together with laser diodes with wavelengths of both 870 and 635 nm.

On the basis of the distribution of V_{OC} across the surfaces of solar cells, measured using incident light of wavelength 870 nm, three basic groups of cells can be defined. The first consists of those with a relatively very homogeneous distribution of V_{OC} (differences in V_{OC} less than 5% of the average value, which is higher than 150 mV). An example of a measured V_{OC} distribution (using incident light of this wavelength) is given in Figure 5.

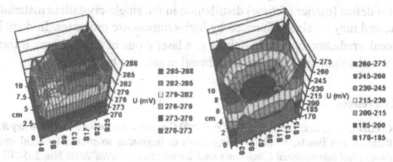

Figure 5. Measured V_{OC} distribution on a homogeneous solar cell (0.27 < V_{OC} < 0.29 V)

Figure 6. Measured non-uniform distribution of V_{OC} (0.17 < V_{OC} < 0.275 V)

292

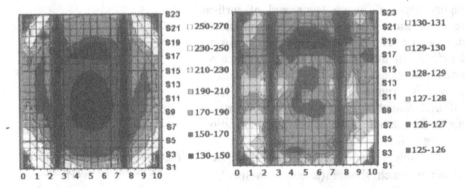

Figure 7. V_{OC} distribution for λ = 870 nm.   *Figure 8.* V_{OC} distribution for λ = 635 nm.

The second group consists of cells with an inhomogeneous distribution of V_{OC}, as in Figure 6 (again obtained with illumination at 870 nm). The non-uniform distribution results in a remarkable decrease in solar cell efficiency (about 14% for the first group and 10% for the second). The distribution of local V_{OC} in the form of concentric circles corresponds to impurity striations [3] (e.g. oxygen, carbon, etc.) in silicon single-crystals grown by the Czochralski technique.

The third group consists of samples with a low V_{OC} (less than 30 mV), indicating contamination during the fabrication process.

The influence of wavelength of the incident light is demonstrated in Figures 7 and 8. In Figure 7 (λ = 870 nm), the areas with a reduced local V_{OC}, in the form of concentric circles, correspond to impurity striations. Figure 8 shows the V_{OC} distribution for the same sample, measured at λ = 635 nm. Similar patterns can be recognised (slightly changed), but the differences in the measured V_{OC} are much smaller. The changes in the pattern can be connected with surface recombination.

4. Conclusions

The measuring method described allows a simplification and speed up of the in-process diagnostics of solar cells. The distribution of the measured V_{OC} induced by local infrared irradiation (e.g. a laser diode with λ = 870 nm) mostly corresponds to the local defect (carrier lifetime) distribution in the single-crystalline material of the p-base, and may be also influenced by high-temperature processes. In combination with local irradiation by visible light (e.g. a laser diode of λ = 635 nm), information about the distribution of surface recombination can also be obtained.

5. References

1. Goetzberger A., Knobloch J. and Voss B. (1998) *Crystalline silicon solar cells*, J.Wiley & Sons.
2. Radil, J. and Benda, V. (2000) Diagnostics of large-area solar cells by local irradiation, *Proc. 22nd International Conference on Microelectronics MIEl'2000*, Nis, 205-207.
3. Ravi, K.V. (1981) *Imperfections and impurities in semiconductor silicon*, J.Wiley & Sons.

DETERMINATION OF TRAP PARAMETERS FROM PHOTOCURRENT DECAY MEASUREMENTS: METAL-FREE PHTHALOCYANINE FILMS

I. ZHIVKOV[1, 2)], S. NEŠPŮREK [2)] and J. SWORAKOWSKI [3)]
[1]Central Laboratory of Photoprocesses, Bulgarian Academy of
Sciences, G. Bonchev Str., bl. 109, 1040 Sofia, Bulgaria
[2]Institute of Macromolecular Chemistry, Academy of Sciences of the
Czech Republic, Heyrovsky Sq. 2, 162 06 Prague 6, Czech Republic
[3]Institute of Physical and Theoretical Chemistry, Technical University
of Wroclaw, Wyb. Wyspianskiego 27, 50-370 Wroclaw, Poland

Abstract

The energy distribution of traps in gap samples of thin vacuum-evaporated metal-free phthalocyanine (H$_2$Pc) films was determined from measurements of photocurrent decays performed at several temperatures, in a nitrogen atmosphere. The parameters of the traps were determined using a straightforward method of the analysis of a first-order process, with distributed parameters.

It was found that the photocurrents are controlled by a narrow distribution of shallow traps peaking at about 0.2 eV, with a distribution width of about 0.015 eV and a very low effective attempt-to-escape factor.

1. Introduction

The photoelectrical properties of phthalocyanines are influenced by the trapping of photogenerated charge carriers [1]. Thus, a correct determination of the energetic and spatial distributions of charge carrier traps is of primary importance. The isothermal decay current method is a simple procedure for determining the parameters of the traps controlling the kinetics of the photocurrent relaxation [2]. Except in the simplest case of a first-order decay controlled by monoenergetic traps, the relaxation processes are usually analyzed using sophisticated Fourier transformations. One can, however, simplify the task using a method of analysis of first-order processes, but with distributed parameters [3].

The present study aimed to investigate the influence of traps on the photoconductivity in metal-free phthalocyanine (H$_2$Pc) thin vacuum-deposited films.

J.M. Marshall and D. Dimova-Malinovska (eds.),
Photovoltaic and Photoactive Materials - Properties, Technology and Applications, 293-296.
© 2002 Kluwer Academic Publishers.

2. Outline of the Method

The occupancy of traps [n(t)] decaying according to first-order kinetics with distributed parameters often follows a so called "stretched exponential" function

$$n(t) = n(0)\exp\left[-\left(\frac{t}{\tau}\right)^{\beta}\right],\tag{1}$$

where β is a parameter describing the deviation from a purely exponential decay ($\beta < 1$) and τ is the time constant of the process

$$\tau = k_1^{-1} = v_0^{-1}\exp\left(\frac{E}{k_B T}\right).\tag{2}$$

Here, k_1 is the rate constant, v_0 the attempt-to-escape factor, E the activation energy, k_B the Boltzmann constant and T the absolute temperature. In most cases, the 'stretching' results from an energy distribution of $k_1 = k_1(E)$. The energy distribution of traps (and hence, their occupancy at $t = 0$) is often approximated by a Gaussian function

$$n(E,0) = \frac{n_{total}}{\sigma\sqrt{2\pi}}\exp\left(-\frac{(E-E_0)^2}{2\sigma^2}\right),\tag{3}$$

where n_{total} is the total concentration of traps, E_0 is the position of the distribution maximum and σ represents the dispersion: for the Gaussian function the full width at half maximum (FWHM) is $\alpha\sigma$, with $\alpha \approx 2.36$. The method advanced in [3] consists of replotting a relaxation curve in $(I \times t)$ vs. $(\ln t)$ coordinates. The maximum of a bell-shaped curve obtained in such a way determines the time constant τ, whereas its width is related to the distribution width

$$\sigma = \sqrt{\frac{(\omega_0^2 - \omega^2)(k_B T)^2}{\alpha^2}},\tag{4}$$

where ω is the FWHM of the transformed relaxation curve $(I \times t)$ vs. $(\ln t)$, and $\omega_0 \approx 2.45$ is the 'intrinsic' FWHM ($\omega \equiv \omega_0$ for $\beta = 1$).

3. Experimental Details

H_2Pc films about 200 nm thick were deposited at a rate of 18 ± 2.5 nm min^{-1} on gold coplanar electrodes (interelectrode gap = 0.7 mm). The deposited samples were heated at 140°C for 90 minutes in a nitrogen atmosphere, in the dark. A biasing voltage of 100 V was then applied to the sample in the dark and, after a steady-state dark current has been attained, the film was exposed to red light (600 - 700 nm) of intensity 2.9×10^{-2} Wcm^{-2}. When the photocurrent reached its steady-state value, the light shutter was closed and the relaxation current was recorded. The measurement of the decay curves was performed at different temperatures in a N_2 atmosphere.

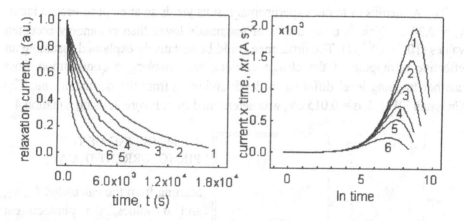

Figure 1. Relaxation currents (normalized to their initial values) measured at: 1 - 40, 2 - 60, 3 - 80, 4 - 90, 5 - 110 and 6 - 130 °C

Figure 2. ($I×t$) vs. ln t curves obtained from the experimental curves shown in Figure 1: 1 - 40, 2 - 60, 3 - 80, 4 - 90, 5 - 110 and 6 - 130 °C.

4. Results and Discussion

4.1. MEASUREMENT OF THE PHOTOCURRENT DECAY

Relaxation currents due to the release of charge carriers from shallow traps, measured at several temperatures, are plotted in linear coordinates in Figure 1. The time constants (τ) of the relaxation process were determined from the experimental data, replotted in ($I × t$) vs. (ln t) coordinates as shown in Figure 2. The relaxation time constants were found to decrease from $1.7 × 10^4$ to $4.3 × 10^3$ s upon increasing the temperature from 40 to 130 °C.

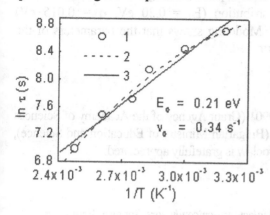

Figure 3. Temperature dependence of the time constant: 1 - experimantal results, 2 - polynomial fit, 3 - linear fit.

The temperature dependence of τ is shown in Figure 3. A fit to Equation (2) yields $E_0 = 0.21$ eV; a result close to the value of 0.25 eV obtained from steady-state photoconductivity measurements [4], and also observed in our earlier studies of thermally modulated space-charge-limited currents in H_2Pc films [5]. Note, however, that the experimental dependence is much better fitted by a polynomial (see Figure 3). The deviation from linearity is probably due to a contribution of side processes whose nature is yet to be established.

The Arrhenius fit to our experimental results yields an attempt-to-escape factor $v_0 = 0.34$ s^{-1}. This is many orders of magnitude lower than commonly accepted values ($10^{10} - 10^{13}$ s^{-1}). The difference could be tentatively explained assuming an effective retrapping of the charge carriers, and invoking a contribution from another trapping level differing in depth and/or v_0 from the dominant one. The Gaussian FWHM, $\sigma = 0.015$ eV, was determined from Figure 2, using equation 4.

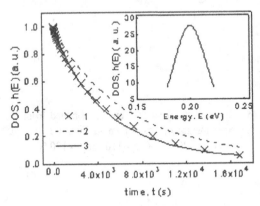

Figure 4. Photocurrent decay at 40 °C: 1 - measured, 2 - modeled, 3 - modeled after correction of E_a. Inset: Gaussian trap distribution corresponding to curve 3.

4.2 MODELING OF THE PHOTOCURRENT DECAY

Starting from the calculated E_0, v_0, and σ values, the photocurrent decay curve was calculated using equations 3, 2 and 1. Figure 4 shows that the modeled decay (curve 2) differs slightly from the measured one (curve 1). The difference results from an error in E_0. A small correction in E_0 (to 0.20 eV) results in a substantial improvement of the fit (curve 3).

5. Conclusions

Photocurrent decay curves were measured on H_2Pc thin film coplanar samples in the Ohmic conductivity region, in the temperature range 40 to 130°C. It was calculated that a Gaussian trap distribution ($E_0 = 0.20$ eV, $\sigma = 0.015$ eV) dominates the photocurrent kinetics. Modeling shows that the parameters of the Gaussian peak are satisfactorily determined.

6. Acknowledgments

Financial support by grants No A1050901 (Grant Agency of the Academy of Sciences of the Czech Republic) and No H806 (Bulgarian Ministry of Education and Science), and by the Technical University of Wroclaw is gratefully appreciated.

7. References

1. Simon, J. J. and Andre, H. J. (1985) *Molecular semiconductors*, Springer, Berlin.
2. Pope, M. and Swenberg, C. (1999) *Electronic Processes in Organic Crystals and Polymers*, Oxford University Press.
3. Sworakowski, J. and Nešpůrek, S. (1998) *Chem. Phys. Lett.* **298**, 21.
4. Schauer, F., Nešpůrek, S. and Zmeškal, O. (1986) *J. Phys. C: Solid State Phys.* **19**, 7231.
5. Zhivkov, I., Nešpůrek, S., Danev, G. and Schauer, F., to be published.

LUMINESCENT PROPERTIES OF $Li_2B_4O_7$ (LTB) POLYCRYSTALS AT THE DEVIATION FROM STOICHIOMETRY

B.M. HUNDA, A.M. SOLOMON, V.M. HOLOVEY, T.V. HUNDA,
P.P. PUGA and R.T. MARIYCHUK[*]
*Institute of Electron Physics, Ukrainian NAS,
21 Universitetska st., Uzhgorod 88000, Ukraine, and*
[*] *Uzhgorod National University, 46 Pidhirna st.,
Uzhgorod 88000, Ukraine*

1. Introduction

An analysis of literature data [1-4], shows that the more widespread use of polycrystalline $Li_2B_4O_7$ (LTB) based dosimetric materials is restricted by irreproducibility of the parameters of samples from different technological batches. In our opinion, one of the most essential causes of this irreproducibility is a deviation from the stoichiometric $Li_2B_4O_7$ composition. This may arise when during synthesis, due to a high hygroscopicity of the initial substances, weighing errors and incongruent evaporation of components during synthesis. However, no studies of the influence of this deviation on the LTB luminescent properties have yet been carried out. In this work, the influence of deviations from stoichiometry on the X-ray luminescent (XL) spectra, and on the intensities and temperatures of maxima in the thermostimulated luminescence (TSL) curves, has been studied.

2. Experimental Details

High-quality $Li_2B_4O_7$ single crystals were the initial materials for polycrystal production. The polycrystal series consisted of samples with no deviation from stoichiometry and those with deviations of 3.0 and 1.0 mol.% towards Li_2O and 1.0, 1.5, 3.0, 5.0, 7.0 and 9.0 mol.% towards B_2O_3.

The experimental TSL curves were measured by an IBM-based automated system, described elsewhere [5]. Samples were excited prior to TSL measurements by emission from an X-ray tube with a copper anode. The temperature was varied within the range 35 to 350°C, at a 2.90 °C/s linear heating rate. The temperature measuring error was 0.1 °C. The integrated luminescence intensity was detected using a FEU-106 photomultiplier in the photon counting mode. The XL spectra were recorded using a MDR-23 monochromator. Data were then smoothed by a standard technique, and averaged over five independent measurements.

J.M. Marshall and D. Dimova-Malinovska (eds.),
Photovoltaic and Photoactive Materials - Properties, Technology and Applications, 297–300.
© 2002 *Kluwer Academic Publishers.*

3. Results and Discussion

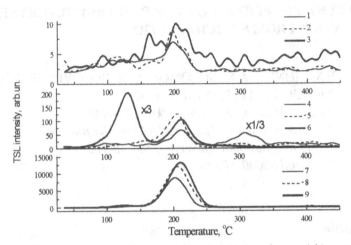

Figure 1. TSL curves of undoped polycrystals with deviations from stoichiometry towards
Li$_2$O by 3 mol.% - (1), 1 mol.% - (2), without deviation - (3), towards B$_2$O$_3$ by 1 mol.% - (4),
1,5 mol.% - (5), 3 mol.% - (6), 5 mol.% -(7), 7 mol.% - (8), 9 mol.% -(9).

TSL curves for isodose-irradiated samples are shown in Figure 1. For polycrystals
with a 3.0 mol.% deviation towards Li$_2$O, a maximum is observed at 200°C.
Reduction of the deviation to 1.0 mol.% causes a slight increase in the intensity of
this maximum, and a change in its shape. For stoichiometric samples, maxima are
observed at 94, 119, 140, 163, 180, 203, 217, 239, 256, 288, 307, 333, 367, 386,
427 and 445°C. A deviation from stoichiometry by 1 mol.% towards B$_2$O$_3$ causes a
twofold increase of the maxima intensities in the 203–217°C range, and a
threefold enhancement for that at 307°C. A further increase of the deviation, to
1.5 mol.% towards B$_2$O$_3$, favors an increase in the maximum close to 200°C
(curve 5). At a 3.0 mol.% B$_2$O$_3$ deviation, the intensity of the maximum at 133°C
increases extremely sharply (up to 30 times), while at larger deviations the
maximum in the 203–217°C range becomes dominant (curves 7–9). Figure 2
presents the dependences of the intensity of this maximum, and of the integrated
XL, on the deviations from stoichiometry. It follows from the above that, contrary
to TSL, the integrated XL intensity depends more weakly on the deviation values.
As our XL spectra show, variations within 3.0 mol.% B$_2$O$_3$ or Li$_2$O have almost no
influence on the spectrum shape and intensity. Above 3.0 mol.% B$_2$O$_3$ deviation, a
reduced emission intensity is observed within the 250–375 nm range, while in the
5–9 mol.% B$_2$O$_3$ range an increasing deviation reduces the XL intensity in the
entire spectral region.

The essential dependence of the TSL on deviations from stoichiometry in LTB
could be explained by the peculiarities of the pseudo-binary phase diagram of the
Li$_2$O–B$_2$O$_3$ system [6,7]. According to this, the LTB homogeneity region does not

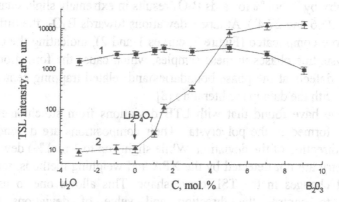

Figure 2. Variations of the XL intensity (1) and the intensities of TSL maxima at 200°C (2)
with LTB stoichiometry deviations towards B_2O_3 and Li_2O.

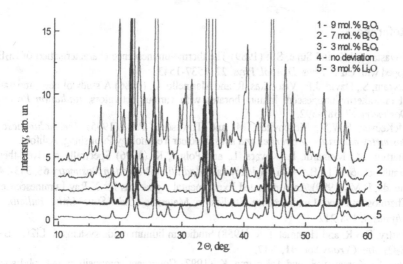

Figure 3. XRD pattern of stoichiometric LTB polycrystals and
polycrystals with stoichiometry deviations.

exceed 0.1 mol.% towards Li_2O and 0.2 mol.% towards B_2O_3. On both sides of
this range, there are regions inside which solid solutions and phase mixtures
($Li_2O \cdot B_2O_3$, $Li_2O \cdot 2B_2O_3$, $Li_2O \cdot 3B_2O_3$, $Li_2O \cdot 4B_2O_3$) exist [7].

To confirm the presence of the above phases in the samples under study, we
carried out X-ray phase studies (Figure 3). For samples with deviations of
composition up to 1 mol.%, no disagreement in the diffractograms is observed,
indicating an insufficient sensitivity of the XPA method to slight deviations from
stoichiometry. For polycrystalline samples with 3 mol.% excess of Li_2O, the
diffractogram shows a redistribution of the intensities of the main maxima, and the
appearance of new maxima at 16.7, 26.6, 27.8 and 29.8°. A deviation from

stoichiometry by 3 mol.% towards B_2O_3 results in extremely slight changes (small maxima at 26.6 and 27.4°). At larger deviations towards B_2O_3, the diffractograms become more complicated (Figure 3, curves 1 and 2), indicating the existence of several crystalline phases in these samples, which cause the formation of a large number of defects at the phase boundaries and related trapping centers, in good agreement with the data in the literature [8].

Thus, we have found that with LTB deviations from stoichiometry, several phases are formed in the polycrystal. Their compositions are dependent on the value and direction of the deviation. While slight (up to 1 mol.%) deviations from stoichiometry are not detected by the XPA and weighing methods, they result in substantial changes in the TSL curve shape. This allows one to use the TSL technique to control the direction and value of deviations from LTB stoichiometry [9]. The high TSL intensity for samples with 7–9 mol.% excess of B_2O_3 opens new possibilities for their use in thermoluminescent dosimetry.

4. References

1. Srivastava, J.K. and Supe, S.J. (1989) The thermoluminescence characterisation of $Li_2B_4O_7$ doped with Cu, *J. Phys. D. Appl. Phys.* **22**, 1537-1543.
2. Lorrain, S., David, J.P., Visocekas, R. and Marinello, G. (1986) A study of new preparations of radiothermoluminescent lithium borates with various activators, *Radiation Protection Dosimetry* **17**, 385-392.
3. McKeever, S.W.S., Moscovitch, M. and Townsend, P.D. (1995) *Thermoluminescence dosimetry materials: properties and uses*, Nuclear Technology Publishing, Ashford.
4. Martini, M., Meinardi, F., Kovacs, L. and Polgar, K. (1996) Spectrally resolved thermo-luminescence of $Li_2B_4O_7$:Cu single crystals, *Radiation Protection Dosimetry* **65**, 343-347.
5. Hunda, B.M. (1999) An Automated Experimental Apparatus for X-Ray Luminescence and Thermostimulated Luminescence Studies in Materials, *Uzh. State Univ. Bulletin. ser. Physics* **5**, 198-212.
6. Sastry, B.S.R. and Hummel, F.A. (1958) Studies in lithium oxide systems: I, Li2O · B_2O_3 - B_2O_3, *Am. Ceram. Soc.* **41**, 7-17.
7. Uda, S., Komatsu, R. and Takayama, K. (1997) Congruent composition and solid solution range of $Li_2B_4O_7$ crystal, *Journal of Crystal Growth* **171**, 458-462.
8. Bube, R.H. (1960) *Photoconductivity of solids*, John Wiley and Sons, inc., New York–London.
9. Hunda, B.M., Holovey, V.M., Solomon, A.M., Turok, I.I. and Puga, P.P. A method of controlling the deviation from the $Li_2B_4O_7$ stoichiometric composition, *Ukrainian patent № 36340A of 16.04.2001, Bul. №3*.

INFLUENCE OF SOLUTION RESISTIVITY AND POSTANODIZING TREATMENTS OF PS FILMS ON THE ELECTRICAL AND OPTICAL PROPERTIES OF METAL/PS/Si PHOTODIODES

K. AIT-HAMOUDA, N. GABOUZE, T. HADJERSI, N. BENREKAA,
R. OUTEMZABET, H. CHERAGA, K. BELDJILALI and B.R. MAHMOUDI
Unité de Développement de la Technologie du Silicium (UDTS),
2 bd. Frantz Fanon, B.P. 399 Alger-Gare, Algiers, Algeria

Abstract

An experimental study of the influence of solution resistivity and the effect of post-anodizing treatments of PS films on the electrical and optical properties of Metal/PS/Si photodiodes was performed. Porous silicon samples were made from *p*-Si in Ethylene-Glycol/HF or Ethanol/HF solutions of different concentrations. The anodised sample was left in the dark for different times in the HF/base electrolyte in which the porous layer had been fabricated, or in methanol. The results show a strong influence of the etching solution parameters and PS surface post-treatments on the kinetic changes of the device's electrical properties.

1. Introduction

Since the discovery of its luminescent properties, porous silicon (PS) has been the subject of much investigation for its use in device technology. Until recently, its electrical properties were poorly studied, and published work presented controversial conclusions about the mechanism of charge carrier transport through a porous matrix of PS [1,2].

In this paper, an experimental study of the influence of solution concentration (resistivity) and the effect of post-anodizing treatments of PS films on the electrical properties of Metal/PS/Si photodiodes has been performed. The I-V characteristics, spectral response and impedance measurements were used to investigate the device's electrical and optical properties.

2. Experimental Procedure

Porous silicon layers were made from *p*-type, (100) oriented boron doped CZ (Czhochralski) silicon (c-Si) 1-100 and 2000 Ω.cm by etching at a constant current density in HF/Ethanoic or HF/Ethylene Glycol solutions of different concentrations. The porous silicon was left for different times in the same

J.M. Marshall and D. Dimova-Malinovska (eds.),
Photovoltaic and Photoactive Materials - Properties, Technology and Applications, 301–304.
© 2002 *Kluwer Academic Publishers.*

electrolyte, or in methanol, in the dark. The photodiode structure was a 4x4 mm cell. Metallic contact was achieved by depositing 99.99% pure Al on the back of the porous silicon sample. The Al contact on the PS layer was in the form of a frame with about 400μm wide sides. The rest of the PS surface was left uncoated.

3. Results and Discussion

3.1. AGEING OF THE POROUS LAYER

The effect of chemical thinning (aging) of the porous silicon structure on the I-V characteristic of an MPS diode is shown in Figure 1. The anodised sample was left in the dark for different times (30, 45, 90 and 120 min) in the HF/base electrolyte in which the porous layer had been fabricated. It is seen that the thinning leads to a significant increase in reverse current and a decrease in the forward regime. The series resistances increase with thinning in the forward regime, indicating a blocking effect of the diode. Therefore, the increase in reverse current for thinner device structures is attributed to a change in the surface defect density.

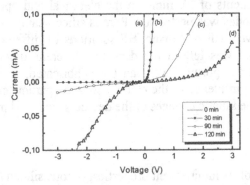

Figure 1. The effect of chemical aging time of the porous silicon in HF/Ethanol (1/1 by volume) on the I-V characteristics of a MPS diode.

3.2. EFFECT OF STORAGE IN METHANOL

The effect of post treatments of porous silicon with organic solvents on its optical properties has been extensively studied [3]. However, the effect on the electrical properties is poorly studied. The influence of methanol treatment on the Al/PS/Si I-V characteristics is illustrated in Figure 2, where data for as prepared (curve a) and 15 min in methanol treated (curve b) samples are presented.

The anodization was performed at 50 mA/cm^2, for 5 min in HF/Ethanoic solution (1:1 by volume). At a constant bias voltage (+3V), the current corresponding to as-anodized material is ten times greater than that observed in samples without methanol treatment. On the contrary, the methanol treated sample exhibits, at 3V reverse bias voltage, a current ten times lower. The behavior in the currents observed for methanol treated device structures is attributed to the formation of SiOCH$_3$ surface groups, reinforcing the surface stability.

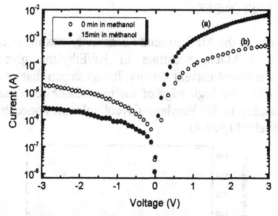

Figure 2. Influence of porous silicon methanol treatment on the I-V characteristics of Al/PS/Si structures: (a) 15 min immersed in methanol and (b) as prepared.

3.3. EFFECT OF SOLUTION RESISTIVITY

The effect of changing the electrolyte on porous silicon formation has recently been studied [4]. It was shown that replacing ethanol by ethylene glycol at different concentrations in solution leads to a large variation of electrolyte resistivity. Current-voltage measurements with devices made on resistive substrates (2 kΩcm) were also shown to be sensitive to the fabrication conditions. The I-V characteristic of the MPS structure indicated that for a given reverse bias, the reverse current strongly depends on the etching solution.

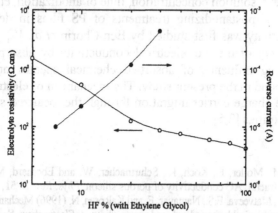

Figure 3. Variations of reverse current, at constant bias voltage (60V), and solution resistivity with HF concentration. The anodization was performed at 20 mA/cm² for 10 s.

The variations of reverse current and solution resistivity with HF concentration are shown in Figure 3. At a constant bias voltage (60V), the reverse current increases with increasing HF concentration, while the electrolyte resistivity decreases. When the potential was scanned in the positive direction, a weak change in forward current was observed.

3.4. SPECTRAL RESPONSES

The spectral response of the MPS structure sensitivity was measured for samples fabricated from a 2 kΩcm substrate in HF/Ethylene glycol of various concentrations, at a constant current density. It was shown that an increase in the HF concentration shifts the high edge of the photodiode response from 0.85 to 0.94 μm, corresponding to HF/Ethylene Glycol solution concentrations of 7/93 and 40/60, respectively (Figure 4).

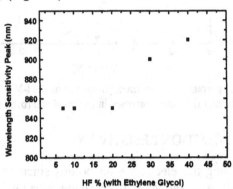

Figure 4. Wavelength sensitivity peak as a function of HF concentration for a MPS structure.

4. Conclusions

We have shown that the reverse current strongly depends on the fabrication parameters (solvent, solution concentration, time of anodization, etc.) of a PS film. The importance of postanodizing treatments of PS films in determining their electrical conductivity was first studied by Ben-Chorin *et al.* [5]. They showed a strong enhancement of the d.c. electrical conductivity by absorption of vapors. Similarly, a strong influence of absorbed chemical species on the electrical properties was found in the present study. The mechanism of electrical conduction is assumed to be charge carrier migration through the localized states associated with surface impurities [2,5].

5. References

1. Ben-Chorin, M., Moller, F., Koch, K., Schirmacher, W. and Eberhard, M. (1995) Hopping transport on a fractal: AC conductivity of porous silicon, *Phys. Rev. B* **51**, 2199-2213.
2. Parkhutik, V.P., Matveeva, E.S., Namavar, F. and Kalcoran, N. (1996) Mechanism of AC electrical transport of carriers in freshly formed and aged porous silicon, *J. Electrochem. Soc.* **143**, 3943-3949.
3. Fellah, S., Wehrspohn, R.B., Gabouze, N., Ozanam, F. and Chazalviel, J.-N. (1999) Photoluminescence quenching of porous silicon in organic solvents: evidence for dielectric effects, *J. Luminescence* **80**, 109-113.
4. Chazalviel, J.-N., Ozanam, F., Gabouze, N., Fellah, S. and Wherspohn, R.B. (2000), Quantitative analysis of the morphology of macropores on low-doped p-Si, 198th Meeting of the electrochemical society (Phoenix).
5. Ben-Chorin, M., Kux, A. and Schechter, I. (1994) Adsorbate effects on photoluminescence and electrical conductivity of porous silicon, *Appl. Phys. Lett.* **64**, 481-483.

STUDY OF SOL-GEL DERIVED VERY THIN FILMS OF MIXED TITANIUM DIOXIDE AND VANADIUM OXIDE

T. IVANOVA, A. HARIZANOVA, M. SURTCHEV[a] and Z. NENOVA
Central Laboratory of Solar Energy and New Energy Sources,
Bulgarian Academy of Sciences, 72 Tzarigradsko chaussee blvd.,
Sofia, Bulgaria, and [a]*Faculty of Physics, Sofia University, 5 James*
Bouchier blvd., Sofia, Bulgaria

1. Introduction

Semiconductor electrodes have gained much attention in the field of solar energy conversion, as electrodes in photoelectrochemical cells [1]. It seems that developing a mixed semiconductor is a way to produce a non-corrosive, narrow bandgap photoelectrode.

TiO_2 and TiO_2-V_2O_5 coatings can be used as anti-reflective films [2]. Nowadays, these are one of the most investigated parts of a solar cell, since they improve the device efficiency. Sol-gel technology, as a method for producing such films, has gained interest as it presents many advantages. These are the use of very simple equipment and a lower required capital investment [3], the high homogeneity of the final thin films, the possibilities of using many different substrates and of controlling the microstructure and density of thin films.

The optimum parameters for an anti-reflective film, to minimize the reflectance of the silicon substrate, are a refractive index ≈ 2.0 and a thickness ≈ 72 nm for the reference wavelength for silicon solar cells, $\lambda = 600$ nm. As TiO_2 and TiO_2-V_2O_5 have refractive indices around 2, they are promising materials for this purpose. They have also been studied as gas sensors [4], catalysts [5], and optoelectronic [6], photocatalytic [7] and electrochromic [8] devices.

In the present work, a sol–gel titanium dioxide and mixed Ti/V oxide system were investigated by XRD measurements and IR analysis. Thin films on Si substrates, prepared by a spin coating method, were characterized by ellipsometry.

2. Experimental Details

The films were prepared by the sol-gel method. The first step [10] is the formation of a titanium dioxide sol. The vanadium component is introduced as V_2O_5 powder dissolved in hydrogen peroxide (30%). The two solutions are mixed and aged for one week at room temperature. Different Ti/V ratios were investigated, and here we use the solution with the highest stability, of up to 1 year ($0.97TiO_2 - 0.03V_2O_5$).

305

J.M. Marshall and D. Dimova-Malinovska (eds.),
Photovoltaic and Photoactive Materials - Properties, Technology and Applications, 305–308.
© 2002 *Kluwer Academic Publishers.*

The vibrational properties were studied in the range of 200 – 1200 cm^{-1}, using a1430 Perkin Elmer Infrared Spectrophotometer. XRD analysis was performed with a URD6 diffractometer with a secondary graphite monochromator, using Cu Kα radiation (35 kV, 20 mA).The refractive indices and film thicknesses were measured with a LEF 3M ellipsometer at λ = 638.2 nm. The refractive index is directly correlated with the porosity. Assuming the film to be composed of a pure anatase phase, the porosity can be estimated as: Porosity $= 1 - (n^2 - 1)/(n_o^2 - 1)$, where **n** is the refractive index of the porous film and n_o is the reference refractive index of dense anatase (2.52) [9]. The porosity was calculated as a function of thermal treatment, for both materials.

3. Results and Discussion

It is of technical importance to obtain precursor solutions with long term stability, since this affects the film thickness in the sol-gel process. The solutions (titanium oxide and mixed sol) have stabilities of more than a year.

XRD measurements of powder samples, treated at 300, 430 and 560°C in air for 1 hour, were made. The 300°C sample shows no peaks, indicating an amorphous phase. The 430 and 560°C samples crystallized, showing the TiO$_2$ anatase phase (JCPDS card 21 – 1272) with no peaks attributable to vanadium oxide. Others also reported that when the concentration of the vanadium component is under 15% there are no significant indications of vanadium oxide [1]. A previous study [10] on the sol–gel titanium oxide system determined that crystallization began at 560°C in the TiO$_2$ anatase phase. In that case, a small amount of vanadium additive decreased the crystallization temperature of the anatase TiO$_2$. Figure 1 presents XRD patterns of TiO$_2$ – V$_2$O$_5$, treated at 560°C, for a powder sample and a thin film.

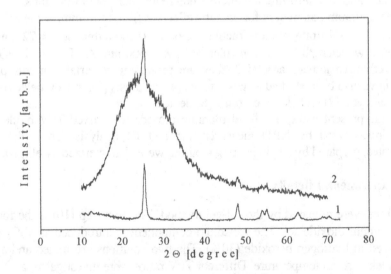

Figure1. XRD patterns of powder (curve 1) and thin film (curve 2) samples treated at 560°C in air for 1 hour.

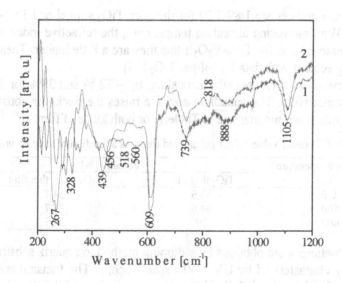

Figure 2. IR spectra of sol-gel TiO₂ - V₂O₅ films treated at 300 (1) and 560°C (2) in air.

IR spectra of the sol-gel films are shown in Figure 2. The absorption band at 1105 cm⁻¹ in both spectra is assigned to the Si-O vibration of the Si substrate, but could also be interpreted as overlapping Ti-O-C vibrations [11]. The main IR band at 609 cm⁻¹ is due to Ti-O stretching vibrations. The weak peaks near 900-980 cm⁻¹, missing for a pure TiO₂ film (not shown), indicate the presence of V=O, vanadyl stretching modes [11]. The absorption bands below 300 cm⁻¹ are due to deformation modes of the crystal lattice. The peak at 739 cm⁻¹ is attributed to Ti-O bonds.

The bands around 560 cm⁻¹ are due to TiO₆ stretching vibrations. The higher annealing temperature yields the appearance of two new absorption bands, at 328 and 439 cm⁻¹. The IR bands at 818 and 439 cm⁻¹ are assigned to vibrations of bound oxygen, shared by two vanadium atoms: the V–O–V bond [12]. The refractive indices and film thickness were obtained by ellipsometry (Figures 3 and 4). The spin-coated films on Si wafers contained 4 layers, and were smooth and uniform.

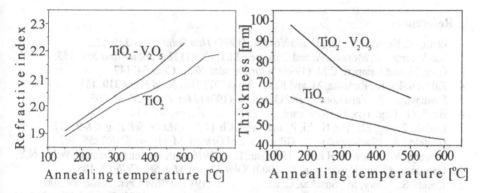

Figure 3. Influence of the annealing temperature on the refractive index.

Figure 4. Influence of the annealing temperature on the film thickness.

The refractive indices are 1.89-2.19 for the pure TiO_2 sol-gel and 1.91-2.28 for mixed films. With increasing annealing temperature, the refractive index increases for the both materials, as for TiO_2-V_2O_5 films they are a little higher. These values are in good agreement with data for sol-gel TiO_2 [13].

The film thickness decreases after annealing, by ~ 32 % and 38% for TiO_2 and TiO_2-V_2O_5, respectively. The vanadium additive raises the thickness considerably. The porosity values are summarized in Table 1 for both kinds of film.

TABLE 1. Porosity values of sol-gel derived films, as a function of temperature.

Annealing temperature [°C]	Porosity [%]	
	TiO_2 thin film	TiO_2-V_2O_5 thin film
400	39.6	34.7
500	34.6	25.7
600	29	21.5

Uniform coatings were obtained by a dipping method, on quartz substrates, and were optically characterized by UV – VIS spectroscopy. The transmittance in the visible range for the sol-gel TiO_2-V_2O_5 thin film is about 70-80%, at different annealing temperatures in the range 300 to 650°C.

The absorption was calculated from transmittance spectra for 560°C samples. Energy gaps were evaluated from the relation $\alpha h \nu \propto (h \nu - E_g)^n$ for the indirect allowed transition. From a linear extrapolation to zero of a plot of $(\alpha h \nu)^{1/2}$ versus $h \nu$, we found E_g for the sol–gel films. The calculated values are $E_g = 3.32$ eV for TiO_2–V_2O_5 and 3.67 eV for TiO_2. These are ascribed to polycrystalline materials.

4. Conclusions

This work demonstrates the possibility of preparing sol–gel TiO_2 and TiO_2–V_2O_5 systems. The sol stabilities are more than one year, making them suitable for industrial applications. The high quality, uniform and smooth thin films have excellent adhesion to various substrates. Film thickness can be controlled by the number of layers. Refractive index values are 1.89-2.19 for a pure TiO_2 sol-gel, and 1.91-2.28 for mixed oxide films. The ability to obtain porous coatings is proved.

5. References

1. Zhao, G., Kozuka, H., Lin, H and Yoko,T. (1999) Thin Solid Films 339, 123.
2. San Vicente, G., Morales, A. and Gutierrez, M.T. (2001) Thin Solid Films 391, 133.
3. Ozer, M. and Lampert, C.M. (1998) Solar En. Mat. Solar Cells 54, 147.
4. Zakrzewska, K., Radeska, M. and Rekas, M. (1997) Thin Solid Films 310, 161.
5. Sawunyama, P., Yasumori, A. and Okada, K. (1998) Mat. Res. Bull. 33, 705.
6. Banfi, G., Degiorgo, V. and Ricard, D. (1999) Adv. Physics 47, 447.
7. Dai, Q., Zhang, Zh., He, N., Li, P. and Yuan, Ch. (1999) Mater. Sci. Eng. C8-9, 417.
8. Hagfeldt, A., Vlachopoulos, N., Gilbert, S. and Grätzel, M. (1994) SPIE 2255, 297.
9. Kingery, W.D., Bowen, H.K. and Uhlmann, D.R. (1976) Introduction to Ceramic, Wiley, N.Y.
10. Harizanov, O. and Harizanova, A. (2000) Solar Energy Mat. Sol. Cells 63, 185.
11. Doeuff, S., Henry, M., Sanchez, C. and Livage, J. (1987) J. Non-Cryst. Solids 89, 206.
12. Benmoussa, M., Ibnouelghazi, E., Bennama, A. and Ameziane, E. (1995) Thin Sol. Films 365, 22.
13. Kumar, P.M., Badrinarayanan, S. and Sastry, M. (2000) Thin Solid Films 358, 122.

PHOTOELECTROCHEMICAL CHARACTERIZATION OF SOME ARGYRODITE-TYPE MATERIALS

Yu. STASYUK, S. KOVACH, A. KOKHAN, V. PANYKO and S. MOTRYA
Chemical Department of Uzhgorod State University, 46, Pidgirna Str., 88000, Uzhgorod, Ukraine

1. Introduction

Materials from the "argyrodite"family (general formula: $A_{(12-n-x)/m}BX_{(6-x)}Y_x$, where A^{m+} = Cu, Ag, Cd, Hg; B^{n+} = P, Ge, Si; X = S, Se; Y = Cl, Br, I) have a wide variety of properties. Among the argyrodites, there are high-efficiency photoconductors with various band-gaps (Cd_4GeSe_6, Cd_4SiSe_6, Cd_4GeS_6) and superionic materials Cu_6PS_5Y; Y – Cl, Br, I). All argyrodites containing Cu^+ and Ag^+ meet the structural requirements for efficient ionic transport - partial occupancy of a sublattice by ions of high mobility in the structurally disordered high-temperature phase. The electrical and ionic conductivities of argyrodites vary in a wide range - from purely electronic conductivity (Cd_4GeSe_6, Cd_4SiSe_6, Cd_4GeS_6), through mixed conductivity (Ag_8GeS_6, Ag_8GeSe_6, Cu_6PS_5I, Cu_6PSe_5I) to nearly purely ionic conductivity (Cu_6PS_5Cl, Cu_6PS_5Br). This vide variety of electrical properties make argyrodites prospective candidates for applications as photoelectrode materials, solid state batteries, high capacity electrochemical devices - ionistors - and others.

On the other hand, the ion mobility in electronically conducting compounds, if they are subjected to electrical or chemical potential gradients, especially under high illumination or at high temperatures, may lead to the degradation and instability of devices. Hence, there is a large interest in investigating the electrochemical processes in mixed conductors, and in developing detailed models for the degradation of commonly used electronic materials.

We present here the results of an electrochemical and photoelectrochemical investigation of Cd_4GeSe_6, Cd_4SiSe_6, Cd_4GeS_6, Ag_8GeS_6, Ag_8GeSe_6 and Cu_6PS_5I electrodes.

2. Experimental Details

Single crystals of Cd_4GeSe_6, Cd_4SiSe_6, Cd_4GeS_6, Cu_6PS_5I were grown by chemical vapour transport (CVT) method, while Ag_8GeS_6 and Ag_8GeSe_6 were grown by Bridgman's method from melts. Electrochemical experiments were made in water solutions with standard electrochemical equipment in a three-electrode cell under white light or monochromatic illumination. The flat-band potentials were

309

J.M. Marshall and D. Dimova-Malinovska (eds.),
Photovoltaic and Photoactive Materials - Properties, Technology and Applications, 309–312.
© 2002 *Kluwer Academic Publishers.*

determined from the sign of the photocurrent change during current-voltage scanning. From the photocurrent quantum efficiency spectral dependence, the band-edge and optical transitions type were determined.

3. Experimental Results and Discussion

Our investigations of the photoelectrochemical properties of Cd_4GeSe_6, Cd_4SiSe_6, and Cd_4GeS_6 electrodes demonstrated that optical transitions in these argyrodites are direct [1]. Manivannan *et al.* [2] determined, for Cd_4GeSe_6 and Cd_4SiSe_6, incorrect values of E_o (1.5 and 1.68 eV respectively) and indirect optical transitions. Quantitative analysis of the spectral dependence of the photocurrent quantum yield (Figure 1) demonstrated the presence of direct optical transitions in the fundamental absorbance range for Cd_4SiSe_6: 1,74±0.02 eV, 1,97 eV, 2.24 eV; and for Cd_4GeSe_6: 1.77±0,02 eV, 2.16 eV. For both materials, a direct impurity transition was established, with energy 1.47 eV for Cd_4SiSe_6 and 1.54 eV for Cd_4GeSe_6. The intensities of the sub-bandgap transitions were 50-100 times lower than the band gap transition intensities. In these experiments, the photocurrent quantum yield registration sensitivity was 10^{-4} [3].

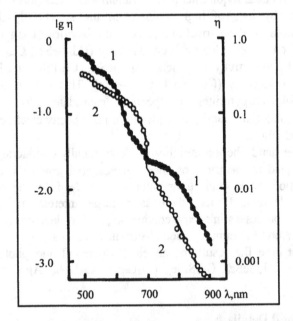

Figure 1. Spectral dependence of the quantum yield of the photocurrent, η, in semilogarithmic coordinates, for Cd_4GeSe_6 (1) and Cd_4SiSe_6 (2) electrodes in 0.05 M H_2SO_4.

In both materials there are direct transitions with Eo≈1.7 eV, as in CdSe. In the case of Cd_4GeS_6 and CdS, the situation is the same. This means that the transitions in argyrodites have the same nature as in the respective binary chalcogenides.

The first electrochemical investigation of Ag-containing argyrodites was made on Ag_8GeS_6 and Ag_8GeSe_6. Ag_8GeS_6 is stable in a 1M KNO_3 electrolyte in the range of -0.3 +1,1 V vs. NHE, and Ag_8GeSe_6 in the range of -0.3 +0.6 V vs. NHE. In the range of stability of Ag_8GeS_6, the current reached only 10 $\mu A/cm^2$ and was independent on the voltage scanning rate. In this potential region, only electronic processes take place, and the electronic conductivity can be determined .The band gap and the specific resistivity of the materials are: Ag_8GeS_6 1.4 eV and 1.2×10^6 ohm.cm, Ag_8GeSe_6 0.9 eV and 5×10^5 ohm.cm. The stability regions correlate with the band gaps of these materials [4].

An anodic photocurrent and a negative shift of the photo-potential were detected under illumination, indicating that these semiconductors are n-type. The flat band potentials of both materials were located at -0.35 V vs. NHE in a 1M KNO_3 electrolyte. They shifted to +0.25 V vs. NHE in 1M KI / 0.05M J_2 and 0.2M $K_3Fe(CN)_6$ / 0.2M $K_4Fe(CN)_6$ electrolytes for Ag_8GeS_6. For Ag_8GeSe_6, the corresponding values were -0.17 V vs.NHE in 1M KI / 0.05M I_2 and +0.76 V vs. NHE in 0.2M $K_3Fe(CN)_6$ / 0.2M $K_4Fe(CN)_6$. At potentials more negative than -0.7 V vs. NHE, cathodic reduction of electrode materials takes place. At positive potentials (outside than the stability region), anodic disolution of the electrodes was observed. The anodic disolution rate increased under illumination.

The additional peak for Ag_8GeSe_6 electrodes, which appeared when the potential was scanned positive +0.5 V vs. NHE corresponded to the Ag^+ ion limiting the current in the electrode. This peak was also detected under illumination. With chopped light, the photocurrent was shunted by ionic transport in the solid state. This means that we registered the interactions of the electronic and ionic subsystems in the solid mixed conductor- Ag_8GeSe_6.

Some results of our investigations are presented in Table 1.

TABLE 1. Parameters of the materials

Material	Electrolyte	Energy transitions	Flat band, V NHE	Band bending
Cd_4GeSe_6	0.05 M H_2SO_4	1.54 ± 0.02	-0.30	~0.35
		1.77 ± 0.02		
		2.16 ± 0.02		
Cd_4SiSe_6	0.05 M H_2SO_4	1.47 ± 0.02	-0.30	~0.1
		1.74 ± 0.02		
		1.97 ± 0.02		
		2.24 ± 0.02		
Cd_4GeS_6	0.05 M H_2SO_4	2.49 ± 0.02	-0.93	0.4
Ag_8GeS_6	1M KNO_3	1.39	-0.35	~0.1
	0.2M $K_3Fe(CN)_6$ / 0.2M $K_4Fe(CN)_6$		-0.25	0.24
	1M KI / 0.05M I_2		-0.25	0.11
Ag_8GeSe_6	1M KNO_3	0.88	-0.35	~0.1
	1M KI / 0.05M I_2		-0.25	0.4

312

Previously, a photoelectrochemical investigation of Cu_6PS_5I in 0.1 M tetraethylammoniumperchlorate acetonitrile solution containing 0.01 M CuCl has been reported [5]. The behavior of this material in aqueous solutions is similar [6,7]. The current-voltage curve of a Cu_6PS_5I-electrode in a 0.1M $CuSO_4$ / 0.1M H_2SO_4 solution is demonstrated in Figure 2.

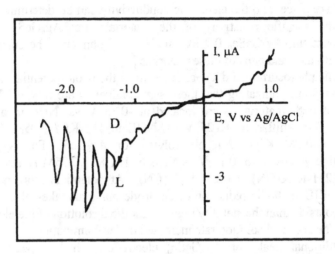

Figure 2. Current-voltage curve of a Cu_6PS_5I-electrode (s = 0.07 cm^2) in 1M H_2NO_3, with V=40 mV/c.

Cu_6PS_5I is a p-type electronic semiconductor with a flat band potential near -0.3 V vs Ag/AgCl. However, we can see that the photocurrent is shunted by ionic current during cathodic polarization. Thus, Cu_6PS_5I is a mixed conductor with a high level of ionic conductivity. This fact limits the possibility of using it as photoelectrode.

4. References

1. Kovach, S.K., Vasko, A.T. *et al.* (1988) Photoelectrochemical properties of semiconductor alloys of $Cd_{4-x}Hg_xGeSe_6$, *Electrochimiya* (Russian) **24**, 1507-1510.
2. Manivanan, A., Fujishima, A., *et al.* (1990) Crystal growth and photoelectrochemical characterization of Cd_4BSe_6 (B=Si or Ge), *Ber. Bunsenges. Phys. Chem.* **94**, 8-12.
3. Kovach, S.K., Motrya, S.F. and Semrad, E.E. (1992) Photoelectrochemical properties of Cd_4SiSe_6 and Cd_4GeSe_6, *Electrochimiya* (Russian) **28**, 1000-1005.
4. Kovach, S.K., Kokhan, A.P. and Voroshilov, Yu.V. (1993) Electrochemical properties of Ag_8GeS_6 and Ag_8GeSe_6, *Ukrainski Khim. Zhurnal* (Ukranian) **59**, 396-398.
5. Betz, G. and Tributsch, H. (1984) Photointercalation and optical information storage using $Cu_{6-x}PS_5I$, J. Electrochem. Soc. **131**, 640-644.
6. Kovach, S.K., Stasyuk, Yu.M. and Panyko, V.V. (1996) Electrochemical processes at the Cu_6PS_5I-electrolyte, *Studia Univ. Babes-Bolyai, chemia* **41**, 252-255.
7. Stasyuk, Yu.M., Kovach, S.K., Panyko, V.V. and Voroshilov, Yu.V. (2000) Electrochemical processes in the volume and at the interface of monocrystalline Cu_6PS_5I, *Ukrainski Khim. Zhurnal* (Ukranian) **9**, 36-39.

OPTICAL ABSORPTION IN APCVD METAL OXIDE THIN FILMS

K. A. GESHEVA, T. IVANOVA and A. SZEKERES[*]

*Central Laboratory of Solar Energy and New Energy Sources,
and [*]Institute of Solid State Physics of the Bulgarian Academy
of Sciences, Tzarigradsko chaussee 72, 1784 Sofia, Bulgaria*

Abstract

The optical absorption in APCVD thin films of MoO_3, WO_3 and mixed MoO_3-WO_3 was studied in the range 300-850 nm, by spectrophotometry. Annealing at $400^{\circ}C$ resulted in an increased absorption in mixed oxide films, while for the MoO_3 and WO_3 films a slight decrease was registered. All films, in the as-deposited and annealed states, have optical band gap energies in the range 2.55-3.2 eV, and refractive indices from 1.8 to 2.6.

1. Introduction

Transition metal oxides are primarily studied in relation to their electrochromic properties, for application as smart windows and display devices. Tungsten oxide is the most investigated material, in respect of its technology, structure and electrochromic properties [1,2]. However, molybdenum trioxide has an optical absorption peak closer to the human eye sensitivity peak, which makes this material interesting [3]. Mixed films based on molybdenum and tungsten oxides offer additional possibilities in this respect. Also, a higher absorption is expected in these films, due to the larger number of metal sites with different valencies in such two-component materials.

This paper examines the optical absorption in MoO_3, WO_3 and mixed MoO_3 - WO_3 oxide thin films, obtained by atmospheric pressure chemical vapor deposition using carbonyl precursors ($W(CO)_6$ and $Mo(CO)_6$).

2. Experimental Details

Films of MoO_3, WO_3 and mixed MoO_3-WO_3 oxides were deposited at atmospheric pressure on glass, by pyrolytic decomposition of the corresponding hexacarbonyls ($W(CO)_6$, $Mo(CO)_6$, or their physical mixture in a ratio of $Mo(CO)_6$:$W(CO)_6$ = 1:4. The deposition was performed in argon-oxygen, in a horizontal cold walled CVD reactor. The precursor powder, placed in a sublimator immersed in a silicon oil bath, was heated to $90 \pm 1^{\circ}C$. This temperature provided a sufficient vapor pressure of the hexacarbonyls. An argon flow through the

J.M. Marshall and D. Dimova-Malinovska (eds.),

Photovoltaic and Photoactive Materials - Properties, Technology and Applications, 313–316.

sublimator carried the vapors to the CVD reactor, with oxygen entering through a separate line. The ratio of the flow rates of Ar and O_2 was 1:32, and the deposition time was kept constant (40 min). Because of the different growth rates, the film thickness, as determined by Talystep, was 300 nm for MoO_3, 400 nm for WO_3 and 120 nm for mixed oxide films.

The deposition temperature was kept at 200°C, because only at this temperature was it possible to deposit the three kinds of film. To improve the film structure and optical properties, post-deposition annealing has previously been conducted in the range 200-500°C [2,4,5]. Annealing at 500°C is a technological limitation for glass substrates, when large-scale production of smart windows is envisaged. Because of this, a 400°C annealing temperature was selected here.

The transmittance and reflectance spectra of the films were measured in the range 300-900 nm, using a Shimadzu UV-190 UV-VIS double-beam spectrophotometer. Measurements of the full (diffusive plus normal) reflectance were carried out in a 100 mm integrating sphere. As a reference, the magnesium oxide reflectance was taken as the 100 % response.

The refractive indices of the films were obtained by spectral-ellipsometic (SE) measurements, using a Rudolph ellipsometer in the spectral region 300-820 nm, at an angle of incidence of 50°. The accuracy of the angles of the polarizer, analyzer and light incidence was ± 0.01°.

3. Results and Discussion

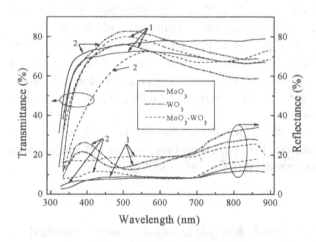

Figure 1. Transmittance and reflectance spectra of CVD metal oxide films: as-deposited (curves 1) and annealed at 400°C (curves 2).

Figure 1 shows the transmission and reflection spectra for as-deposited and annealed metal oxide films, in the visible wavelength range. The transmittance for all films is within the range 60-80 %, while the reflectances are around 20 %. After annealing, tungsten and molybdenum oxide films became more transparent, and a slight change of the reflectance values was detected. The transparency of the mixed oxide films decreased, together with a noticeable decrease in the reflectance. This suggests a considerable increase in the absorption.

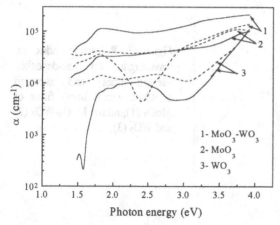

Figure 2. Absorption coefficient, α, vs. photon energy of as-deposited (dashed lines) and annealed at 400°C (solid lines) films of WO_3 (3), MoO_3 (2) and mixed MoO_3-WO_3 (1).

The absorption coefficient, α, was determined from the spectrophotometry data, using the equation $\alpha = 1/t_f \ln[(1-T)/R]$, where t_f is the film thickness. The absorption spectra of as-deposited and annealed films are presented in Figure 2. In the as-deposited state, the WO_3 spectrum is almost flat up to ~ 3 eV, while there is a small maximum at 1.7 eV for MoO_3 and a broad and a deep maximum at 1.8 eV for the mixed films. In the same energy range, Faughnan *et al.* [3] found an absorption peak for tungsten-molybdenum mixed oxide films. After annealing, the absorption in the pure metal oxides decreased, while in mixed oxide films it increased significantly.

Approaching the absorption edge, the absorption rapidly increases due to band-to band electronic transitions. The absorption is expressed as $\alpha h\nu \sim (h\nu - E_g)^{\eta}$, where $\eta = 2$ and E_g is the optical band gap energy. Values of E_g can be evaluated by extrapolating the linear part of a plot of $(\alpha h\nu)^{1/2}$ versus $h\nu$ to zero absorption. Assuming indirect and allowed electron transitions in MoO_3 and WO_3 films ($\eta = 2$), the band gap energies were determined. For mixed MoO_3-WO_3 films, direct and allowed electron transitions were assumed, ($\eta = 1/2$), since the extrapolation gave real values of E_g only if such transitions were assumed to prevail.

It is established empirically that a structural transformation from an amorphous to a crystalline phase causes an E_g increase for MoO_3 [1,4,6,] and a decrease for WO_3 [1,7]. The explanation of the optical band gap narrowing as a result of crystallization is still unclear. For MO_3, the E_g value is 2.55 eV and after annealing it increases to 2.7 eV, suggesting crystallization. For as-deposited WO_3 films, E_g is 3.1 eV, both before and after annealing. This indicates that 400°C annealing does not affect the film structure. For as-deposited mixed MoO_3-WO_3 films, the E_g value is 3.23 eV, falling to 2.93 eV after annealing. This could be explained in terms of the coexistence of crystalline MoO_3 inclusions in the host amorphous WO_3 matrix. Such inclusions probably define defects, the energy levels of which contribute to the band gap narrowing. The observed increase of the absorption after annealing in mixed oxides could be explained via the same defects.

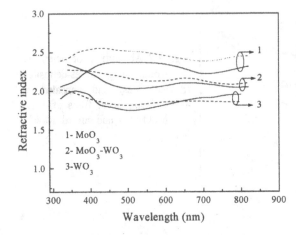

Figure 3. Refractive index vs. wavelength for as-deposited (dashed lines) and annealed at 400°C (solid lines) films of MoO_3 (1), mixed MoO_3-WO_3 (2) and WO_3 (3).

The spectral dependences of the refractive indices obtained from the SE data are presented in Figure 3. The refractive index values are in the range 1.8-2.6, which are characteristic for these films prepared by different techniques [8,9]. The peculiarities observed could arise from an inhomogeneous film density, due to the mixture of amorphous and crystalline phases with different modifications. The coexistence of amorphous and crystalline phases with different modifications is proved by our Raman, IR and SE investigations [4,7,8].

4. Conclusions

CVD films of MoO_3, WO_3, and mixed MoO_3-WO_3 possess transmittances of about 80 % and reflectances of 20 %. The optical absorption coefficient in all films in the as-deposited state is below 5×10^4 cm^{-1}. It is unchanged after annealing for MoO_3 and WO_3 films, but for mixed films a strong increase is observed. The optical band gap values can be related to the crystalline structure of the MoO_3 films, and the amorphous structure of the WO_3 and mixed MoO_3-WO_3 films.

5. References

1. Granqvist, C.G. (2000) *Solar Energy Mat. and Solar Cells* **60**, 201-262 and references therein.
2. Gogova, D., Gesheva, K., Szekeres, A. and Sendova-Vassileva, M. (1999) *Phys. Stat. Sol. (a)* **176**, 969-984.
3. Faughnan, B.W. and Crandall, R.S. (1977) *App.Phys. Lett.* **31**, 834.
4. Gesheva, K.A., Ivanova, T., Szekeres, A., Maksimov, A. and Zaitzev, S. (2001) *J. Phys. IV France* **11**, Pr3-1023-1028.
5. Ivanova, T., Gesheva, K.A., Szekeres, A., Maksimov, A. and Zaitzev, S. (2001) *J. Phys. IV France* **11**, Pr3-385-390.
6. Julien, C., Khelfa, A., Hussain, O.M. and Nazri, G.A. (1995) *J. Crystal Growth* **156**, 235-244.
7. Szekeres, A., Gogova, D. and Gesheva, K. (1999) *J. Crystal Growth* **198/199**, 1235-1239.
8. Abdellaoui, A., Leveque, G., Donnadieu, A., Bath, A. and Bouchikhi, B. (1997) *Thin Solid Films* **304**, 39-44.
9. Miyata, N. and Akiyshi, S. (1985) *J. Appl. Phys.* **58**, 1651-1655.

EXTRINSIC SURFACE PHOTOVOLTAGE SPECTROSCOPY – AN ALTERNATIVE APPROACH TO DEEP LEVEL CHARACTERISATION IN SEMICONDUCTORS

K GERMANOVA, V. DONCHEV, CH. HARDALOV[*] and
M. SARAYDAROV

*Faculty of Physics, Sofia University, 5 blvd. J Bourchier,
Sofia - 1164, Bulgaria, and*
[*] *Dept. Appl. Physics, Technical University of Sofia, 8, Kl.Ohridski
Str. Sofia -1000, Bulgaria.*

1. Introduction

Surface photovoltage (SPV) spectroscopy in the intrinsic absorption energy range has been successfully used to study semiconductor surface and interface states [1-6]. However, simultaneously with the surface states, deep levels in the semiconductor bulk can also be optically excited, thus changing the surface potential [6]. This makes feasible the use of extrinsic SPV spectroscopy for the assessment of deep bulk semiconductor levels, which are a subject of increasing interest from both a fundamental and a technological viewpoint. At the same time, a further improvement of the SPV spectra analyses is necessary, in order to take into account the contributions to the surface potential of both surface and bulk electron transitions stimulated by extrinsic illumination. Theoretical and experimental studies supporting the concept depicted above have been already performed [1,2]. The comparison of the extrinsic SPV and photoconductivity (PC) spectra is of prime interest in studying the validity and the applicability of this concept.

In this paper, we present a comparative study of the extrinsic SPV and PC spectra of Cr-doped semi-insulating (SI) GaAs bulk crystals with real (100) surfaces. Both SPV and PC are measured on the same samples under identical experimental conditions. Both kinds of spectra reveal similar structures, which can be explained by the photoionization mainly of the bulk deep centers: Cr^{2+} acceptors and EL2 donors. Some optical parameters of these centers are obtained from the SPV and the PC spectra, and the results agree very well. Thus, evidence is given that the concept discussed above is valid in the investigated material.

2. Experimental Details

The samples were LEC grown Cr-doped SI GaAs bulk crystals ($[Cr]=10^{16}$ cm^{-3}) with a dark room-temperature resistivity of 2.8×10^{8} $\Omega.cm$. They were of

J.M. Marshall and D. Dimova-Malinovska (eds.),
Photovoltaic and Photoactive Materials - Properties, Technology and Applications, 317–320.
© 2002 *Kluwer Academic Publishers.*

parallelepiped form: 10 x 3 x 0.5 mm^3. The illuminated face (10 x 3 mm^2) was chemico-mechanically polished. Ohmic contacts were made on both of the smallest faces, by In alloying and annealing.

The experimental system included a LH optical cryostat with He exchange gas, a 100 W halogen tungsten lamp, a single SPEX monochromator (f = 0.25 m) and a filter system to cut off the high order diffraction. The scanning was from high to low wavelengths, with a resolution of ~10 nm. The photon flux was kept constant at each wavelength, at a value of ~10^{14} cm^{-2}s^{-1} which was sufficiently low to prevent EL2 quenching. The SPV spectra were measured at 293K, by an improved Kelvin method. The vibrating probe was a Pt boss (diameter 1.2 mm), driven by a piezoelectric plate at 330 Hz. The signal was recorded by a PARC 9305 lock-in amplifier, with a sensitivity of 0.1mV. The PC spectra were measured at 293K as well as at 70K, in dc operation mode, using a HP dc source/picoammeter.

3. Results and Discussion

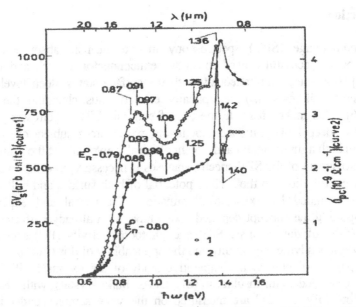

Figure 1. SPV (curve 1) and PC (curve 2) spectra of Cr-doped SI GaAs at 293 K

A typical SPV spectrum at 293 K is shown in Figure 1 (curve 1). It exhibits two main structures, ranging from 0.66 to 0.97 eV (I) and from 0.97 to 1.32 eV (II). Structure I is characterised by a large peak at 0.91 eV and a small peak (sometimes a shoulder) at 0.87 eV superimposed on an increasing background. Structure II includes a step (sometimes a small bump) around 0.97 eV, and two step-like bands with thresholds at 1.08 and 1.25 eV, respectively. Between 1.08 and 1.25 eV, the spectrum reveals an increasing background and an extra band located around 1.18 eV. Sharp peaks at 1.36 and 1.4 eV can be also seen. For larger photon energies, the SPV response is constant.

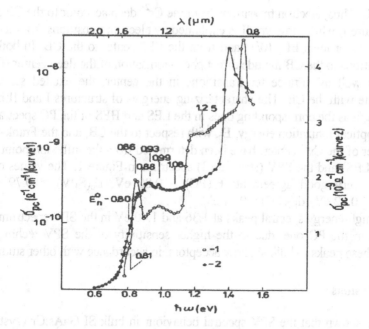

Figure 2. PC spectra of Cr-doped SI GaAs at 70K (curve 1) and 293 K (curve 2)

Curve 1 in Figure 2 is a 70K PC spectrum. It reveals a low-energy structure (LES), ranging from 0.74 to 0.90 eV and a high-energy one (HES), from 0.90 to 1.40 eV. The dc PC spectral behaviour in SI GaAs:Cr has been explained in terms of the optical cross sections of the main deep levels [7,8]. The sharp peak at 0.86 eV is due to Cr^{2+} internal electron transitions $^5T_2$ - ^2E, with excited state ^2E being degenerate with the conduction band (CB). The low-energy shoulder (sometimes a small peak) at 0.81 eV is attributed to direct photoionization from the Cr^{2+} centers to the CB. Electronic transitions from the valence band (VB) to the EL2 (0/+) and the Cr^{2+} - Cr^{3+} deep levels account for the hump around 0.99 eV. The extra band in the range 1.08-1.25 eV is ascribed to free electron generation via EL2 intra-center transitions T_2 – A_1. The background in this range, and the 1.25 eV threshold band, correspond to EL2 photoionization to the L and X CB minima, respectively. The origin of the HES has been further confirmed by measuring a 70K PC spectrum, after intentional photo-induced transfer of EL2 into its metastable state [7,8]. The 293 K PC spectrum (Figure 2, curve 2) reveals the same LES and HES, but they are not so clearly separated as at 70K. The low-energy threshold of the LES is located at 0.60 eV. The shoulder and the sharp peak are shifted to higher energies: 0.88 and 0.93 eV respectively, with the energy spacing between them remaining the same. The energies characterizing the HES are not changed.

The SPV and PC spectra, measured at 293K, are compared in Figure 1. The SPV structures I and II correlate perfectly with the LES and the HES in the PC spectrum. They have virtually identical shapes and characteristic energies. This implies that in the range 0.66 – 1.32 eV, the SPV is wholly determined by the same bulk deep levels as the PC. Hence, it can be explained in terms of the model employed above for the PC. Thus, electron transitions from the Cr^{2+} deep acceptor to the CB account for structure I, while structure II is explained by electron transitions from the VB to the deep donor level EL2 (0/+) and from the EL2 center to the CB. In both cases, the transitions to the CB include direct photo-ionisation of the deep center (Cr^{2+} and EL2), as well as intra-center transitions in the center; the excited states being degenerate with the CB. The characterizing energies of structures I and II have the same origin as the corresponding ones in the LES and HES of the PC spectra.

The optical ionization energy, E_o, with respect to the CB, and the Frank-Condon parameter of the Cr^{2+} center, have been determined from the inflexion point of both the PC (LES) and the SPV (structure I) spectrum in Figure 1. The values obtained are in a very good agreement: $E_o(PC) = 0.80$ eV; $E_o(SPV) = 0.79$ eV and $d_{FC}(PC) = 0.12$ eV , $d_{FC}(PC) = 0.11$ eV.

The high-energy spectral peaks at 1.36 and 1.40 eV in the SPV spectrum are not detected in the PC one, due to the higher sensitivity of the SPV technique. We ascribe these peaks to bulk shallow acceptors, in accordance with other studies [9].

4. Conclusions

We have shown that the SPV spectral behaviour in bulk SI GaAs:Cr crystals with real (100) surfaces is entirely determined by the photo-ionisation of the main deep centers (Cr^{2+} acceptors, EL2 donors) and shallow levels in the sample bulk. This model is consistent with our earlier investigations, which have shown that in SI GaAs:Cr the surface states are optically inactive in the extrinsic spectral range up to the band-gap energy [10]. This work gives an alternative approach for deep level characterisation in semiconductors, based on extrinsic SPV spectroscopy.

Financial support from the Bulgarian National Science Fund and the Sofia University Research Fund is gratefully acknowledged.

5. References

1. Kronik, L. and Shapira, Y. (1999) *Surface photovoltage phenomena: theory, experiment and application*, Elsevier.
2. Timoshenko, V.Yu., Kashkarov, P.K., Matveeva, A.B., Konstantinova, E.A., Flietner, H. and Ditrich, Th. (1996) *Thin Solid Films* **276**, 216-218.
3. Lagowski, J., Balestra, C. and Gatos, H. (1972) *Surface Sci.* **29**, 213.
4. Gatos, H. and Lagowski, J. (1973), *J. Vac. Sci. Technol.* **10**, 130
5. Czekala-Mukalled, Z. *et al.* (1995) *Vacuum* **46**, 489.
6. Monch, W., Clemens, H., Gorlich, S. Enninghrst, R. and Gatos, H. (1981) *J. Vac. Sci. Technol.* **19**, 525.
7. Germanova, K., Donchev, V., Hardalov, Ch. and Nikolov, L. (1987) *J. Phys. D: Appl.Phys.* **20**, 1507.
8. Donchev, V (1991), PhD Thesis, Sofia University.
9. Tsukada, N., Kikuta, T. and Ishida, K. (1986), *Jap. J. Appl. Phys.* **25**, L196.
10. Germanova, K. and Hardalov, Ch. (1987) *Appl. Phys. A* **43**, 117.

CELLO: AN ADVANCED LBIC MEASUREMENT TECHNIQUE FOR SOLAR CELL LOCAL CHARACTERISATION

J. CARSTENSEN, G. POPKIROV, J. BAHR and H. FÖLL
Faculty of Engineering, Christian-Albrechts University of Kiel,
Kaiserstr. 2, D-24143 Kiel, Germany

1. Introduction

A solar cell is a large area device, and thus its global IV-characteristic and efficiency depend strongly upon its local properties. Local defects, such as a locally reduced diffusion length, a strong local shunt resistance or a high local series resistance will adversely influence the cell's global properties. Experimental techniques suitable for mapping the spatial distribution of such local parameters can provide valuable information, and thus help to improve the technology for the production of efficient and reproducible solar cells. The LBIC (Light Beam Induced Current) technique [1] allows calculation of the local diffusion length of the solar cell material from local photocurrent data obtained under short-circuit conditions. As well as allowing LBIC measurements, the PVScan 5000 mapping analyzer by NREL [2] can be used to map defects and grain boundaries using reflectivity data and special surface etching. Localized shunts can be mapped by CCD cameras [3,4] or nematic liquid crystals [5]. The electron-beam-induced current (EBIC) is an alternative technique for the investigation of defects in solar cells [6]. Destructive techniques, MASC [7] for local IV-characterisation and RAMP [8] using a scanning (by scratching) tungsten electrode, have also recently been developed.

This paper describes a new advanced LBIC measurement technique, 'CELLO' (solar cell local characterisation), which allows the determination of all local parameters, especially the local series- and shunt-resistances, $R_s(x,y)$ and $R_{sh}(x,y)$, and thus the identification of all material- and process-induced defects relevant to the efficiency. In principle, the data obtained could also be used to simulate the behaviour of a complete solar cell, for any set of technological parameters.

2. The Measurement Technique

A simplified schematic diagram of CELLO is shown in Figure 1. The solar cell is illuminated homogeneously by a set of halogen lamps, at near AM1.5 intensity. Additionally, an intensity-modulated infrared laser beam focussed to about 100µm diameter scans the sample using piezo-controlled mirrors, allowing for high resolution maps of the local linear response. A custom-made, computer controlled, low noise potentiostat/galvanostat is used for voltage/current control and

J.M. Marshall and D. Dimova-Malinovska (eds.),
Photovoltaic and Photoactive Materials - Properties, Technology and Applications, 321–324.
© 2002 *Kluwer Academic Publishers.*

322

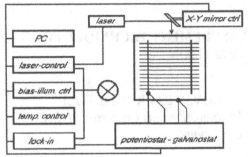

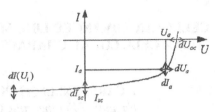

Figure 2. The response of the current (potentiostatic) or of the voltage (galvanostatic) is measured at several points along the IV curve.

Figure 1. Block scheme of the CELLO system

measurement, with a high signal to noise ratio. A lock-in amplifier, synchronised to the laser beam modulation signal, is used to measure the solar cell response to the laser beam perturbation. The CELLO technique measures the solar cell global response, $dI(U_{cell},x,y)$ and $dU(I_{cell},x,y)$, to local perturbations, as shown in Figure 2. The data obtained are fitted to a complete (and partially novel) model of the solar cell which allows the: i) generation of solar cell surface maps of the measured data, ii) calculation of maps of the local series and shunt resistances, diffusion length and back-surface field, and iii) construction of a complete local IV-curve for each point on the cell. In this context, the LBIC mode is just the measurement at $U_{cell} = 0$.

3. Modelling and Calculations

If sufficient data sets have been obtained experimentally, the raw data can be converted into the local parameters of the solar cell. This is done with the help of the equivalent circuit shown in Figure 3. The solar cell is divided in a global part and a local part, which are described by different sets of parameters. The global part is just the constant IV-curve of the complete solar cell. In a linear response approach, any small local illumination can be added with a proper local set of solar cell parameters without changing the global values, by assuming that the interconnect (the grid) between the complete solar cell and the additionally illuminated area serves as an equipotential layer.

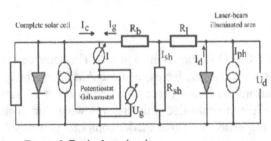

Figure 3. Equivalent circuit

There are two important modifications in the equivalent circuit shown in Figure 3, with respect to conventional diagrams: i) the two parts of the cell are connected via two resistors, as shown, ii) the combination of the local diode and the local current source (laser induced photo current) is not described by the usual

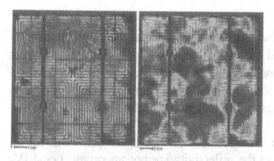

Figure 4. Maps of *dI*, *U*=300mV (left), *dU*, *I*=300mA (right)

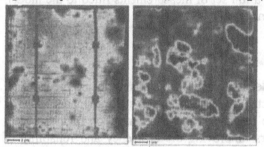

Figure 5. Maps of R_{sh} (left) and R_{ser} (right)

equations, but couples the diode current I_d and the generated and collected photo currents I_{ph} and $I_{ph,0}$ for any given point on the solar cell surface, via $I_{ph} = I_{ph,0} - C_{rec}I_d^3$. Here, $C_{rec}I_d^3$ represents the current due to local recombination losses, with C_{rec} as a fitting parameter. This equation describes the lateral diffusion and recombination of minority carriers, and expresses the sensitivity of the solar cell to gradients of the diffusion length. In fact, not all carriers photo-generated close to the surface are collected and contribute to the current at finite voltages. Some recombine in areas laterally removed from their points of origin. Thus, if the local diffusion length varies significantly in the lateral direction, this will influence the local properties of the point on the cell under consideration. Thus, the average diffusion length of the cell material is not a sufficient measure of the local recombination activity.

The gradients in the diffusion length distribution are just as important. The complete set of equations (the grid is an equipotential surface) is

$$I_g = I_d + I_{sh} - I_{ph} \qquad (1)$$

$$I_d = I_{d1}\left(e^{\frac{qU_d}{n_1kT}} - 1\right) + I_{d2}\left(e^{\frac{qU_d}{n_2kT}} - 1\right) \qquad (2)$$

$$U_g = R_{sh}\,I_{sh} + R_b I_g \qquad (3)$$

Here, k is Boltzmann's constant, T the temperature, q the electronic charge, I_{d1}, n_1, and I_{d2}, n_2 the saturation currents and ideality factors of the two diodes. R_b is the lateral resistance of the emitter between the illuminated 'pixel' and the grid. Photocarriers generated at one point can be collected by the grid, or be lost by recombination or via a local shunt. Using linear order approximations in the local laser beam induced photocurrent, this set of equations is generally sufficient to extract all required quantities, by fitting the experimental data. The complete equivalent circuit diagram, and therefore the complete set of local characteristics, can be calculated from a finite set of experimental data. Local IV-curves can be calculated using data for the local shunt and series resistances, and the local diffusion length.

4. Experimental Results

The operation of CELLO will be illustrated by maps, obtained from one measurement on a solar cell formed on a μc-Si wafer. Maps of the linear current response, dI, at $U = 300\text{mV}$ and of the voltage response, dU, measured galvanostatically at $I = 300\text{mA}$, are shown in Figure 4. Figure 5 presents calculated maps of R_{sh} and R_{ser}. We do not measure the reflectivity of the wafer with CELLO, and thus do not get information about the quantum efficiency. All other measurements are analysed relative to (e.g.) the short circuit current. Thus, for all other parameters, local differences in the reflectivity are not important. Local IV-characteristics can be obtained from a full set of measurements fitted to the model, as will be shown in future publications.

5. Conclusions

CELLO is a universal method for detecting and characterising local defects in all solar cells, since it is not restricted to silicon or crystalline materials. The measurements can provide valuable feedback for the optimisation of the process technology and materials parameters, improving the efficiency and reliability of solar cells.

6. References

1. Kress, A., Pernau, T., Fath, P. and Bucher, E. (2000) LBIC measurements on low cost contact solar cells, *Proc. 16th European Photovoltaic Conference on Solar Energy Conversion, Glasgow*, VA1.39, (File D339.pdf).
2. NREL – Technology brief (Document: NREL/MK-336-21116, 8/96)
3. King, D.L., Kratochvil, J.A., Quintana., M.A. and McMahon, T.J. (2000) Applications for infrared imaging equipment in photovoltaic cells, modules and system testing, *WWW-Link: http://www.sandia.gov/pv/ieee2000/kingquin.pdf.*
4. Langenkamp, M., Breitenstein, O., Nell, M.E., Wagemann, H.-G. and Estner, L. (2000) Microscopic localisation and analysis of leakage currents in thin film silicon solar cells, *Proc. 16th European Conference on Photovoltaic Solar Energy Conversion, Glasgow*, VD3.39 (File D411.pdf) .
5. Färber, G., Bardos, R.A., McIntosh, K.R., Honsberg, C.B. and Sproul, A.B. (1998) Detection of shunt resistance in silicon solar cells using liquid crystals, *Proc. 2nd World Conf. and Exhibition on Photovoltaic Solar Energy Conversion, Vienna.*
6. Boudaden, J., Riviere, A., Ballutaud, D., Muller, J.-C. and Monna, R. (2000) EBIC characterization of multicrystalline silicon solar cell emitters, *Proc. 16th European Conference on Photovoltaic Solar Energy Conversion, Glasgow*, VD3.62.
7. Häßler, B. Thurm, S., Koch, W., Karg, D. and Pensl, G. (1995) *Proc. 13th European Photovoltaic Solar Energy Conference*, Nice, 1364.
8. van der Heide, A.S.H., Schönecker, A., Wyers, G.P. and Sinke, W.C. (2000) Mapping of contact resistance and locating shunts on solar cells using resistance analysis by mapping of potential (RAMP) technique, *Proc. 16th European Conference on Photovoltaic Solar Energy Conversion, Glasgow*, VA1.60 (File D359.pdf).

THE BEHAVIOR OF PV MODULE PARAMETERS AS A FUNCTION OF SOLAR CELL TEMPERATURE IN HOT CLIMATES

P. VITANOV, K. IVANOVA, D. VELKOV, Y.G. KUDDAN and N. TYUTYUNDZHIEV
Central Lab. of Solar Energy & New Energy Sources, Bulgarian Academy of Sciences, 72, Tzarigradsko chausse Blvd., 1784 Sofia, Bulgaria

1. Introduction

In hot climates, the use of PV modules is sometimes problematic, since their efficiency normally decreases with increasing solar cell temperature. The temperature behavior of PV modules is thus important when considering their application for large-scale economical power generation.

In the present work, the electrical parameters of a PV module were investigated as a function of solar cell temperature, under real exploitation conditions. For this purpose, a database was used to store measurements systematically recorded in Sofia during the period June to October. The accumulated information permitted the dependence of the photovoltaic parameters to be determined as a function of solar cell temperature in the module. The results were compared to laboratory solar cell tests. The thermal dissipation coefficient of the PV module was assessed. This characterizes the relation between the temperature of solar cells inside the module and parameters such as the ambient air temperature and solar radiation intensity.

2. Measuring Procedure and Experimental Results

Single solar cells and a PV module were studied. The cells were realized on 530 μm p-type <100> oriented 4" monocrystalline silicon wafers, of 1 Ω.cm resistivity. A low cost multilayer (Ni-Cu-Sn) metallization method [1] was used for easy soldering and interconnecting. Cell front faces were chemically treated to form random pyramids, 1-2 μm in height. As antireflection films, a SiO_2 + Si_3N_4 system was used.

A solar simulator delivering the equivalent of 1 kW/m^2 of solar power radiation was used for the measurements. The cells were heated by a "Temptronics" thermochuck system, providing $\pm1°C$ temperature control. The current-voltage (I-V) load characteristics of the cells were recorded, allowing determinations of the maximum generated electric power P_{max}, the fill factor FF, the short circuit current

325

J.M. Marshall and D. Dimova-Malinovska (eds.),
Photovoltaic and Photoactive Materials - Properties, Technology and Applications, 325–328.
© 2002 *Kluwer Academic Publishers.*

I_{sc}, the open circuit voltage U_{oc}, and the conversion efficiency η. The important material parameters which determine the temperature dependence of solar cell behavior are the intrinsic carrier density, n_i, the diffusion length during the lifetime, and the absorption coefficient.

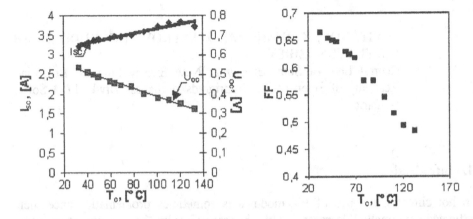

Figure 1. Short circuit current, I_{sc}, and open circuit voltage, U_{oc}, with temperature for a solar cell illuminated by a simulator supplying the equivalent of solar radiation of 1 kW/m². *Figure 2.* Dependence of the fill factor, FF, on temperature, T_c, for a solar cell illuminated by a simulator supplying the equivalent of solar radiation of 1 kW/m².

Figure 1 shows that I_{sc} increases slightly with temperature, partly due to an improvement in the base diffusion length and partly to an absorption edge shift to lower energies. Both of these improve the long wavelength spectral response. As also shown, U_{oc} decreases with temperature, at a rate of about 2.2 mV/°C for our Si cells. This is due to the strongly increasing dark current ($\propto n_i^2$): $U_{oc} \approx kT/q \ln(I_{sc}/I_o)$.

The fill factor decreases with increasing temperature, as seen in Figure 2. This is partly due to the lower open circuit voltage and partly to the increasing "softness" (roundness) in the knee of the I-V curve, as temperature increases via the $\exp(qV/kT)$ term. FF decreases with temperature at a rate of about 1.25×10^{-3} per °C. As a result of the falls in U_{oc} and FF, the maximum generated, P_{max} and η normally decrease with increasing temperature. The rate of change of efficiency, $\Delta\eta/\Delta T$, is around 0.07% per °C. If PV modules with Si solar cells are used on earth, then the cell temperature is conditioned by the ambient air temperature, the solar radiation intensity and the module's thermal characteristics, which provide efficient convective and radiative cooling. For simplicity, the PV module thermal behavior can be expressed empirically as $T_c = T_a + K.E_s$, where T_c is the cell temperature [°C], T_a the ambient air temperature [°C], E_s the solar radiation power [W/m²] and K the thermal dissipation coefficient [°C.m².W⁻¹]. Measurements were carried out for a 18W PV module, with an implanted Pt resistor on the back of a cell and a second one on its front to check the cell temperature. The I-V characteristics for the module were automatically recorded [2]. In the period July to October, each day from 7 a.m. to 7 p.m., measurements of E_s, T_a, T_c and the I-V characteristics of the module were measured at one-hour intervals. From these data, P_{max}, FF, I_{sc}, U_{oc}, and η were determined. A database stored the results obtained (about 10^5 data), allowing various functional dependences to be identified. Also, the total solar radiation per day, the average daily air and cell temperatures, the average solar energy per day and the amounts of generated electricity per day could be calculated.

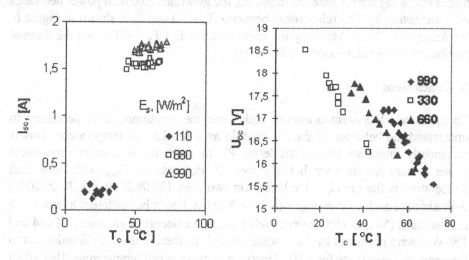

Figure 3. Dependence of the short circuit current on temperature for various intensities of solar power radiation, E_s (W/m^2).

Figure 4. Dependence of the open circuit voltage on temperature for various intensities of solar power radiation, E_s (W/m^2).

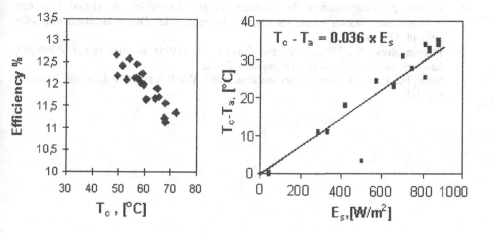

Figure 5. Dependence of conversion efficiency on T_c for an intensity of solar power radiation of E_s = 990 W/m^2.

Figure 6. (T_c -T_a) vs. E_s. The slope of the linear fit is equal to thermal coefficient, K. T_c is the solar cell temperature and T_a the ambient air temperature.

The dependence of I_{sc} on the temperature of the solar modules is shown in Figure 3, for several solar radiation values. The rate of change in I_{sc} is similar to that for a single cell. At low solar radiation intensities, I_{sc} does not change significantly, in agreement with prior data [3]. Figure 4 shows that U_{oc} falls with increasing cell temperature, while the dependence of η on T_c is shown in Figure 5. The dependence of P_{max} on T_c looks similar. The efficiency tends to drop due to the higher operating temperature. As expected, the generated electrical power decreases with increasing T_c. The relationship between $(T_c - T_a)$ and E_s is shown in Figure 6, for August 31st 2000. Assuming the linear relation $T_c - T_a = K.E_s$, then the thermal coefficient of the solar module is $K = 0.036$.

3. Conclusions

To optimize PV system operation under extreme conditions, it is necessary to understand the behavior of the PV module as a function of temperature. This is particularly important when considering PV modules for large-scale economical power generation in very hot climates. A decrease in P_{max} with solar cell temperature is the principal tendency. On two days (30.08.2000 and 26.09.2000) with almost equal solar energy values (≈ 8400 W.h.m^{-2}) but different average cell temperatures (54 and 45°C, respectively), average amounts of electricity of 144 and 158 W.h were generated by the module studied; ie. there was a $\sim 10\%$ reduction in the generated electricity for a 10°C increase in average cell temperature. This effect should be kept in mind when PV systems are designed for use in hot climates.

4. References

1. Vitanov, P., Tyutyundzhiev, N., Stefchev, P. and Karamfilov, B. (1996) Low cost multilayer metallization system for silicon solar cells, *Solar Energy Materials and Solar Cells* 44, 471-484.
2. Tyutyundzhiev, N., Vitanov, P. and Peneva, M. (1995) *Internal report IC/95/383*, *International center for theoretical physics* - Trieste.
3. Hovel, H.J. (1975) *Semiconductors and semimetals*, Vol. II Solar cells, Academic Press, ch. 8, p. 169.

APPLICATION OF SiO$_2$:Re LAYERS FOR IMPROVEMENT OF THE UV SENSITIVITY OF a-Si:H SOLAR CELLS

M. SENDOVA-VASSILEVA, M. NIKOLAEVA, O. ANGELOV,
A. VUTCHKOV[1], D. DIMOVA-MALINOVSKA and J. C. PIVIN[2]
CL SENES, BAS, 72 Tzarigradsko Chaussee, 1784, Sofia, Bulgaria
[1]UCTM, 8 Kl. Ohridski bul., 1756 Sofia, Bulgaria
[2]CSNSM Bâtiment 108, 91405 Orsay Campus, France

1. Introduction

Rare earth ions (Re) have potential applications in solar cells, because of the fact that they absorb light at shorter wavelengths and emit at longer ones. Thus, light which is ineffectively used by solar cells can be transformed into radiation with an energy at which the cell material absorbs well. Hence, a wider spectrum of light can be utilized for photovoltaic conversion. This idea was suggested and investigated in [1], for the example of Eu^{2+}.

It is well known that the optical spectra of rare earth ions consist of narrow 4f-4f transitions and broader 4f-5d transitions. The former are parity forbidden in the free ions and only partially allowed when the ions are inserted in a matrix. The 4f levels are shielded by the outer 6s and 6p electronic shells, and this is why their energy position does not vary much with the matrix.

In this paper, we demonstrate the UV absorption and visible luminescence of rare earth ions, for the example of Tb^{3+} and Sm^{3+} in SiO$_2$ thin films deposited by magnetron co-sputtering.

2. Experimental Details

SiO$_2$ thin films containing Tb or Sm were deposited by magnetron co-sputtering of a complex target consisting of a SiO$_2$ plate with rare earth chips placed on it. The sputtering was performed in Ar gas without intentional heating of the substrates, which were silicon wafers and fused silica plates. The thickness of the films was 1000 nm. The transmission of the samples was measured on fused silica substrates using a CARY UV-VIS-NIR spectrophotometer. The PL was excited with the 351 nm and 488 nm lines of an Ar$^+$ ion laser, and measured using a double monochromator or a Fourier transform spectrometer and a photomultiplier with a GaAs photocathode. The rare earth concentration was measured by RBS.

329

J.M. Marshall and D. Dimova-Malinovska (eds.),
Photovoltaic and Photoactive Materials - Properties, Technology and Applications, 329–332.
© 2002 *Kluwer Academic Publishers.*

3. Results and Discussion

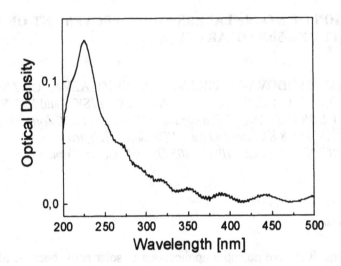

Figure 1. Optical absorption spectrum of a SiO₂:Tb sample containing 2.5 at.% Tb.

Figure 1 shows the transmission spectrum of a SiO₂ sample containing 2.5% Tb. A strong absorption band with a maximum at 225 nm is observed. It is most probably due to an allowed 4f-5d transition in the Tb^{3+} ion.

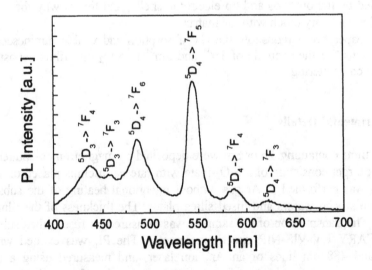

Figure 2. Photoluminescence of Tb^{3+} in a SiO₂ thin film excited with the 351 nm line of an Ar^+ laser.

A typical room temperature photoluminescence (PL) spectrum of the SiO$_2$:Tb thin films excited with the 351 nm line is shown in Figure 2. It consists of six characteristic bands, which are assigned to the following 4f - 4f transitions of the Tb^{3+} ion: $^5D_3 \rightarrow {}^7F_4$ (~ 439 nm), $^5D_3 \rightarrow {}^7F_3$ (~ 458 nm), $^5D_3 \rightarrow {}^7F_6$ (~ 487 nm), $^5D_4 \rightarrow {}^7F_5$ (~ 545 nm), $^5D_4 \rightarrow {}^7F_4$ (~ 590 nm), $^5D_4 \rightarrow {}^7F_3$ (~ 620 nm). The 351 nm laser line excites an electron from the ground $^7F_6$ state to the $^5L_9$ and $^5G_5$ multiplets of the Tb^{3+} ion. It then moves down non-radiatively to the $^5D_3$ and $^5D_4$ levels, from which we observe the radiative transitions to the levels lying just above the ground state [2]. In our experiment these emission lines appear on a broad background, which we attribute to defect luminescence of the SiO$_2$ matrix. A decay of the intensity of this background with time under the laser light illumination was detected. We found no such decrease in the Tb^{3+} luminescence. As a measure of the intensity of the Tb^{3+} PL, the area under the strongest $^5D_4 \rightarrow {}^7F_5$ band after subtraction of the background is taken. For Tb concentrations up to 2.5 at.%, the PL intensity increases with Tb content (Figure 3).

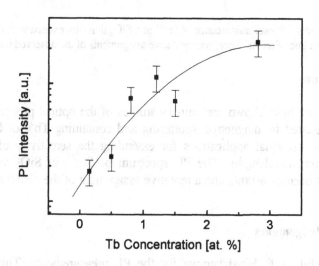

Figure 3. Dependence of the PL intensity in SiO$_2$:Tb thin films on Tb concentration.

Another ion which is a candidate for improving the conversion efficiency of solar cells is Sm^{3+}. It has absorption bands at 340, 370 and 400 nm [1]. A typical photoluminescence spectrum of this ion excited with the 488 nm line of an Ar$^+$ ion laser is shown in Figure 4. Five narrow lines in the visible are clearly observed. Most of them we tentatively assign to transitions from the $^4G_{5/2}$ 4f level, as it is the lowest of a number of closely lying levels at an energy just below that of the exciting phonons. Below this level, there is a large energy gap to the next lower $^6F_{11/2}$ level. In such cases, electrons usually thermalize to the level above the gap and start radiative transitions from there.

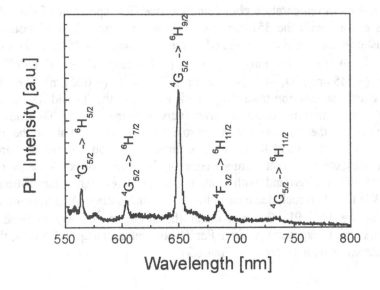

Figure 4. Photoluminescence of Sm^{3+} in a SiO$_2$ thin film excited with the 488 nm line of an Ar$^+$ laser, with tentative assignments of the observed bands.

4. Conclusions

In this paper, we have shown preliminary studies of the optical properties of SiO$_2$ thin films prepared by magnetron sputtering and containing Tb^{3+} or Sm^{3+}. These materials have potential applications for extending the sensitivity of solar cells towards shorter wavelengths. The PL spectrum of Sm^{3+} in SiO$_2$ with 488 nm excitation has been measured, and a tentative assignment of the observed bands has been made.

5. Acknowledgements

We are grateful to T. Merdzhanova for the PL measurements. This study was supported by the Bulgarian National Scientific Fund under contract X-903, and by a UK Royal Society Joint Project Grant between the Bulgarian Academy of Sciences and the University of Wales Swansea.

6. References

1. Kawano, K., Sado, T., Nishikawa, M. and Nakata, R. (1998) Conversion efficiency of solar cell with ion-implanted rare earth into anti-reflection film, in J. Schmid, H.A. Ossenbrink, P. Helm, H. Ehmann and E.D. Dunlop (eds.) *Proc. 2nd Second World Conference and Exhibition on Photovoltaic Solar Energy Conversion, Vienna, Austria,* European Communities, Luxemberg, pp.334-337.
2. Amekura, H., Eckau, A., Carius, R. and Buchal, Ch. (1998) Room-temperature photoluminescence from Tb ions implanted in SiO$_2$ on Si, *J. Appl. Phys.* **84**, 3867-3871.

COST-EFFECTIVE POROUS SILICON TECHNOLOGY FOR SOLAR CELL INDUSTRIAL APPLICATIONS

V. YEROKHOV, M. LIPINSKI*, A. MYLYANYCH and
YU. BOGDANOVSKY

University "Lviv Polytechnic", PO Box 1050, 79045 Lviv, Ukraine, and
*Inst. Metallurgy and Mat. Sci., Reymonta Str. 25, 30-059 Cracow, Poland

1. Introduction

For porous silicon (PS) layer preparation, only the electrochemical method of DC-anodizing in HF-based electrolytes and the chemical method using HF/HNO$_3$ electrolytes are widely used. In solar cell applications, the former possesses major disadvantages, such as the need for a rear contact (a metal or high conductivity diffused layer) and for other components of the electrode system. Also, additional operations are needed to prepare the rear face for the final product. Moreover, very high currents are required for large areas, limiting the industrial use of the method.

Chemical etching avoids these problems. Fabrication simply involves immersion in an electrolyte, rinsing and drying. Thus, it is a promising technological direction for porous structure fabrication in photovoltaic and optoelectronic applications.

Films made from HF:HNO$_3$ based electrolytes are similar in appearance to those made by traditional anodizing [1]. Porous films can be obtained using solutions of NaNO$_2$:HF and CrO$_3$:HF. The PS layers are externally similar to HF:HNO$_3$ or electrochemically etched films [2], indicating a similarity in the mechanisms of formation. Thus, to study the processes occurring at the electrolyte/Si interface, the mechanisms considered are usually those for HF:HNO$_3$ based electrolyte systems. The speed of the etching (dissolution) processes, which depends on both the properties of the substrate (Si) material being used and the etchant itself, is thus of considerable applied and theoretical importance.

2. Modelling

There still is no agreement on the underlying mechanism of crystalline Si self-dissolution in concentrated HF and HNO$_3$ in general, or etch pitch formation in particular. In our opinion, the limiting factor is the velocity of supply of the solution components to the cathode areas formed on the semiconductor surface.

For solar cell applications, the texture developed on the crystalline Si surface must feature new important functional properties. It must have a low reflection coefficient, allowing the creation of an optical system for the efficient capture of

J.M. Marshall and D. Dimova-Malinovska (eds.),

Photovoltaic and Photoactive Materials - Properties, Technology and Applications, 333–336.

light. At the same time, it must be can readily suitable for further technological processing (p/n junction formation, passivation).

The above advantages require the presence of etch pits with sharp edges. The main criterion for such surface formation is the ratio of the velocities of dissolution in directions perpendicular and parallel to the surface $V_\perp/V_{II}$, with $V_\perp \gg V_{II}$. The chemical composition of the reagent has a decisive influence on the shapes and dimensions of the etch pits for a polished silicon substrate, because $V_\perp$ depends on the phase-boundary energy and V_{II} can be lowered by appropriate inhibitors.

3. Results

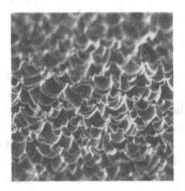 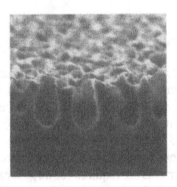

Figure 1. Morphology of a chemically textured Si surface (width of the picture: 203 μm)

Figure 2. Morphology of an electrochemically textured Si surface (width of the picture: 10.1 μm)

If we take the reduction of the reflection level, plus the requirements for high manufacturability and low price as the main goals, then one can propose a structure comprising two layers of macro-texture [3]. The first (coarse) macro-porous layer can be prepared by chemical etching (Figure 1) and the second (fine) one by electrochemical etching (Figure 2). Such a double structure is interesting from several points of view: obtaining a low reflection coefficient [4], preparing a surface which enables the use of well proven methods of antireflection layer deposition and passivation, and developing a rapid, low cost process to improve the profitability of the final solar cell (Figure 3). This involves creation of (a) the chemical, (b) the electrochemical, and (c) the double layer texture. The sequence is controlled by the dimensions of each separate texture. The size of the elementary cell of the chemical texture is tens of microns, while that of the electrochemical texture is microns. Therefore, the latter can be laid over the former without changing its shape.

For a macro-porous layer created by chemical etching, the dimensions of the elementary hole can be varied over a wide range. Depending on the silicon substrate type and the technology being used, it is possible to obtain textures with elementary holes of crater-like shape, with dimensions from 10 to 50 μm. This allows the

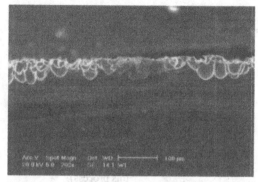

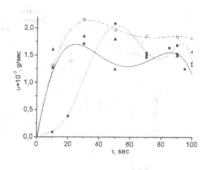

Figure 3. SEM cross-sectional image of a solar cell with macro PS texture formed in p-Si.

Figure 4. Process kinetics of the etching of crystal silicon in various solutions: 1 (◊) - initial solution; 2 (●) - solution with different additives of inhibitors; 3 (▲), 4 (■) - solution with additives of inhibitors and a surface activator.

parameters to be varied to suit the end requirement. Varying the hole size does not significantly influence the reflection coefficient, which in principle can oscillate near a certain average value typical for the type of structure in question. However, the so-called subdimension influences the reflection coefficient much more strongly. That is, if we regard the crater-like elementary hole as a semi-ellipsoid, then it can be described in terms of its diameter and depth.

With this approach, we can develop a theoretical model for the final appearance of a texture, and its reflection coefficient. The ellipsoid diameter and depth can be set by the etching conditions. However, this problem has several aspects that are contrary to the simultaneous end goals of high efficiency and profitability. There are two conditions that should be satisfied simultaneously in one technological process. As noted above, the diameter can be varied over a broad range. However, the condition for obtaining the necessary depth is related to so-called "wear" of the material during etching. Substrates for future solar cells will have thicknesses of only a few hundreds of microns. Therefore, a large amount of material should not be removed, because this would affect the mechanical properties of the wafer, and also increase the fabrication cost of the solar cell.

Our studies demonstrated that acid solutions without additives were not sufficiently efficient to form a geometrical surface profile with a low reflection coefficient. They did not allow achievement of the required etch pit size and, because of the high values of etching velocity (Figure 4, Curve 1), they led to a large loss of sample mass.

A number of organic additives (Figures 4 and 5) with different ratios of etching velocities (Figure 4, Curves 3 and 4) were examined, giving elementary cells from 10 to 50 μm in diameter. The porous surface of a silicon wafer with a cell diameter near 25 μm (Figure 2b) was found to be the optimal one from the point of view of fabrication and practical usage (etchant composition, processing time,

336

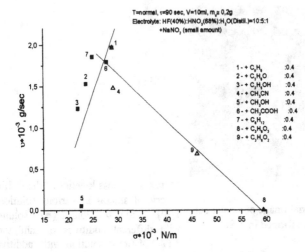

T=normal, τ=90 sec, V=10ml, m₂ 0,2g
Electrolyte: HF(40%):HNO₃(68%):H₂O(Distil.)=10:5:1
+NaNO₂ (small amount)

1 - + C₂H₃   :0.4
2 - + C₂H₆O   :0.4
3 - + C₂H₅OH   :0.4
4 - + CH₃CN   :0.4
5 - + CH₃OH   :0.4
6 - + CH₃COOH   :0.4
7 - + C₆H₁₂   :0.4
8 - + C₃H₆O₃   :0.4
9 - + C₃H₆O₂   :0.4

Figure 5. Etching velocity from surface strain of the different additives in an etching solution (T-normal, at room temperature in this process).

suitability for subsequent technological operations to form p/n junctions and the collecting grid, etc.). Note, that the etching average velocity initially increases and then decreases in the interval of surface strain values from 20 to 30 N/m. Obviously, it is related to the presence of CH_3 groups in the organic substances structure positively influencing on the etching velocity, whereas the OH-groups are unwanted.

4. Conclusions

The technologies of isotropic chemical etching of silicon surfaces and of chemical growth of PS have been outlined and applied. Additionally, experiments on the electrochemical growth of textures have been carried out. We succeeded in determining the optimal conditions for the fabrication of a new macro-porous (crater-like or honeycomb) texture on a silicon substrate. This features several considerable advantages. It gives a geometrical relief that can easily be subjected to further processing (p/n junction formation, passivation). Moreover, together with a second electrochemically etched layer, it can create a high quality optical system for the efficient capture of light into the surface.

5. References

1. Archer, R.J. (1960) *J. Phys. Chem. Solids* **14**, 104.
2. Aoyagi, H., Motohashi, A., Kinoshita, A., Aono, T. and Satou, A. (1993) A comparative study of visible photoluminescence from anodized and from chemically stained silicon wafers, *Jpn. Appl. Phys. Lett.* **32**, L1.
3. Yerokhov, V., Hezel R., Lipinski M., Nagel H., Mylyanych A. and Panek P. (2002) Cost-effective methods of texturing for silicon solar cells, *Solar Energy Materials & Solar Cells*, 2002, v. 72 (1-4), P. 291-298.
4. Yerokhov V., Hezel R., Nagel H., Melnyk I. and Semochko I. (2000) Development of profitable methods of texturing for silicon solar cells, VA2/15, *Proc. 16th European Photovoltaic Solar Energy Conference and Exhibition, Glasgow, UK*

INFLUENCE OF VACUUM ANNEALING ON THE COMPOSITION OF CuInSe₂

M. KAUK, M. ALTOSAAR and J. RAUDOJA
Institute of Materials Technology, Tallinn Technical University,
Ehitajate tee 5, 19086 Tallinn, Estonia.

Abstract

In-rich CuInSe₂ powder, synthesized in the liquid phase of a CuSe-Se flux, was annealed for 15 minutes in dynamic vacuum, at different temperatures from 100 to 530°C. The chemical compositions of the annealed materials were determined polarographically, and changes in composition were characterised by deviations from molecularity, Δm, and stoichiometry, Δs. The results revealed that in the vacuum annealing process, the molecularity stayed constant up to 450°C, but excess selenium started to be removed at temperatures higher than 100°C. The grain resistivity increased up to 400°C, and tended to decrease at higher temperatures.

1. Introduction

The chalcopyrite semiconductor CuInSe₂ (CIS) is a promising material for high-efficiency photovoltaic applications, because of its favourable electrical and optical properties, durability and cost effectiveness [1]. CIS based absorber materials have been prepared by various methods, including monograin powder technology. It is well known that the electrical and optical properties of CIS are very sensitive to compositional deviations from stoichiometry, and to intrinsic or impurity defects. It is also well known that the optical and electrical properties can be modified by post-growth annealing in vacuum or in selenium vapor. The composition of CuInSe₂ can be varied over a wide range. Deviations from the stoichiometric composition are compensated by intrinsic point defects. The deviation from the ideal chemical formula of CuInSe₂ can be described by two parameters, the non-molecularity Δm and the non-stoichiometry Δs, which are defined [2] as:

$$\Delta m = ([Cu]/[In]) - 1, \qquad (1)$$

$$\Delta s = 2([Se]/([Cu]+3[In])) - 1. \qquad (2)$$

337

J.M. Marshall and D. Dimova-Malinovska (eds.),
Photovoltaic and Photoactive Materials - Properties, Technology and Applications, 337-339.
© 2002 *Kluwer Academic Publishers.*

In this study, a monograin powder of CuInSe$_2$ was used to investigate the relationships between thermal treatment (vacuum annealing) and composition. The annealed materials were characterized by grain resistivity and deviations from molecularity and stoichiometry.

2. Experimental Details

p-type In-rich CuInSe$_2$ powder material was synthesized from a Cu-In alloy and Se, in the liquid phase of a CuSe-Se flux in an evacuated quartz ampoule, by technology described elsewhere [3]. Heat-treatments of the powder were made under dynamic vacuum (continuous pumping), at different temperatures in the range 100 to 530°C, for a constant time (15 min).

The compositions of the materials were determined polarographically, using a Metrohm 746 VA Trace Analyzer. Monograin powder was dissolved in HClO$_4$. All three elements were determined in the same solution in a citrate buffer (pH=6,0±0,2) using EDTA-Na$_2$ for complexation. In the first run, copper and selenium were determined. After acidification, In was determined in a second run.

The grain resistivities were measured by pressing single powder crystals between two In contacts. The type of conductivity was determined by the hot probe method.

3. Results

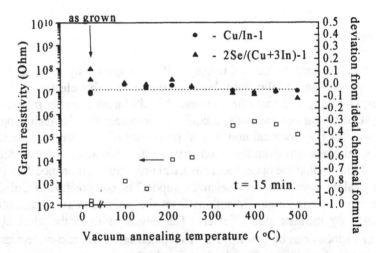

Figure 1. Grain resistivity and deviations from the ideal chemical composition of In-rich CuInSe$_2$ powder, as a function of vacuum annealing temperature. Annealing time = 15 minutes.

The results of the post-growth annealings of initially In-rich CuInSe$_2$ powder, in conditions of continuous vacuum pumping, are plotted in Figure 1. The as-grown powder, and all the vacuum annealed materials, had a *p-type* conductivity. Figure 1 shows increasing grain resistivities with increasing treatment temperatures up to 400°C. At higher temperatures, the grain resistivity tends to decrease. The results of polarographical analyses reveal that the deviation from molecularity stays constant in a 15 minute vacuum annealing process. Excess selenium starts to be removed even at 100°C.

We believe that the annealing time in the applied conditions was not long enough to change the *p-type* as-grown powder to an *n-type* one, as predicted by the thermal dissociation of $CuInSe_2$ [4]:

$$2CuInSe_2 \rightarrow Cu_2Se \text{ (solid)} + In_2Se \text{ (gas)} + Se_2 \text{ (gas)} \quad K_{CuInSe2} = P_{In2Se}P_{Se2}. \quad (3)$$

4. Conclusions

The annealing of $CuInSe_2$ monograin powder in vacuum, at different temperatures, resulted in clearly detectable changes in the composition of the material. An increase in the vacuum annealing temperature reduced the selenium content in the materials, but the ratio of Cu to In stayed practically constant. The conductivity of the material decreased with increasing vacuum annealing temperature, but the material's conductivity type did not change.

5. Acknowledgements

Financial support by the Estonian Scientific Foundation under contract No.3409 is gratefully acknowledged.

6. References

1. Akl, A.A.S., Ashour, A., Ramadan, A.A. and El-Hady, K. (2001) Structural study of flash evaporated $CuInSe_2$ thin films, *Vacuum* **61**, 75-84.
2. Neumann, H. and Tomlinson, R.D. (1990) Relation between electrical properties and composition in $CuInSe_2$ single crystals, *Solar Cells* **28**, 301-313.
3. Altosaar, M. and Mellikov, E. (2000) $CuInSe_2$ Monograin Growth in CuSe-Se Liquid Phase, *Jpn. J. Appl. Phys.* **39**, Suppl.39-1, 65-66.
4. Neumann, H. and Kühn, G. (1989) Thermal decomposition of $CuInSe_2$, *J. Less-Common Metals* **155**, 13-17.

NATO ADVANCED STUDY INSTITUTE
Sozopol, Bulgaria, September 2001

NATO Advanced Study Institute
Photovoltaic and Photoactive Materials - Properties, Technology and Applications
Sozopol, Bulgaria, September 2001

Participants, Observers and Administrative Staff

Deniz AKDAS, Viacheslav ANDREEV, Christo ANGELOV, Bogdan ATANASIU, Mirela ATANASIU, Csaba BALAZSI, Vitezslav BENDA, Rémi De BETTIGNIES, Tom BETTS, Monica BRINZA, Özlem CANKURTARAN, Reinhard CARIUS, Krassimir DIMITROV, Nicholas DMITRUK, Doriana DIMOVA-MALINOVSKA, Kazimierz DRABCZYK, Aynur ERAY, Sermet ERAY, Gokhan ERDOGAN, Thierry Langlois d'ESTAINTOT, Noureddine GABOUZE, Maxim GANCHEV, Kostadinka GESHEVA, Antoaneta HARIZANOVA, Tatyana IVANOVA, Marko JANKOVEC, Vida JANUSONIENE, Tomas KALETA, Marijka KAMENOVA, Stefan KANEV, Marit KAUK, Suzi KAVLAKOVA, Alp Osman KODOLBAŞ, Yalin Gueorgui KUDDAN, Vera KUDOYAROVA, Ivo KURITKA, Carl LAMPERT, Weng-Mui (May) LEE, Sergiy LYTVYNENKO, Vladislav MALAKHOV, Ruslan MARIYCHUK, Joe MARSHALL, John MAUD, Geoffrey MUNYEME, Ani NEDIALKOVA, Ágoston NEMETH, Miglena NIKOLAEVA, Giuseppe NOBILE, Jean-Michel NUNZI, Ortac ONMUS, Miklos PALFY, Piotr PANEK, Margorzata POCIASK, Georgi POPKIROV, Jatindra Kumar RATH, Michal RUŽINSKÝ, Avgustina RACHKOVA, Julia ŠABATAITYTE, Yucel SAHIN, Vladimir ŠALY, Miroslav SARAYDAROV, Marushka SENDOVA-VASSILEVA, Merih SERIN, Petro SMERTENKO, Yuriy STASYUK, Natalia STRATIEVA, Martin STUTZMANN, Jozef SZLUFCIK, Alek TAMEEV, Maria TEODOREANU, Veronika TIMAR-HORVATH, Yasemin UDUM, Ruben VARDANYAN, Franz WITTMANN, Chenyang XUE, Valery YEROKHOV, Joanna ZELAZNY, Ivaylo ZHIVKOV, and (top left): Debbie (Ph. D.), the ASI cat.

PARTICIPANTS

AKDAS, Deniz
Izmir Institute of Technology
Science Faculty, Physics Department
Gulbahce/Izmir
TURKEY
denizak@likya.iyte.edu.tr, or
denizak@photon.iyte.edu.tr

ANGELOV, Christo
Institute for Nuclear Research and
Nuclear Energy
Bulgarian Academy of Sciences
72, Tzarigradsko Chaussee Blvd.
1784 Sofia
BULGARIA
hangelov@inrne.bas.bg

ATANASIU, Georgeta Mirela
Research Inst. for Electrical Engineering
New Energy Sources Laboratory
313, Splaiul Unirii Str., 74204
Sector 3, Bucharest
ROMANIA
femopet@icpe.ro

BENDA, Vitezslav
Czech Technical University in Prague
Faculty of Electrical Engineering
Department of Electrotechnology
Technicka 2, 166 27 Praha 6
CZECH REPUBLIC
benda@fel.cvut.cz

BETTS, Tom
Centre for Renewable Energy Systems
Technology, AMREL
Loughborough University
Loughborough, LE11 3TU
U.K.
T.R.Betts@lboro.ac.uk

ANDREEV, Viacheslav
Physical Technical Institute
Russian Academy of Sciences
Polytechnicheskaya 26
194 021 St.Petersbourg
RUSSIA
Andreev@scell.ioffe.rssi.ru

ATANASIU, Constantin Bogdan
Research Institute for Electrical
Engineering
New Energy Sources Laboratory
313, Splaiul Unirii Str., 74204
Sector 3, Bucharest
ROMANIA
femopet@icpe.ro

BALAZSI, Csaba
Hungarian Academy of Sciences
Research Institute of Technical Physics
and Materials Science, Ceramics
Department
HUNGARY
balazsi@mfa.kfki.hu

De BETTIGNIES, Rémi
ERT Cellules Solaires
PhotoVolotaïques Plastiques
Laboratoire POMA, UMR-CNRS 6136
Université d'Angers
2 Blvd. Lavoisier, F49045 Angers cedex
FRANCE
remi.debettig@univ-angers.fr

BRINZA, Monica
K.U. Leuven, Afd.Halfgeleiderfysica
Celestijnenlaan 200D
3001 Heverlee-Leuven
BELGIUM
monica.brinza@.fys.kuleuven.ac.be

CANKURTARAN, Özlem
Yildiz Technical University
Department of Chemistry
Davutpasa Kampusu, Davutpasa
cad.No:127, 34210 Esenler-Istambul
TURKEY
kurtaran@yildis.edu.tr

CARIUS, Reinhard
Forschungszentrum Julich
ISI-PV
52425 Julich
GERMANY
R.Carius@kfa-juelich.de

DIMITROV, Krassimir
Institute of Solid State Physics
Bulgarian Academy of Sciences
72, Tzarigradska Chaussee Blvd.
1784 Sofia
BULGARIA

DMITRUK, Nicholas
Institute for Physics of Semiconductors
National Acad. of Sciences of Ukraine
UKRAINE
nicola@dep39.semicond.kiev.ua

DIMOVA-MALINOVSKA, Doriana
Central Lab. for Solar Energy and New
Energy Sources, Bulgarian Acad. of Sci.
72, Blvd. Tzarigradsko Chaussee
1784 Sofia
BULGARIA
doriana@phys.bas.bg

DRABCZYK, Kazimierz
Silesian Technical University
Gliwice Akademicka Str. 16
POLAND
bilbo@zeus.polsl.gliwice.pl

ERAY, Aynur
Dept. of Physics Engineering
Hacettepe University
06532 Beytepe, Ankara
TURKEY
feray@hacettepe.edu.tr

ERAY, Sermet
Dept. of Physics Engineering
Hacettepe University
06532 Beytepe, Ankara
TURKEY
feray@hacettepe.edu.tr

ERDOGAN, Gokhan
Izmir Institute of Technology
Department of Physics
Urla Campus Gulbahce, Izmir
TURKEY
gerdogan@likya.iyte.edu.tr, or
gokhaner@yahoo.com

d'ESTAINTOT, Thierry Langlois
Principal Scientific Officer
European Commission
Research Directorate-General
200, Rue de la Loi
B-1049 Bruxelles
BELGIUM
thierry.d'estaintot@cec.eu.int

GABOUZE, Noureddine
Unit of Silicon Developpement
Technology (UDTS)
2, Boulevard Frantz-Fanon, B.P. 399
Alger-Gare, Algiers
ALGERIA
ngabouze@usa.net

GANCHEV, Maxim
Central Lab. of Solar Energy & New
Energy Sources, Bulgarian Acad. of Sci.
72, Tzarigradsko Chaussee Blvd.
1784 Sofia
BULGARIA
ganchev@phys.bas.bg

345

GESHEVA, Kostadinka
Central Lab. of Solar Energy and New
Energy Sources, Bulgarian Acad. of Sci.
72, Tzarigradsko Chaussee Blvd.
1784 Sofia
BULGARIA
kagesh@phys.bas.bg

IVANOVA, Tatyana
Central Lab. of Solar Energy and New
Energy Sources, Bulgarian Acad. of Sci.
72, Tzarigradsko Chaussee Blvd.
1784 Sofia
BULGARIA
tativan@phys.bas.bg

JANUSONIENE, Vida
Institute of Lithuanian Scientific Society
A. Goštauto 11, LT-2600 Vilnius
LITHUANIA
stepas.@uj.pfi.lt

KAMENOVA, Marijka
Central Lab. of Solar Energy and New
Energy Sources, Bulgarian Acad. of Sci.
72, Tzarigradsko Chaussee Blvd.
1784 Sofia
BULGARIA
photomat@phys.bas.bg

KAUK, Marit
Tallinn Technical University
Ehitajate tee 5, 19086 Tallinn
ESTONIA
kmarit@stuff.ttu.ee

KUDDAN, Yalin Gueorgui
Central Lab. of Solar Energy and New
Energy Sources, Bulgarian Acad. of Sci.
72, Tzarigradsko Chaussee Blvd.
1784 Sofia
BULGARIA
vitanov@phys.bas.bg

HARIZANOVA, Antoaneta
Central Lab. of Solar Energy and New
Energy Sources, Bulgarian Acad. of Sci.
72, Tzarigradsko Chaussee Blvd.
1784 Sofia
BULGARIA
tonyhari@phys.bas.bg

JANKOVEC, Marko
Faculty of Electrical Engineering
University of Ljubljana
Trzaska 25, SI-1000 Ljubljana
SLOVENIA
Marko.Jankovec@fe.uni-lj.si

KALETA, Tomasz
Foundation for Materials Science
Development
24 Warszawska Str., 31-155 Krakow
POLAND
rciach@pk.edu.pl

KANEV, Stefan
Central Lab. of Solar Energy and New
Energy Sources, Bulgarian Acad. of Sci.
72, Tzarigradsko Chaussee Blvd.
1784 Sofia
BULGARIA
vitanov@phys.bas.bg

KODOLBAŞ, Alp Osman
Hacettepe University
Department of Physics Engineering
TR-06532, Beytepe, Ankara
TURKEY
kodolbas@hacettepe.edu.tr

KUDOYAROVA, Vera
A.F.Ioffe Physical-Technical Institute of
Russian Acad. of Sci.
26, Politechnicheskaya ul.
194021 St.-Petersburg
RUSSIA
kudoyarova@pop.ioffe.rssi.ru

346

346

KURITKA, Ivo
Brno University of Technology
Institute of Physical and Applied Chemistry
Purkynova 118, Brno 612 00
CZECH REPUBLIC
kuritka@fch.vutbr.cz

LEE, Weng-Mui (May)
Centre for Renewable Energy Systems Technology, AMREL
Loughborough University
Loughborough LE11 3TU
U.K.
W.M.Lee@lboro.ac.uk

MALAKHOV, Vladislav
Institute for Problems of Materials Science
National Acad. of Sci. of Ukraine
3, Krzhizhanovsky St., 03142 Kiev
UKRAINE
malakhov@ipms.kiev.ua

MARSHALL, Joe
Department of Materials Engineering
University of Wales Swansea
Swansea SA2 8PP
U.K.
joe.marshall@killay9.freeserve.co.uk

MUNYEME, Geoffrey
University of Utrecht, Debye Institute
Department of Interface Physics
P.O.Box 80.000 NL-3508 TA, Utrecht
THE NETHERLANDS
Gmunyeme@phys.uu.nl

NIKOLAEVA, Miglena
Central Lab. of Solar Energy and New Energy Sources, Bulgarian Acad. of Sci.
72, Tzarigradsko Chaussee Blvd.
1784 Sofia
BULGARIA
Miglena@phys.bas.bg

LAMPERT, Carl
6114 LaSalle Avenue Suit 612
Piedmont, CA 94611-2802
U.S.A.
cmlstar@hooked.net, or
cmlstar@juno.com

LYTVYNENKO, Sergiy
Kiev National Taras Shevchenko University, Radiophysical Faculty
64, Volodimirska str., 01033 Kiev
UKRAINE
litvin@uninet.kiev.ua, or
litvin@rpd.univ.kiev.ua

MARIYCHUK, Ruslan
Faculty of Chemical Technology and Industrial Ecology, Dept. of Chemistry
Uzhgorod National University
46, Pidgirna str., 88000, Uzhgorod
UKRAINE
rusmar@chem.univ.uzhgorod.ua

MAUD, John
Department of Chemistry
University of Wales Swansea
Swansea SA2 8PP
U.K.

NEMETH, Ágoston
Budapest University of Technology & Economics (BUTE), Department of Electron Devices
Goldmann Gy. tér 3, 1521-H Budapest
HUNGARY
nemeth.agoston@freemail

NOBILE, Giuseppe
ENEA (Agency for New Technologies, Energy and Environment)
Portici Research Center
Loc. Granatello, 80055 Portici (NA)
ITALY
nobile@portici.enea.it

NUNZI, Jean-Michel
Lab POMA
2, Boulevard Lavoisier
49045 Anger
FRANCE
Jean-michel.nunzi@univ-anger.fr

PALFY, Miklos
SOLART SYSTEM LTD.
Gulyas u. 20
H-1112 Budapest
HUNGARY
Solartsy@elender.hu

POCIASK, Margorzata
Institute of Physics
Pedagogical University if Rzeszow
16 A Rejtana Str., 35-310 Rzeszow
POLAND
pociask@atena.univ.rzeszow.pl

RATH, Jatindra Kumar
Utrecht University, Debye Institute
Interface Physics (Physics of Devices)
RvdG Laboratory
Princetoplein 4, P.O. BOX 80000
NL-3508 TA Utrecht
THE NETHERLANDS
J.K.Rath@phys.uu.nl

RACHKOVA, Avgustina
Institute of Solid State Physics
Bulgarian Academy of Sciences
72, Tzarigradsko Chaussee Blivd.
1784 Sofia
BULGARIA

SAHIN, Yucel
Anadolu University, Faculty of Science
Department of Chemistry
26470 Eskisehir
TURKEY
ysahin@anadolu.edu.tr

ONMUS, Ortac
Izmir Institute of Technology
Department of Physics, Izmir Yüksek
Teknoloji Enstitüsü, Fen Fakültesi Fizik
Bölümü, Gülbahce – Izmir
TURKEY
onmus@photon.iyte.edu.tr, or
onmus@likya.iyte.edu.tr

PANEK, Piotr
Inst. of Metallurgy & Materials Science
Polisch Academy of Sciences
Reymonta 25 Str.
POLAND
nmpanek@imim-pan.krakow.pl

POPKIROV, Georgi
Central Lab. of Solar Energy and New
Energy Sources, Bulgarian Acad. of Sci.
72, Tzarigradsko Chaussee Blvd.
1784 Sofia
BULGARIA
popkirov@yahoo.com

RUŽINSKÝ, Michal
Slovak University of Technology (SUT)
Faculty of Electrical Engineering and
Information Technology
Ilkovičova 3, SK-81219 Bratislava
SLOVAK REPUBLIC
ruzinskm@elf.stuba.sk

ŠABATAITYTE, Julia
Semiconductor Physics Institute
A. Goštauto 11, LT-2600 Vilnius
LITHUANIA
Julija@uj.pfi.lt

ŠALY, Vladimír
Slovak University of Technology (SUT)
Faculty of Electrical Engineering and
Information Technology
Ilkovičova 3, SK-81219 Bratislava
SLOVAK REPUBLIC
vsaly@elf.stuba.sk

SARAYDAROV, Miroslav
Faculty of Physics, Department of Solid
State of Physics and Microelectronics
Sofia University
5, blvd. J Bourchier, 1164 Sofia
BULGARIA
Saraydar@Phys.uni-sofia.bg

SERIN, Merih
Yildiz Technical University, Physics
Department
Davutpasa Campus, 34210, Topkapi
Istanbul
TURKEY
serin@yildiz.edu.tr, or
smerih@hotmail.com

STASYUK, Yuriy
Inst. of Physics and Chemistry of Solids
Uzhgorod National University
46, Pidgirna str., 88000 Uzhgorod
UKRAINE
skovach@chem.univ.uzhgorod.ua, or
mls@uzh.ukrtel.net

STUTZMANN, Martin
W. Schottky Institute, TU Munich
Am Coulombwall
D-85747 Garching
GERMANY
stutz@WSI.tu-muenchen.de

TAMEEV, Alek
Frumkin Institute of Electrochemistry
Russian Academy of Sciences
31, Leninsky prosp., build. 5
117071, Moscow
RUSSIA
altam@online.ru

TIMAR-HORVATH, Veronika
Budapest University of Technology &
Economics (BUTE)
Department of Electron Devices
Goldmann Gy. tér 3, 1521-H Budapest
HUNGARY
timarne@eet.bme.hu

SENDOVA-VASSILEVA, Marushka
Central Lab. of Solar Energy and New
Energy Sources, Bulgarian Acad. of Sci.
72, Tzarigradsko Chaussee Blvd.
1784 Sofia
BULGARIA
marushka@phys.bas.bg

SMERTENKO, Petro
Institute of Semiconductor Physics
National Academy of Sciences of
Ukraine
45, prospect Nauki, 03028 Kyiv
UKRAINE
eureka@irva.kiev.ua

STRATIEVA, Natalia
Central Lab. of Solar Energy and New
Energy Sources, Bulgarian Acad. of Sci.
72, Tzarigradsko Chaussee Blvd.
1784 Sofia
BULGARIA
soleil@phys.bas.bg

SZLUFCIK, Jozef
Inter-University Microelectronics Centre
Kapeldreef 75
3001 Leuven
BELGIUM
szlufcik@imec.be

TEODOREANU, Maria
Research Inst for Electrical Engineering
New Energy Sources Laboratory
313, Splaiul Unirii, 74204 Bucharest
ROMANIA
mariateo@icpe.ro

UDUM, Yasemin
Hacettepe University, Faculty of
Science
Department of Chemistry
06 532 Beytepe/ Ankara
TURKEY
yarslan@hacettepe.edu.tr

VARDANYAN, Ruben
State Engineering University of
Armenia
105 Teryan St.
Yerevan, 375009
ARMENIA
rvaradn@seua.am

XUE, Chenyang
National Technical University of Athens
Institute of Applied Sciences
Department of Physics
Zografou Campus, Athens 157 73
GREECE
xuechina@central.ntua.gr

ZELAZNY, Joanna
Foundation for Materials Science
Development
24 Warszawska Str., 31-155 Krakow
POLAND
rciach@pk.edu.pl

WITTMANN, Franz
Lehrstuch fuer Technische
Elektrophysik
Technische Universitaet Muenchen
Arcisstrasse 21, D-80290 Muenchen
GERMANY
wittmann@tep.ei.tum.de

YEROKHOV, Valery
National university "Lviv Polytechnic"
P.O. Box 1050, 79045, Lviv-45
UKRAINE
verohov@polynet.lviv.ua

ZHIVKOV, Ivaylo
Central Laboratory of Photoprocesses
Bulgarian Academy of Sciences
G. Bonchev Str., bl. 109, 1113 Sofia
BULGARIA
izhiv@clf.bas.bg

INDEX